Race and Ethnicity in Society

The Changing Landscape

ELIZABETH HIGGINBOTHAM
University of Delaware

MARGARET L. ANDERSEN
University of Delaware

THOMSON
™
WADSWORTH

Australia • Brazil • Canada • Mexico • Singapore • Spain
United Kingdom • United States

D0019030

THOMSON
WADSWORTH

Sociology Editor: *Robert Jucha*
Assistant Editor: *Elise Smith*
Editorial Assistant: *Christina Cha*
Technology Project Manager: *Dee Dee Zobian*
Marketing Manager: *Wendy Gordon*
Marketing Assistant: *Gregory Hughes*
Marketing Communications Manager:
  Linda Yip
Signing Representative: *Adrienne Krysiuk*

Project Manager, Editorial Production:
  Cheri Palmer
Creative Director: *Rob Hugel*
Print Buyer: *Judy Inouye*
Permissions Editor: *Sarah Harkrader*
Production Service: *Ginny Somma, Stratford
  Publishing Services*
Cover Designer: *Yvo Riezebos*
Compositor: *Cadmus*
Printer: *Webcom Limited*

Cover art by James Little, "Countdown"
James Little received a BFA from the Memphis Academy of Art and an MFA from Syracuse University. His work has appeared in numerous national and international exhibitions. He won the Pollock-Kranswer Foundation Award for his work and has been recognized by numerous organizations for his work. "Countdown" is part of the Paul Jones Collection of African-American Art at the University of Delaware—a collection of over 1500 pieces of art. You can get more information about James Little and the Paul Jones Collection at www.museums.udel.edu/jones/.

Thomson Higher Education
10 Davis Drive
Belmont, CA 94002-3098
USA

For more information about our products,
contact us at:
**Thomson Learning Academic Resource Center
1-800-423-0563**

For permission to use material from this text or product, submit a request online at
http://www.thomsonrights.com.

Any additional questions about permissions can be submitted by email to
thomsonrights@thomson.com.

Library of Congress Control Number: 2005921946
ISBN 0-534-57648-6

Contents

PART IV

Race, Relationships, and Identity   123
Introduction by Elizabeth Higginbotham and Margaret L. Andersen

PART V

The Political Economy of Race   167
Introduction by Elizabeth Higginbotham and Margaret L. Andersen

A. Race, Citizenship, and Labor

PART VI

Institutional Segregation and Inequality 269
Introduction by Elizabeth Higginbotham and Margaret L. Andersen

Preface

The study of race and ethnicity is changing, just as the character of race and ethnic relations is changing. The United States is now a multiracial society. New patterns of immigration since the late 1960s have brought new populations to the United States—populations that are changing the racial and ethnic composition of the nation. Whereas the focus of study about race and ethnicity has long assumed a "Black/White" framework, there is now more attention to the different racial and ethnic groups that make up the United States.

In addition, the current generation of college students has come of age not just in a more diverse nation, but at a time when race appears to have lost some of its significance in organizing relationships and social institutions. Popular culture makes it appear that interracial relationships are common and that race no longer matters in shaping social relations. The dominant ideology is one of "color blindness," as if recognizing race is the same as being racist.

But more accurate observation and, indeed, volumes of scholarship indicate that race still matters—and it matters a lot. Racial segregation, not integration, is the norm, despite the illusion of inclusion portrayed in popular culture and in the few numbers of people of color who occupy highly visible places in social, economic, and political institutions. As the articles in this book will show, race is still a fundamental part of social structures. Relationships, resources, and identities continue to be shaped by race and ethnicity.

This anthology is intended to introduce students to the study of race by engaging them in the major topics and themes now framing the study of race in the United States. Because society is changing in its racial and ethnic makeup, current scholarship on race is also changing, focusing more on the social construction of race and the diversity of group experiences. But the available textbooks have not kept up with these new realities. The dominant framework in most texts on the sociology of race and ethnicity is still a smorgasbord approach, wherein the book documents the different histories of an array of groups, but where ethnicity is seen as only applying to Whites of European ancestry. Within this model, ethnicity has been perceived as mostly "white," race as "color." Such an approach has led people to ignore diverse ethnic experiences *within* all groups, not just those among people of European ancestry.

We offer this anthology so that those teaching courses on race and ethnicity will have material to use that reflects the current state of scholarship on race and ethnicity in the United States. This book is intended primarily for courses on the sociology of race and ethnicity, although it can be used for courses in other departments and interdisciplinary programs where courses on race are being taught (such as education, political science, ethnic studies, and some humanities departments).

We have organized this book to reflect the different themes that underlie the study of race and ethnicity—consciously doing so instead of organizing it around particular groups. The major themes of this book include:

- showing the diversity of experiences that now constitute "race" in the United States;
- teaching students the significance of race as a socially constructed system of social relations;
- showing the connection between different racial identities and the social structure of race;
- understanding how racism works as a belief system rooted in societal institutions;
- providing a historical perspective on how the racial order has emerged and how it is maintained;
- examining how people have contested the dominant racial order;
- exploring current strategies for building a just multiracial society.

ORGANIZATION OF THIS BOOK

This book is organized in seven parts, beginning with a short introductory section that shows students the significance of understanding race by opening the book with articles that build from the direct experiences of those from diverse racial-ethnic backgrounds. Thus, Part I ("Race: Why It Matters") establishes the importance of examining race as a contemporary social issue. Articles are selected to grab students' attention by showing the importance of changes in the U.S. population, presenting the importance of race in under-standing the plight of different groups. We include some personal narratives to engage students' empathy and understanding.

Part II ("The Social Construction of Race and Ethnicity") establishes the analytical frameworks that are now being used to think race in society. The section examines the social construction of race as a concept and experi-ence. Together, the articles show the grounding of racial categories in specific historical and social contexts and their fluidity over time.

Part III ("Representations of Race and Group Beliefs: Prejudice and Racism") examines the most immediately experienced dimensions of race: beliefs and ideology. We think that students learn best about race by examin-ing the manifestations of a racially stratified society in people's beliefs and in the images and representations of popular culture. This section also includes material on the current ideology of colorblind thinking and how it perpetu-ates racial inequality.

Part IV ("Race, Relationships and Identity") examines racial identity and interracial relationships—topics that we think are especially interesting to stu-dents. In a racially diverse and changing society, many people now form racial identities that cross racial and ethnic boundaries. Many also have had interra-cial relationships that spark their curiosity about race and racism. We examine these issues in the section by looking at interracial relationships and racial identity.

Part V ("The Political Economy of Race") analyzes the importance of the political economy of race, showing how the economic exploitation of racial groups is buttressed by political arrangements in the state. In particular, the racial division of labor is supported by concepts of citizenship that deny full rights of citizenship to certain groups. Part A ("Race, Citizenship, and Labor") provides articles that examine these issues. In light of current discussions about citizenship, national security, immigration, and racial profiling, we think the analysis provided in this section is especially timely. Part B ("Immigration, Ethnicity, and Migration") examines the increasing significance of immigra-tion and how it has changed over time. Part C of this section ("Opportunity Structure, Class Formation, and Social Mobility") furthers this discussion of racial stratification by focusing on opportunity structures and the relationships between race and class. Several of the articles also examine race and social mobility, such as in the formation of the Black middle class.

Part VI ("Institutional Segregation and Inequality") details the conse-quences of race and racism as manifested in different social institutions. Within this part, we include subsections on different institutional sites where racial stratification can be seen, including work, family, health, housing, education, and social justice. Each section includes articles examining the outcomes within social institutions that stem from the reality of racial inequality in soci-ety. We include articles on work and labor markets, health care, the criminal justice system, housing, education, and social welfare.

Part VII ("Mobilizing for Change: Looking Forward and Learning from the Past") focuses on social movements and social change. One purpose of this section is to teach students how the past matters in understanding how people have mobilized for change (for example, in the Civil Rights Movement). But this section studies activism of the past with an eye toward guiding future social policy, social activism, and social movements. We include articles in this section (and throughout the book) that show how people have resisted the oppression of race.

PEDAGOGICAL FEATURES

We have included **Discussion Questions** for each article in this book, with the goal of helping students grasp the major points of each argument. These questions can also be used for student paper assignments, research exercises, and class discussion.

In addition to the articles published in this book, we utilize Wadsworth's InfoTrac® College Edition feature to include a **Bonus Reading** at the end of each major section, with a brief annotation by the editors. This gives students access to more information than we can include in this book and provides a Web-based exercise for students to explore online articles.

ACKNOWLEDGEMENTS

We have benefited from the support and encouragement of many people who have either discussed the contents of the book with us or provided clerical and computer assistance, or other forms of help that enabled us to complete/work on this book even in the midst of many other commitments. We thank Maxine Baca Zinn, Victoria Baynes Becker, Bonnie Thornton Dill, Ben Fleury-Steiner, Charles Gallagher, Valerie Hans, Linda Keen, Richard Rosenfeld, and Judy Watson for all they have done to help us. Special thanks go to Bethany Brown for her research assistance. We also appreciate the support provided by the University of Delaware and the Center for the History of Business, Technology, and Society at the Hagley Museum and Library. The editors at Wadsworth have also been enthusiastic about this project, so we thank Bob Jucha for his good judgment and sound advice, as well as Eve Howard for supporting this project. The suggestions of those who carefully reviewed the first drafts of this book were extremely valuable and have helped make this a stronger anthology, thus we thank Harriett Romo, University of Texas San Antonio; Eileen Diaz McConnell, University of Illinois at Urbana–Champaign; William Egelman, Iona College; Sandra Woodside, Modesto Junior College; Joseph Carroll, Colby–Sawyer College; Youlanda Gibbons, Georgetown University; Pamela Williams-Paez, College of the Canyons; Elizabeth J. Clifford, Towson University; and Emily Ignacio, Loyola University Chicago.

About the Editors

Elizabeth Higginbotham
(B.A., City College of the City University of New York; M.A., Ph.D., Brandeis University) is Professor of Sociology and Women's Studies at the University of Delaware. She is author of *Too Much to Ask: Black Women in the Era of Integration* (University of North Carolina Press, 2001) and co-editor of *Women and Work: Exploring Race, Ethnicity, and Class* (Sage Publications, 1997; with Mary Romero). She has also authored many articles in journals and anthologies. While teaching at the University of Memphis, she received the Superior Performance in University Research Award for 1991–92 and 1992–93. Along with colleagues Bonnie Thornton Dill and Lynn Weber, she is a recipient of the American Sociological Association Jessie Bernard Award (1993) and Distinguished Contributions to Teaching Award (1993) for the work of the Center for Research on Women at the University of Memphis. She also received the 2003–2004 Robin M. Williams Jr. Award from the Eastern Sociological Society, given annually to one distinguished sociologist.

Margaret L. Andersen
(B.A., Georgia State University; M.A., Ph.D., University of Massachusetts, Amherst) is Professor of Sociology and Women's Studies at the University of Delaware. She is the co-editor of the best-selling anthology, *Race, Class, and Gender* (Wadsworth, 2004; with Patricia Hill Collins) and author of *Thinking about Women: Sociological Perspectives on Sex and Gender* (Allyn & Bacon, 2006); *Sociology: Understanding a Diverse Society* (Wadsworth, 2006; co-authored with

Howard F. Taylor); *Sociology: The Essentials* (Wadsworth, 2004; also co-authored with Howard F. Taylor), *Understanding Society: An Introductory Reader* (Wadsworth, 2004; co-edited with Kim Logio and Howard F. Taylor), and *Social Problems* (Addison Wesley Longman, 1997; co-authored with Frank R. Scarpitti and Laura L. O'Toole). She is a recipient of the University of Delaware's Excellence in Teaching Award and the College of Arts and Sciences Outstanding Teacher Award, former President of the Eastern Sociological Society, and Chair of the National Advisory Board for the Center for Comparative Studies in Race and Ethnicity at Stanford University, where she has been a Visiting Professor. She was the 2004–05 Sociologists for Women in Society (SWS) Feminist Lecturer, an award given annually to a social scientist who has made significant contributions to the study of women in society.

Introduction

BY ELIZABETH HIGGINBOTHAM
AND MARGARET L. ANDERSEN

Race in the United States is changing. Whereas once there were laws that formally segregated people from one another, now the United States is supposed to be an open society where people have equal opportunities, are free to mix with one another, and enjoy freedoms unfettered by their racial or ethnic background. Formal, state-sanctioned segregation is illegal. There are laws that protect civil rights based on race, sex, national origin, and religion. Most people believe that there should be equal opportunity for all and that no one should be barred from full participation in society because of their race or ethnic background.

Yet, racial and ethnic inequality continue. Communities and schools are racially segregated. Police, shopping clerks, airport security officials, and others engage in racial profiling, defined as the singling out of individuals because of their presumed identification with a particular group, such as Arab Americans or Black Americans. Disparities in levels of income and wealth persist, despite several decades of equal employment legislation. Poverty rates are highest among African American and Latina single-parent families. Many people, including people of color, hold racial and ethnic stereotypes about other groups. Indeed, a "we/they" dichotomy characterizes much of the world conflict in which the United States is now engaged.

At the same time, the U.S. population is becoming more diverse. Population projections indicate that White people will soon be a numerical minority. In some states (notably, California, Texas, and Hawaii), that is nearly the case now. Immigration to the United States is also bringing more diversity to the nation. Whereas in the past most immigrants came from western Europe, today's immigrants are most likely to be from Asia, Latin America, and the Caribbean. This increased diversity is bringing new forms of culture and new relationships to this nation. Young people, regardless of their race or ethnic identity, listen to music and buy clothing inspired by urban African American, Latina/Latino, and Asian cultures. People meet and fall in love across different racial and ethnic identities, resulting in an increase of so-called bi-racial or multiracial people. And, whereas in the past, the nation's largest minority group was African Americans, now Hispanics are the largest "minority" group.

All of these changes are transforming the social landscape of what it means to be American. While some people in America might not think much about their racial identity, others proudly assert they are from Korea or El Salvador, that they are American-born Chinese, fifth generation African American, or a member of the Choctaw nation. How does this population diversity change how people think about race and ethnicity in the United States?

Thinking about race was once framed by a presumed "Black-White" dichotomy. That way of thinking about race is no longer adequate. Certainly African American experience is a central part of how U.S. social institutions emerged, but the landscape of race in America is more complex than thinking in Black-White terms can capture. Diversity both within and across racial groups cannot be understood in simple or one-dimensional frameworks. The multiracial character of the United States demands that we think more broadly and more inclusively about the meaning of race and its sister concept, ethnicity.

Race and ethnicity are social categories that define a complex set of social relationships. They are ingredients in the social institutions of our nation. Social scientists have pondered the meaning and significance of race and ethnicity for years. Their analyses have had to change as the society itself changes and as groups shift in their positions relative to each other. Fundamentally, U.S. society is marked by racial hierarchy, where groups are ranked and assigned different rights and values based on their race and ethnicity, just as other social hierarchies exist based on gender, social class, sexuality, and age. These hierarchies are based on power differences between groups. These power differences and the structures in which they play out are supported by ideas that define race as a "natural" trait of groups—an idea we examine in Part II. Who decides what racial categories are important and what they mean? What are the consequences for various individuals in the society? These are the kinds of questions you'll begin to ask once you no longer take race for granted and begin to see its underlying social basis.

At different points in history "race" has taken on different meanings. Many of the people currently considered White and thought of as the majority group are descendents of immigrants who at one time were believed to be racially distinct from native-born White Americans, the majority of whom were Protestants. Protestants were hesitant to welcome German Catholics and Jews from western Europe in the early nineteenth century. They were even more distrusting of Chinese immigrants in the 1850s and 1860s and, later in the 1880s, of the Europeans from eastern and southern Europe, including Poland, Russia, and Italy. Powerful White leaders at the time used the law and other means to push for greater control of the national borders, attempting to shape the racial composition of the nation. Asians, Jewish people, and Italians were viewed as strangers who were not as "worthy" as the descendants of the presumed founders of the nation.

Race has now shifted in meaning, but racial borders between groups are still contested and changing. How we think about race and ethnicity now can help one prepare for the continuing changes in society. Thus, this book is about *race and ethnicity in society.* It is not just about people of color, though surely understanding the experience of people of color—and more dominant groups—is critical to understanding the social reality of race. Even though many are reluctant to acknowledge the social reality of race, race and ethnicity are social facts. They are fundamental to how we think about ourselves. We all walk around in bodies that are assigned racial meanings—meanings we give to ourselves and that others give to us. Race shapes basic social institutions, and racial politics are a major dimension of social change—even when not explicitly known as such. Even though many deny that race is "real," it is a lens through which we often view our lives and those of others.

Our task in this book is to examine the changing landscape of race and ethnicity in the United States. We do so by examining the *social meaning of race and ethnicity*—how they form our identities and beliefs about each other; how they become the basis for the distribution of social and economic resources, and how they structure social institutions including work, families, communities, education, and the law. But we also emphasize how people have organized to resist the inequities that exist in a racially divided society. Underlying this book is the idea that racial hierarchies, like other social hierarchies, are unstable, fluid, and constantly changing. Systems of inequality are usually challenged by those who are disadvantaged by them, adding to the changes in society that define race and ethnic relations.

We also note that the study of race has changed dramatically in recent years. It has moved from a focus on the Black-White dichotomy to an understanding or acknowledgement that many different groups that make up the United States. The current generation of college students has come of age in a more diverse nation, but also at a time when issues of race continue to be contested. Current scholarship on race has also changed. It now focuses on how race is socially constructed and is a fundamental part of social structure. This anthology is intended to introduce students to the study of race by engaging them in the major topics and themes now framing the study of race in the United States.

Several themes guide the articles selected for this book:

- the diversity of experiences that now constitute race and ethnicity in the United States;
- the significance of race and ethnicity as socially constructed systems of social relations;
- the connection between diverse racial identities and the social structure of race and ethnicity;
- how racism works as a belief system rooted in societal institutions;
- how the racial order has emerged historically and how it is maintained;
- how people have contested the dominant racial order; and
- current strategies for building a more just multiracial society.

ORGANIZATION OF THE BOOK

The book is divided into seven parts, beginning with this introductory section by the editors discussing just how much race matters in U.S. society. Part I establishes the importance of examining race as a contemporary social issue and introduces you to some of the diverse experiences that people have as the result of their different racial-ethnic backgrounds. The articles in this part show how the U.S. population is becoming more racially and ethnically diverse; some of the articles in this part use personal narratives to help you gain empathy and understanding.

Part II ("The Social Construction of Race") examines the social construction of race. Several articles in this part show how thinking about race and racism has emerged over time in western thought. In this part, you will learn to be critical of purely biological constructions of race, learning instead how race is defined through society, history, and culture. Together, the articles show both the fluidity of racial categories and their grounding in specific historical and social contexts.

Part III ("Representations of Race and Group Beliefs: Prejudice and Racism") examines the most immediately experienced dimensions of race: racial beliefs and the representation of race in popular culture. We think you will learn best about race by examining how people's beliefs about race emerge in a racially stratified society. Images in popular culture are central in this process. Because society is so strongly influenced by the media, we include articles on film and music to help you see from the beginning how much thinking about race is influenced by popular culture. This part also includes material on the current ideology of colorblind thinking and how it cloaks the persistence of institutional racism.

In Part IV ("Race, Relationships, and Identity") we examine racial identity and interracial relationships, topics that we think are especially interesting to you as students. In a racially diverse and changing society, many people now form racial identities that cross racial boundaries. Many also have had interracial relationships that spark their curiosity about race and racism. We examine these issues in the section by looking at interracial relationships and racial identity.

Part V ("The Political Economy of Race") analyzes the importance of the political economy of race, showing how the economic exploitation of racial groups is buttressed by political arrangements in the state. We have broken this part into multiple sections, each reflecting different dimensions of the political economy of race. Historically, the racial division of labor has been supported by concepts of citizenship that deny the full rights of citizenship to certain groups. Section A, "Race, Citizenship, and Labor," includes articles that examine these issues. In light of current discussions about citizenship, national security, and racial profiling, we think the analysis provided in this section is especially timely. Section B of this part, "Immigration, Ethnicity, and Migration," is devoted to discussion of contemporary immigration. The articles also compare contemporary trends in immigrant experience to those of past immigrant groups, showing not only how U.S. society is transformed by immigration, but also how immigration is linked to patterns of global migration. Section C, "Opportunity Structure, Class Formation, and Social Mobility," focuses on opportunity structures and the relationships between race and class. Several of the articles also examine race and social mobility, such as in the formation of the Black middle class.

In Part VI ("Institutional Segregation and Inequality") we detail the consequences of race and racism as manifested in different social institutions, including work, family, health, housing, education, and social justice. As with Part V, this part is longer than others because we wanted to examine in detail the different social institutions that reflect racial stratification. Institutional racism means that different groups have different access to institutional resources. In this part we also emphasize what groups have done to withstand institutional exploitation. The subsections include: work and labor markets; families, communities, and welfare; residential segregation and education; and social justice and social control.

Part VII ("Mobilizing for Change: Looking Forward and Learning from the Past") focuses on social movements and social change. One purpose of this part is to illustrate how the past matters in understanding how people have mobilized for change (for example in the Civil Rights Movement). We also include articles in this part that show the different ways that people have resisted racial oppression.

Throughout the book we include the experiences of diverse groups, showing the complexity, pervasiveness, and importance of race and ethnicity in people's lives. As the United States moves through the twenty-first century, we know that the racial landscape will continue to change, but we hope that it will do so for the better as people become better informed about how race and ethnicity continue to shape the experience of diverse groups in society.

PART I

Race: Why It Matters

Supreme Court Justice Ruth Bader Ginsburg wrote:

> In the wake of a system of racial caste only recently ended, large dispar-
> ities endure. Unemployment, poverty, and access to health care vary
> disproportionately by race. Neighborhoods and schools remain racially
> divided. . . . Irrational prejudice is still encountered in real estate markets
> and consumer transactions. Bias both conscious and unconscious, reflect-
> ing traditional and unexamined habits of thought, keeps up barriers
> that must come down if equal opportunity and nondiscrimination are
> even genuinely to become this country's law and practice.[1]

In other words, race matters, as philosopher Cornel West reminded the
public in his best-selling book entitled *Race Matters* (1993).

Race matters because it is one of the most significant ways that U.S. society distributes economic, social, political, and cultural resources. Race matters
because it is one of the ways we define ourselves and other people. Race matters because it segregates our neighborhoods, our schools, our churches, and
our relationships. Race matters because it is often a matter of heated political

---

[1] *Gratz v. Bollinger*, 123 S. Ct. 2411.

debate, and the dynamics of race lie at the heart of the systems of justice and social welfare.

Yet, many people claim that the United States is now a colorblind society where race no longer matters. Personal attitudes, not societal factors, are seen as causes for success or failure. People are often blamed when they do not succeed, and that blame is often associated with race.

Some think that as a society we can overcome racial problems by abolishing race as a category of identification. Evidence from the biological sciences might seem to support that belief because biologists have concluded that there is no gene for race (as we discuss in Part II). But race is still "real." It is real in that people attribute meaning and significance to it. Its meaning is embedded in social institutions and in social relationships. In other words, race is real because it has real social and historical consequences.

Consider these facts:

- There is a 65 percent gap between Black family and White family income, compared to 56 percent in 1954 (DeNavas-Walt, Proctor, and Mills 2004).

- U.S. schools are now more segregated than thirty years ago, and the most successful plans for desegregation are being dismantled (Frankenberg, Lee, and Orfield 2003).

- Despite high levels of poverty, there is also a substantial growth of the Black, Asian American, Native American and Latino middle class (Pattillo-McCoy 1999).

- Interracial marriage, although increasing, is still rare—constituting only 3 percent of all marriages (U.S. Census Bureau 2004).

The official statement of the American Sociological Association reprinted in Part I shows the many ways that race matters. The statement was collectively developed by a group of sociological experts on race who reviewed the many social scientific studies of race. The Association concluded that, despite beliefs to the contrary, race is still a mechanism for sorting people in society, in their words—a "stratifying practice." Disparities in jobs, housing, health, education, and other facets of life remain, and these can be attributed to race, along with its relationship to other social factors such as social class, gender, age, nationality, and so on. Because of the strong sociological significance of race, we cannot just make it go away by denying that it is there.

The social reality of race is becoming even more evident as the nation becomes more racially and ethnically diverse. Sociologists often frame discussions of race and ethnicity in terms of "minority/majority group." These terms are used not in a numerical sense but as a way to represent who holds power in society, namely the "majority." Although the nomenclature is intended to refer to the power relations associated with race and ethnicity, it can be misleading in the context of greater diversity. Indeed, it may seem to neutralize the relations of power that exist between dominant and subordinate groups. "Dominant and subordinate" are stronger in their connotation and that better represent how power shapes relations between race and ethnic groups in society. As society changes, so does the language that social scientists use to describe and analyze it.

How much society is changing is reflected in Farai Chideya's essay ("A Nation of Minorities: America in 2050"), in which she describes the population shifts that are transforming the character of the U.S. population. The United States has long been considered a "White people's society," but that understanding is shifting. As Chideya shows, the idea of "Whiteness" has historically relied on a bipolar construction of race as either "White" or "Black." With a more diverse population emerging, this binary system of thinking— "either you're a Black person or you're a White person"—is undergoing profound change. Young people are now coming of age at a time when they see a different mix of racial identities and racial groups. How this trend will transform society is yet to be known, but it is fascinating to watch it unfold.

Anyone who thinks that race does not matter much might want to step into the shoes of those who know it does. Robert Blauner ("Talking Past One Another") shows how a racial fault line has developed with White and Black people having different world views about the reality of race. This racial fault line inhibits cross-race communication and understanding, especially if people deny that racism still matters. And clearly it does. Something as simple as the act of shopping elicits different experiences for White people and people of color. Blauner also points out that racism can sometimes be subtle, sometimes not; it is not always overt, nor is it always intentional. Racism is institutionalized—that is, built into the very fabric of society. As the nation is becoming more racially and ethnically diverse, the racial fault lines may be unfolding into multiple worldviews—each of them founded in a different racial-ethnic social location.

As U.S. society becomes more racially and ethnically diverse, what race and ethnicity mean also changes. Clara Rodriquez ("Changing Race") shows us that "race" and "ethnicity" have shifting, context-dependent meanings. Thus, some Latinos/Latinas identify as "Black," others as "White" and some only by their specific ethnic identification, say, as Puerto Rican or Mexican or Guatemalan or Cuban. The complexity of the definitions of race and ethnicity stem both from how society perceives groups and how groups perceive themselves.

In a racially stratified society, people of color become marked with suspicion. This social reality has become familiar to Arab Americans in the aftermath of the devastating events of September 11, 2001. Many Arab Americans are now experiencing racial profiling in their daily life. They join other people of color who appear "foreign" and who are subjected to racial and ethnic stereotyping. Once you realize that profiling happens, you can see how resentment, hostility, and suspicion characterize human interaction under conditions of racial inequality. But, we can learn to challenge these ways of interacting by becoming more aware of the reality of race in everyday life and working to change it—in our personal lives, as well as within larger societal structures.

Moustafa Bayoumi ("What Does It Feel Like to Be a Problem?") asks us to think about race as something other than a social problem. Yes, race does generate social problems, but seeing people only in the framework of a problem diminishes their humanity and makes us lose sight of the human creativity and adaptation that takes place even when people face harsh conditions. Reflecting on 9/11, Bayoumi wonders if this event will further divide people or provide an opportunity through which people can challenge the commission of hate crimes, the growth of exclusionary nationalism, and "we/they" thinking. The tragedy of 9/11 may be that this was the first time that many White Americans may have felt someone hated them simply because of the group they belonged to—a feeling that many people of color have repeatedly experienced.

In this first Part, we examine some of the ways that race matters, thus setting the stage for understanding the many dimensions of race in society explored in subsequent parts.

REFERENCES

DeNavas-Walt, Carmen, Bernadette D. Proctor, and Rovert J. Mills. 2004. *Income, Poverty, and Health Insurance Coverage in the United States: 2003*. Washington, DC: U.S. Census Bureau.

Frankenberg, Erica, Chungmei Lei, and Gary Orfield. 2003. "A Multiracial Society with Segregated Schools: Are We Losing the Dream?" The Civil Rights Project, Harvard University, Web site: www.civilrightsproject.harvard.edu

Pattillo-McCoy, Mary. 1999. *Black Picket Fences: Privilege and Peril among the Black Middle Class*. Chicago: University of Chicago Press.

U.S. Census Bureau, 2004. *Statistical Abstract of the United States 2003*. Washington, DC: U.S. Census Bureau.

West, Cornel. 1993. *Race Matters*. Boston: Beacon Press.

1

A Nation of Minorities

America in 2050

BY FARAI CHIDEYA

merica is facing the largest cultural shift in its history. Around the year
2050, whites will become a "minority." This is uncharted territory for
this country, and this demographic change will affect everything.
Alliances between the races are bound to shift. Political and social power will
be re-apportioned. Our neighborhoods, our schools and workplaces, even
racial categories themselves will be altered. Any massive social change is bound
to bring uncertainty, even fear. But the worst crisis we face today is not in our
cities or neighborhoods, but in our minds. We have grown up with a fixed
idea of what and who America is, and how race relations in this nation work.
We live by two assumptions: that "race" is a black and white issue, and, that
America is a "white" society. Neither has ever been strictly true, and today
these ideas are rapidly becoming obsolete.

Just examine the demographic trends. In 1950, America was nearly 85
percent non-Hispanic white. Today, this nation is 75 percent non-Hispanic
white, 12 percent black, 12.5 percent Hispanic, 3.6 percent Asian, 1 percent
Native American, and 5.6 percent other groups.[1] (To put it another way, we're
about three-quarters "white" and one-quarter "minority.") But America's racial
composition is changing more rapidly than ever. The number of immigrants in
America is the largest in any post–World War II period. Nearly one-tenth of
the U.S. population is foreign born. Asian Americans, the fastest-growing group

From *Civil Rights Journal* 4 (Fall 1999): 34. Published by the U.S. Commission on Civil Rights.

[1]Note: Figures have been updated by the editors to reflect the 2000 U.S. census data.

in America, have begun to come of age politically in California and the Pacific Northwest. [By the 2000 census, Hispanics outnumbered African Americans in the U.S. population. ED.]

Yet our idea of "Americanness" has always been linked with "whiteness," from tales of the Pilgrims forward. We still see the equation of white=American every day in movies and on television. We witness it in the making of social policy. (The U.S. Senate is only 4 percent nonwhite—though over 25 percent of the country is.) We make casual assumptions about who belongs in this society and who is an outsider. (Just ask the countless American-born Asians and Latinos who've been complimented on how well they speak English.)

"Whiteness" would not exist, of course, without something against which to define itself. That thing is "blackness." Slavery was the forging crucible of American racial identity, setting up the black/white dichotomy we have never broken free from. The landmarks of American history are intimately intertwined with these racial conflicts—the Civil War, Jim Crow, the Civil Rights movement. But today, even as America becomes more diverse, the media still depicts the world largely in black and white. The dramas and sitcoms we watch are so segregated that the top-10 shows in black households and the top-10 shows in white households barely overlap. . . . Race is almost always framed as bipolar—the children of slaves vs. the children of slaveowners—even when the issues impact Asians, Latinos and Native Americans as well. School segregation, job integration—they're covered in black and white. Political rivalries, dating trends, income inequalities—they're covered as two-sided dilemmas as well.

Everyone gets exposed to media images of race. Kids who have never met an African American will learn about slavery in school, listen to rap or R & B, and read an article on welfare reform or the NBA. It's only human nature to put together those pieces and try to synthesize an idea of what it means to be "black." The media and pop culture have such a tremendous power in our society because we use them to tell us what the rest of the society is like, and how we should react to it. The problem is that, too often, the picture we're getting is out of kilter.

If you're not black and not white, you're not very likely to be seen. According to a study by the Center for Media and Public Affairs, the proportion of Latino characters on prime-time television actually dropped from 3 percent in the 1950s to 1 percent in the 1980s, even as the Latino population rapidly grew. Asian Americans are even harder to find in entertainment, news, or on the national agenda, and Native Americans rarer still. How we perceive race, and how it's depicted in print and on television, has less to do with demographic reality than our mindset. National opinion polls reveal that, in the basest and most stereotypic terms, white Americans are considered "true" Americans; black Americans are considered inferior Americans; Asians and Latinos are too often considered foreigners; and Native Americans are rarely thought of at all.

The media's stereotypic images of race affect all of us, but especially the young Americans who are just beginning to form their racial attitudes. I call the young Americans coming of age today the Millennium Generation. These 15–25 year olds are the most racially mixed generation this nation has ever seen—the face of the new America. As a group, they are 60 percent more likely to be non-white than their parent and grandparent generations. . . . No less than one-third of young Americans aged 15 to 25 are black, Latino, Asian or Native American. While the older generations largely rely on the media to provide them with images of a multi-ethnic America, this generation is already living in it.

THE MILLENNIUM GENERATION

The teens and twenty-somethings of the Millennium Generation are the true experts on the future of race, because they're re-creating America's racial identity every single day. They're more likely to interact with people of other races and backgrounds than other generations, and they've grown up seeing multi-ethnic images. Critically important, a third of this generation is non-white, not just black but Asian, Latino, Native American and multi-racial. Yet the rhetoric which they hear about race clashes abruptly with the realities of their lives. . . . Politicians (and parents) of every political persuasion tend to cast the race debate in black and white, but the truth of this generation's lives is far more complex and colorful.

The members of the Millennium Generation defy the easy racial stereotypes. Take an issue as heated as illegal immigration—and the life of an Oakland teen named Diana. Serious and thoughtful, with hopes of going on to college, the Mexican immigrant has lived most of her life in California. She's more familiar with American culture (not to mention more articulate in English) than most teens. But she doesn't have a green card, and her chances of pursuing her college dreams seem slim. Her dad has a green card and two of her four siblings are U.S. citizens because they were born in the United States. Diana was born in Mexico. So, even though she came to the U.S. at the age of two, Diana will have a nearly impossible time getting citizenship unless she finds the money to hire an immigration lawyer to fight her case. It would be easy to think of Diana as some kind of anomaly, but she's not. Countless undocumented immigrants have spent the majority of their lives in this country. And in California alone, there are over a million residents who belong to families of mixed immigration status. Another flashpoint is the battle over affirmative action. Berkeley student LaShunda Prescott could be portrayed as a case of affirmative action gone awry, a black student admitted to a school she wasn't ready for. An engineering student, LaShunda dropped out of Berkeley twice before graduating. But during that time she looked out for a drug-addicted sister, took care of one of the sister's children, and dealt with the death of one family member and the shooting of another. In context, her circuitous route through college is not a failure but a triumph. . . .

A SPLINTERING DIVIDE

Young Americans like these illustrate a fault line in the race debates that most of us don't even think about: a massive generation gap. On the one hand, America is led by Baby Boomers and people from the generations that came before them. These movers and shakers in government and industry came of age before and during the Civil Rights era, while America was dealing with (and reeling from) the struggles of blacks to gain legal equality with whites. When they grew up, America was much whiter, both demographically and culturally. The most powerful images of the era show the divide. The top movies and television shows excluded blacks, and our archives are filled with photographs of black and white youth during the Civil Rights Era, such as the stormy desegregation of [Central High School in Little Rock].

On the other hand, Americans in their teens and twenties are coming of age at a time which seems less momentous than the Civil Rights Era, but is even more complex. This generation sees firsthand evidence in their own schools and neighborhoods that America is becoming less white and more racially mixed. Yet the court battles of today aren't over providing legal equality for African Americans; they're about whether to keep or end programs like affirmative action, which were set up to achieve civil rights goals. The cultural battles loom even larger than the legal ones, from the debate over multiculturalism on campus to issues like inter-racial dating. America's pop culture today is infinitely more likely to show blacks as well as whites (though other races often remain unseen). The billion-dollar hip hop industry, produced by blacks but driven by sales to young fans of all races, is one indicator of the cultural shift. Even more significant, eighty percent of teens have a close friend of another race. . . .

The very idea that America will become "majority-minority" scares the hell out of some people. That's why we find ourselves not only at a point of incredible change, but of incredible fear. The 1990s have seen a full-scale backlash against immigrants and non-whites, both in word and in deed. As the visibility of non-whites has been rising, hate crimes have too—with attacks on increasingly visible Latinos and Asian Americans rising the fastest. Over the 1999 Fourth of July weekend, a white supremacist named Benjamin Nathaniel Smith went on a shooting spree in Illinois, killing an African American and an Asian American, and wounding another Asian American and six Orthodox Jews. But extremists like Smith are not the only Americans clinging to prejudices. A study by the National Opinion Research Center found that the majority of whites still believe blacks to be inferior (with smaller numbers holding the same views of Southern whites and Hispanics). . . .

The halls of power in America are still segregated. Many corporations and even government agencies look much like they did half a century ago, before Martin Luther King, Jr. marched to Selma. Ninety-five percent of corporate management—the presidents, vice presidents, and CEOs who run America— are white males. Or as Newsweek's article put it: "White males make up just 39.2 percent of the population, yet they account for 82.5 percent of the Forbes 400 (folks worth at least $265 million), 77 percent of Congress, 92 percent of state governors, 70 percent of tenured college faculty, almost 90 percent of daily-newspaper editors, 77 percent of TV news directors." The image of a hostile takeover of America by non-white guerrilla forces is patently a lie.

What remains a sad truth is the racial divide in resources and opportunity. The unemployment rate is one good indicator. For decades, the black unemployment rate has been approximately twice that of whites. In 1995, the unemployment rate was 3.3 percent for whites, 6.6 percent for blacks, 5.1 percent for Hispanics, and 3.2 percent for Asian Americans.

Recent polls indicate that most Americans know little about the profound differences separating the income, health and educational opportunities of Americans of different races. This makes a profound difference in how we think of racial issues. In a series of polls, Americans who believed that the opportunities and incomes of blacks and whites were equal were much less likely to support programs to end racial discrepancies. Too many of us try to wish the problem of race away instead of confronting it. Instead of attacking the problems of race, we seem intent on attacking non-white races, including those members of the next generation who belong to "minority" ethnic groups.

PATHS FOR THE FUTURE

We have better options than tearing each other apart. Instead of fearing the change in American society, we can prepare for it. Here are some simple suggestions:

- Know the Facts About America's Diversity. Evaluate how much you know about race in America. According to an array of surveys, white Americans—who at this moment in time make up over three-quarters of the adult population—have an inaccurate view of the racial opportunity gap. Those misperceptions then contribute to their views on issues like the need for the government to address racial inequality.

- Demand Better Media Coverage of Race. One study which tracked a year's worth of network news coverage found that sixty percent of images of blacks were negative, portraying victims, welfare dependents and criminals. That is a far cry from the reality about the black community. The news and even the entertainment we read, listen to and watch has a tremendous influence on our perception of societal problems.

- Foster Coalitions Between Non-White Groups. Particularly in urban areas, it's becoming increasingly likely that various non-white groups will share the same community. For example, South Asians and Latinos live next to each other in parts of Queens, New York, and Blacks, Latinos, and Asians share the same neighborhood in Oakland, California. But even though blacks, Latinos, Asians and Native Americans often share common issues, they don't have a good track record of joining together. Every city has groups trying to make a difference. One example is Los

Angeles's MultiCultural Collaborative, a group of Korean, Latino and black grassroots organizers formed in response to the destruction following the Rodney King verdict.

- Foster Coalitions Between Whites and Non-Whites. Just as important as forming coalitions between different nonwhite groups is changing the often antagonistic politics between the racial majority (whites) and racial "minorities." One way of doing this is to bring together like-minded groups from different communities. For example, the Parent-Teacher Association from a majority-black school could meet with the PTA from a mostly-Asian school, to discuss their common goals, specific challenges, and how they might press government officials to improve education in their district.

- Demand "Color Equality" Before "Color Blindness." Segregation is still a pervasive problem in American society, most of all for blacks but for virtually every other race as well. But does that mean we should attempt to overcome segregation and bias by demanding a "color blind" society—one where we talk less, think less, and certainly act without regard to race. The term "color blind" has become increasingly popular, but it avoids a couple of fundamental truths. If racial inequality is a problem, it's terribly difficult to deal with the problem by simply declaring we're all the same. Moreover, do we want to be the same, or equal? Who, for example, could envision New York without a Chinatown and a Little Italy?

- Re-desegregate the School System. Four decades after the Brown v. Board of Education ruling, over sixty percent of black students still attend segregated schools. In many municipalities, the statistics are getting worse, not better. The Supreme Court has consistently ruled in the past decade that even strategies like creating magnet programs in mostly-minority schools could not be used as a desegregation strategy. It would be nothing less than a tragedy if at the precise moment we are becoming a more diverse country, we are steering children and teens into increasingly segregated schools.

The changes the next millennium brings will at the very least surpass and quite possibly will shatter our current understanding of race, ethnicity, culture and community. The real test of our strength will be how willing we are to go beyond the narrowness of our expectations, seek knowledge about the lives of those around us—and move forward with eagerness, not fear.

DISCUSSION QUESTIONS

1. What evidence of a generation gap in race relations does Chideya see? Do you see similar evidence in your own life?

2. What do you think will be some consequences of so-called minorities becoming a larger proportion of the U.S. population? What suggestions does Chideya give for approaching this change without racial division?

2

Talking Past One Another

Black and White Languages of Race

BY ROBERT BLAUNER

I want to advance the proposition that there are two languages of race in America. I am not talking about black English and standard English, . . . which refer to different structures of grammar and dialect. "Language" here signifies a system of implicit understandings about social reality, and a racial language encompasses a worldview.

Blacks and whites differ on their interpretations of social change . . . because their racial languages define the central terms, especially "racism," differently. Their racial languages incorporate different views of American society itself, especially the question of how central race and racism are to America's very existence, past and present. Blacks believe in this centrality, while most whites, except for the more race-conscious extremists, see race as a peripheral reality. Even successful, middle-class black professionals experience slights and humiliations—incidents when they are stopped by police, regarded suspiciously by clerks while shopping, or mistaken for messengers, drivers, or aides at work—that remind them they have not escaped racism's reach. For whites, race becomes central on exceptional occasions: collective, public moments . . . when the veil is lifted, and private ones, such as a family's decision to escape urban problems with a move to the suburbs. But most of the time European-Americans are able to view racial issues as aberrations in American life, much as Los Angeles Police Chief Daryl Gates used the term "aberration" to explain his officers' beating of Rodney King in March 1991.

From *The American Prospect* 3 (June 1992). Reprinted by permission.

Because of these differences in language and worldview, blacks and whites often talk past one another, just as men and women sometimes do. I first noticed this in my classes, particularly during discussions of racism. Whites locate racism in color consciousness and its absence in color blindness. They regard it as a kind of racism when students of color insistently underscore their sense of differences, their affirmation of ethnic and racial membership, which minority students have increasingly asserted. Many black, and increasingly also Latino and Asian, students cannot understand this reaction. It seems to them misinformed, even ignorant. They in turn sense a kind of racism in the whites' assumption that minorities must assimilate to mainstream values and styles. Then African Americans will posit an idea that many whites find preposterous: Black people, they argue, cannot be racist, because racism is a system of power, and black people as a group do not have power.

In this and many other arenas, a contest rages over the meaning of racism. Racism has become the central term in the language of race. From the 1940s through the 1980s new and multiple meanings of racism have been added to the social science lexicon and public discourse. The 1960s were especially critical for what the English sociologist Robert Miles has called the "inflation" of the term "racism." Blacks tended to embrace the enlarged definitions, whites to resist them. This conflict, in my view, has been at the very center of the racial struggle during the past decade.

THE WIDENING CONCEPTION OF RACISM

The term "racism" was not commonly used in social science or American public life until the 1960s. "Racism" does not appear, for example, in the Swedish economist Gunnar Myrdal's classic 1944 study of American race relations, *An American Dilemma*. But even when the term was not directly used, it is still possible to determine the prevailing understandings of racial oppression.

In the 1940s racism referred to an ideology, an explicit system of beliefs postulating the superiority of whites based on the inherent, biological inferiority of the colored races. Ideological racism was particularly associated with the belief systems of the Deep South and was originally devised as a rationale for slavery. Theories of white supremacy, particularly in their biological versions, lost much of their legitimacy after the Second World War due to their association with Nazism. In recent years cultural explanations of "inferiority" are heard more commonly than biological ones, which today are associated with such extremist "hate groups" as the Ku Klux Klan and the White Aryan Brotherhood.

By the 1950s and early 1960s, with ideological racism discredited, the focus shifted to a more discrete approach to racially invidious attitudes and behavior, expressed in the model of prejudice and discrimination. "Prejudice" referred (and still does) to hostile feelings and beliefs about racial minorities and the web of stereotypes justifying such negative attitudes. "Discrimination" referred to actions meant to harm the members of a racial minority group. The logic of this model was that racism implied a double standard, that is, treating a person of color differently—in mind or action—than one would a member of the majority group.

By the mid-1960s the terms "prejudice" and "discrimination" and the implicit model of racial causation implied by them were seen as too weak to explain the sweep of racial conflict and change, too limited in their analytical power, and for some critics too individualistic in their assumptions. Their original meanings tended to be absorbed by a new, more encompassing idea of racism. During the 1960s the referents of racial oppression moved from individual actions and beliefs to group and institutional processes, from subjective ideas to "objective" structures or results. Instead of intent, there was now an emphasis on process: those more objective social processes of exclusion, exploitation, and discrimination that led to a racially stratified society.

The most notable to these new definitions was "institutional racism." In their 1967 book *Black Power*, Stokely Carmichael and Charles Hamilton stressed how institutional racism was different and more fundamental than individual racism. Racism, in this view, was built into society and scarcely required prejudicial attitudes to maintain racial oppression.

This understanding of racism as pervasive and institutionalized spread from relatively narrow "movement" and academic circles to the larger public with the appearance in 1968 of the report of the commission on the urban riots appointed by President Lyndon Johnson and chaired by Illinois governor Otto Kerner. The Kerner Commission identified "white racism" as a prime reality of American society and the major underlying cause of ghetto unrest. America, in this view, was moving toward two societies, one white and one black (it is not clear where other racial minorities fit in). Although its recommendations were never acted upon politically, the report legitimated the term "white racism" among politicians and opinion leaders as a key to analyzing racial inequality in America.

Another definition of racism, which I would call "racism as atmosphere," also emerged in the 1960s and 1970s. This is the idea that an organization or an environment might be racist because its implicit, unconscious structures were devised for the use and comfort of white people, with the result that people of other races will not feel at home in such settings. Acting on this understanding of racism, many schools and universities, corporations, and other institutions have changed their teaching practices or work environments to encourage a greater diversity in their clientele, students, or workforce.

Perhaps the most radical definition of all was the concept of "racism as result." In this sense, an institution or an occupation is racist simply because racial minorities are underrepresented in numbers or in positions of prestige and authority.

Seizing on different conceptions of racism, the blacks and whites I talked to in the late 1970s had come to different conclusions about how far America had moved toward racial justice. Whites tended to adhere to earlier, more limited notions of racism. Blacks for the most part saw the newer meanings as more basic. Thus African Americans did not think racism had been put to rest by civil rights laws, even by the dramatic changes in the South. They felt that it still pervaded American life, indeed, had become more insidious because the subtle forms were harder to combat than old-fashioned exclusion and persecution.

Whites saw racism largely as a thing of the past. They defined it in terms of segregation and lynching, explicit white supremacist beliefs, or double standards in hiring, promotion, and admissions to colleges or other institutions. Except for affirmative action, which seemed the most blatant expression of such double standards, they were positively impressed by racial change. Many saw the relaxed and comfortable relations between whites and blacks as the heart of the matter. More crucial to blacks, on the other hand, were the underlying structures of power and position that continued to provide them with unequal portions of economic opportunity and other possibilities for the good life.

The newer, expanded definitions of racism just do not make much sense to most whites. I have experienced their frustrations directly when I try to explain the concept of institutional racism to white students and popular audiences. The idea of racism as an "impersonal force" loses all but the most theoretically inclined. Whites are more likely than blacks to view racism as a personal issue. Both sensitive to their own possible culpability (if only unconsciously) and angry at the use of the concept of racism by angry minorities, they do not differentiate well between the racism of social structures and the accusation that they as participants in that structure are personally racist.

The new meanings make sense to blacks, who live such experiences in their bones. But by 1979 many of the African Americans in my study, particularly the older activists, were critical of the use of racism as a blanket explanation for all manifestations of racial inequality. Long before similar ideas were voiced by the black conservatives, many blacks sensed that too heavy an emphasis on racism led to the false conclusion that blacks could only progress through a conventional civil rights strategy of fighting prejudice and discrimination. (This strategy, while necessary, had proved very limited.) Overemphasizing racism, they feared, was interfering with the black community's ability to achieve greater self-determination through the politics of self-help. In addition, they told me that the prevailing rhetoric of the 1960s had affected many young blacks. Rather than taking responsibility for their own difficulties, they were now using racism as a "cop-out."

In public life today this analysis is seen as part of the conservative discourse on race. Yet I believe that this position originally was a progressive one, developed out of self-critical reflections on the relative failure of 1960s movements. But perhaps because it did not seem to be "politically correct," the left-liberal community, black as well as white, academic as well as political, has been afraid of embracing such a critique. As a result, the neoconservatives had a clear field to pick up this grass-roots sentiment and to use it to further their view that racism is no longer significant in American life. This is the last thing that my informants and other savvy African Americans close to the pulse of their communities believe.

By the late 1970s the main usage of racism in the mind of the white public had undoubtedly become that of "reverse racism." The primacy of "reverse racism" as "the really important racism" suggests that the conservatives and the liberal-center have, in effect, won the battle over the meaning of racism.

Perhaps this was inevitable because of the long period of backlash against all the progressive movements of the 1960s. But part of the problem may have been the inflation of the idea of racism. While institutional racism exists, such a concept loses practical utility if every thing and every place is racist. In that case, there is effectively nothing to be done about it. And without conceptual tools to distinguish what is important from what is not, we are lost in the confusion of multiple meanings. . . .

The question then becomes what to do about these multiple and confusing meanings of racism and their extraordinary personal and political charge. I would begin by honoring both the black and white readings of the term. Such an attitude might help facilitate the interracial dialogue so badly needed and yet so rare today.

Communication can only start from the understandings that people have. While the black understanding of racism is, in some sense, the deeper one, the white views of racism (ideology, double standard) refer to more specific and recognizable beliefs and practices. Since there is also a cross-racial consensus on the immorality of racist ideology and racial discrimination, it makes sense whenever possible to use such a concrete referent as discrimination rather than the more global concept of racism. And reemphasizing discrimination may help remind the public that racial discrimination is not just a legacy of the past.

The intellectual power of the African American understanding lies in its more critical and encompassing perspective.

DISCUSSION QUESTIONS

1. What different definitions of racism does Blauner identify as having emerged at different points in time in this country? How does he now define racism?

2. What different views of racism does Blauner say White people and Black people hold? How does this result in Black people and White people "talking past one another?"

3

Changing Race

BY CLARA E. RODRÍGUEZ

According to definitions common in the United States, I am a light-skinned Latina with European features and hair texture. I was born and raised in New York City; my first language was Spanish; and I am today bilingual. I cannot remember when I first realized that the color of one's skin, the texture of one's hair, or the cast of one's features determined how one was treated in both my Spanish-language and English-language worlds. I do know that it was before I understood that accents, surnames, residence, class, and clothing also determined how one was treated.

Looking back on my childhood, I recall many instances when the lighter skin color and European features of some persons were admired and terms such as *pelo malo* (bad hair) were commonly used to refer to "tightly curled" hair. It was much later that I came to see that this Eurocentric bias, which favors European characteristics above all others, was part of our history and cultures. In both Americas and the Caribbean, we have inherited and continue to favor this Eurocentrism, which grew out of our history of indigenous conquest and slavery.

I also remember a richer, more complex sense of color than the simple dichotomy of black and white would suggest, a genuine aesthetic appreciation of people with some color and an equally genuine valuation of people as people,

From Clara E. Rodriguez, *Changing Race: Latinos, the Census, and the History of Ethnicity in the United States* (New York: New York University Press, 2000), p. 3–6. Reprinted by permission of the author.

regardless of color. Also, people sometimes disagreed about an individual's color and "racial" classification, especially if the person in question was in the middle range, not just with regard to color, but also with regard to class or political position.

As I grew older, I came to see that many of these cues or clues to status—skin color, physical features, accents, surnames, residence, and other class characteristics—changed according to place or situation. For example, a natural "tan" in my South Bronx neighborhood was attractive whereas downtown, in the business area, it was "otherizing." I also recall that the same color was perceived differently in different areas. Even in Latino contexts, I saw some people as lighter or darker, depending on certain factors such as their clothes, occupation, and families. I suspect that others saw me similarly, so that in some contexts, I was very light, in others darker, and in still others about the same as everyone else. Even though my color stayed the same, the perception and sometimes its valuation changed.

I also realize now that some Latinos' experiences were different from mine and that our experiences affect the way we view the world. I know that not all Latinos have multiple or fluctuating identities. For a view, social context is irrelevant. Regardless of the context, they see themselves, and/or are seen, in only one way. They are what the Census Bureau refers to as *consistent*; that is, they consistently answer in the same way when asked about their "race." Often, but not always, they are at one or the other end of the color spectrum.

My everyday experiences as a Latina, supplemented by years of scholarly work, have taught me that certain dimensions of race are fundamental to Latino life in the United States and raise questions about the nature of "race" in this country. This does not mean that all Latinos have the same experiences but that for most, these experiences are not surprising. For example, although some Latinos are consistently seen as having the same color or "race," many Latinos are assigned a multiplicity of "racial" classifications, sometimes in one day! I am reminded if the student who told me after class one day, "When people first meet the, they think I'm Italian, then when they find out my last name is Mendez, they think I'm Spanish, then when I tell them my mother is Puerto Rican, they think I'm nonwhite or black." Although he had not changed his identity, the perception of it changed with each additional bit of information.

Latino students have also told me that non-Latinos sometimes assume they are African American. When they assert they are not "black" but Latino, they are either reproved for denying their "race" or told they are out of touch with reality. Other Latinos, who see whites as other-than-me, are told by non-Latinos, "But you're white." Although not all Latinos have such dramatic experiences, almost all know (and are often related to) others who have.

In addition to being reclassified by others (without their consent), some Latinos shift their own self-classification during their lifetime. I have known Latinos who became "black," then "white," then "human beings," and finally again "Latino"—all in a relatively short time. I have also known Latinos for whom the sequence was quite different and the time period longer. Some

Latinos who altered their identities came to be viewed by others as legitimate members of their new identity group. I also saw the simultaneously tricultural, sometimes trilingual, abilities of many Latinos who manifested or projected a different self as they acclimated themselves to Latino, African American, or white context.

I have come to understand that this shifting, context-dependent experience is at the core of many Latinos' life in the United States. Even in the nuclear family, parents, children, and siblings often have a wide range of physical types. For many Latinos, race is primarily cultural; multiple identities are a normal state of affairs; and "racial mixture" is subject to many different, sometimes fluctuating, definitions.

Some regard *racial mixture* as an unfortunate or embarrassing term, but others consider the affirmation of mixture to be empowering. Lugones (1994) subscribes to this latter view and affirms "mixture," *mestizaje*, as a way of resisting a world in which purity and separation are emphasized and one's identities are controlled: "Mestizaje defies control through simultaneously asserting the impure, curdled multiple state and rejecting fragmentation into pure parts . . . the mestiza . . . has no pure parts to be 'had,' controlled" (p. 460). Also prevalent in the upper classes is the hegemonic view that rejects or denies "mixture" and claims a "pure" European ancestry. This view also is common among middle- and upper-class Latinos, regardless of their skin color or place of origin. In some areas, people rarely claim a European ancestry, such as in indigenous sectors of Latin America, in parts of Brazil, and in the coastal areas of Colombia, Venezuela, Honduras, and Panama. Recently, some Latinos have encouraged another view in which those historical components that were previously denied and denigrated, such as indigenous and African ancestry, were privileged.

Many people, however—mostly non-Latinos—are not acquainted with these basic elements of Latino life. They do not think much about them, and when they do, they tend to see race as a "given," an ascribed characteristic that does not change for anyone, at any time. One is either white or not white. They also believe that "race" is based on genetic inheritance, a perspective that is just another construct of race.

Whereas many Latinos regard their "race" as primarily cultural, others, when asked about their race, offer standard U.S. race terms, saying that they are white, black, or Indian. Still others see themselves as Latinos, Hispanics, or members of a particular national-origin group *and* as belonging to a particular race group. For example, they may identify themselves as Afro-Latinos or white Hispanics. In some cases, these identities vary according to context, but in others they do not.

I have therefore come to see that the concept of "race" can be constructed in several ways and that the Latino experience in the United States provides many illustrations of this. My personal experiences have suggested to me that for many Latinos, "racial" classification is immediate, provisional, contextually dependent, and sometimes contested. But because these experiences apply to many non-Latinos as well, it is evident to me that the Latino construction of race and the racial reading of Latinos are not isolated phenomena. Rather, the government's recent deliberations on racial and ethnic classification standards reflect the experiences and complexities of many groups and individuals who are similarly involved in issues pertaining to how they see themselves and one another.

Throughout my life, I have considered racism to be evil, and I oppose it with every fiber of my being. I study race to understand its influence on the lives of individuals and nations because I hope that honest, open, and well-meaning discussions of race and ethnicity and their social dynamics can help us appreciate diversity and value all people, not for their appearance, but for their character.

REFERENCES

Lugones, María. 1994. "Purity, Impurity, and Separation." *Signs: Journal of Women and Culture in Society.* 19(2) Winter, 458–479.

DISCUSSION QUESTIONS

1. Why do you think that some Latinos/Latinas identify themselves as Black and some as White?

2. Would you describe Latinos/Latinas as an ethnic or a racial group? Why? Do you think Latinos are in the process of becoming "racialized?" If so, why? If not, why not?

4

How Does It Feel
to Be a Problem?

BY MOUSTAFA BAYOUMI

(NEW YORK CITY, SEPTEMBER 25, 2001)

Thankfully, I was spared any personal loss. Like so many others in the city which I love I have spent much of the past two weeks reeling from the devastation. Mostly this has meant getting back in touch with friends, frantically calling them on the phone, rushing around the city to meet with them to give them a consoling hug, but knowing that really it was me looking for the hug. I dash off simple one-line emails, "let me know you're okay, okay?"

Old friends from around the world responded immediately. An email from Canada asks simply if I am all right. Another arrives from friends in Germany telling me how they remember, during their last visit to see me, the view from the top of the towers. A cousin in Egypt states in awkward English, "I hope this attack will not affect you. We hear that some of the Americans attack Arabs and Muslims. I will feel happy if you be in contact with me."

I am all right, of course, but I am devastated. In the first days, I scoured the lists of the dead and missing hoping not to find any recognizable names, but I come across the name of a three-year-old child, and my heart collapses. I hear my neighbor, who works downtown, arrive home, and I knock on her door. She tells me how she was chased by a cloud of debris into a building, locked in there for over an hour, and then, like thousands of others, walked home. I can picture her with the masses in the streets, trudging bewildered like

From *Amerasia Journal* 27:3 (2001)/28:1 (2002): 69–77. Reprinted by permission.

refugees, covered in concrete and human dust. Later, I ride the subway and see a full-page picture of the towers on fire with tiny figures in the frame silently diving to their deaths, and I start to cry.

In the following days, I cried a lot. Then, with friends, I attended a somber peace march in Brooklyn, sponsored by the Arab community. Thousands, overwhelmingly non-Arab and non-Muslim, show up, and I feel buoyed by the support. A reporter from Chile notices my Arab appearance and asks if she can interview me. I talk to her but am inwardly frightened by her locating me so easily among the thousands. Many people are wearing stickers reading: "We Support Our Arab neighbors," which leaves me both happy and, strangely, crushed. Has it really come to this? Now it has become not just a question of whether we—New Yorkers —are so vulnerable as a city but whether we—in the Arab and Muslim communities—are so vulnerable by our appearances. Is our existence so precarious here? I want to show solidarity with the people wearing the stickers, so how can I possibly explain to them how those stickers scare me?

Before September 11, I used to be fond of saying that the relations between the Muslim world and the West have never been at a lower point since the crusades. They have now sunk lower. The English language lexicon is, once again, degraded by war. President George W. Bush's ignorant use of the word "crusade" is but a manifestation. Why don't we ask the Apache what they think of the Apache helicopter? Is there any phrase more disingenuous in the English language than "collateral damage"? . . .

For the first four weeks after the attacks, I felt a bubble of hope in the dank air of New York. The blunt smell of smoke and death that hung in the atmosphere slowed the city down like I had never experienced it before. New York was solemn, lugubrious, and, for once, without a quick comeback. For a moment, it felt that the trauma of suffering—not the exercise of reason, not the belief in any God, not the universal consumption of a fizzy drink, but the simple and tragic reality that it hurts when we feel pain—was understood as the thread that connects all of humanity. From this point, I had hope that a lesson was being learned, that inflicting more misery cannot alleviate the ache of collective pain.

When the bombing began, the bubble burst. Where there was apprehension, now there was relief in the air. It felt like the city was taking a collective sigh, saying to itself that finally, with the bombing, we can get back to our own lives again. With a perverted logic, dropping munitions meant all's right with the world again.

Television, the great mediator, allows the public to feel violence or to abstract it. New Yorkers qualify as human interest. Afghans if they are lucky, get the long shot. In late October, CNN issued a directive to its reporters, for it seems that even a little bit of detached compassion is too much in the media world. "It seems perverse to focus too much on the casualties or hardship of Afghanistan," their leadership explains. "We must talk about how the Taliban is using civilian shields and how the Taliban has harbored the terrorists responsible for killing close to 5,000 innocent people." God forbid, we viewers see the pain ordinary Afghans are forced to endure. CNN must instead issue policy like a nervous state, rather than investigate how cluster bombs, freely dropped in the tens of thousands from the skies, metamorphose into land mines since about 7 percent of these soda-can-sized bombs don't explode on contact. In Canada, in the U.K., across the Arab world, this is becoming an issue. But in the United States, a cluster bomb sounds like a new kind of candy bar.

This is not to say that people in the United States are foolish, but they are by and large woefully underinformed. A study taken during the course of the Gulf War revealed that the more TV one watched, the less one actually knew about the region. In the crash course on Islam that the American public is now receiving, I actually heard a group of well-suited pundits on MSNBC (or its equivalent, I can no longer separate the lame from the loony) ask questions of a Muslim about the basics of the faith, questions like "Now, is there a difference between Moslem and Muslim?" No lie. The USA-Patriot Act, the end of the world as we know it (or at least of judicial review), actually includes the expression "Muslim descent," as if Islam is a chromosome to be marked by the human genome project. About Islam, most people in the United States still know nothing, unlike professional sports, where many are encyclopedias. . . .

For years, the organized Arab American community has been lobbying to be recognized with minority status. The check boxes on application forms have always stared defiantly out at me. Go ahead, try to find yourself, they seem to be taunting me. I search and find that, in the eyes of the government, I am a white man.

It is a strange thing, to be brown in reality and white in bureaucracy. Now, however, it is stranger than ever. Since 1909, when the government began questioning whether Syrian immigrants were of "white" stock (desirable) or Asian stock (excludable), Arab immigrants in this country have had to contend with fitting their mixed hues into the primary colors of the state. As subjects of the Ottoman empire, and thus somehow comingled with Turkish stock (who themselves claim descent from the Caucuses, birthplace of the original white people in nineteenth century thinking, even though the location is Asia Minor), Arabs, Armenians, and other Western Asians caused a good deal of consternation among the legislators of race in this country. Syrians and Palestinians were in 1899 classified as white, but by 1910 they were reclassified as "Asiatics." A South Carolina judge in 1914 wrote that even though Syrians may be white people, they were not "that particular free white person to whom the Act of Congress had donated the privilege of citizenship" (that being reserved exclusively for people of European descent). What is it Du Bois wrote: "How does it feel to be a problem?"

In the twenty-first century, we are back to being white on paper and brown in reality. After the attacks of September 11, the flood to classify Arabs in this country was drowning our community like a break in a dam. This impact? Hundreds of hate crimes, many directed at South Asians and Iranians, whom the perpetrators misidentified as Arab (or, more confusingly, as "Muslim": again as if that were a racial category). In the days following the attack (September 14/15), a Gallup Poll revealed that 49 percent of Americans supported "requiring Arabs, including those who are U.S. citizens, to carry a special ID." Fifty-eight percent also supported "requiring Arabs, including those who are U.S. citizens, to undergo special, more intensive security checks before boarding airplanes in the U.S." Debate rages across the nation as to the legitimacy of using "racial profiling" in these times (overwhelmingly pro). The irony, delicious if it were not so tragic, is that they are racially profiling a people whom they don't even recognize as a race. . . .

From the fall to the fallout, I have been living these days in some kind of limbo. The horrific attacks of September 11 have damaged everyone's sense of security, a principle enshrined in the Universal Declaration of Human Rights, and I wonder if for the first [time] that I can remember in the United States, we can start to reflect on that notion more carefully. All the innocents who have perished in this horrendous crime deserve to be mourned, whether they be the rescue workers, the financiers, the tourists, or the service employees in the buildings. An imam in the city has told me how a local union requested his services for a September 11 memorial of their loss since a quarter of their membership was Muslim. Foreign nationals from over eighty countries lost their lives, and the spectacular nature of the attacks meant that the world could witness the United States' own sense of security crumble with the towers. The tragedy of September 11 is truly of heartbreaking proportions. The question remains whether the United States will understand its feelings of stolen security as an unique circumstance, woven into the familiar narrative of American exceptionalism, or whether the people of this country will begin to see how security of person must be guaranteed for all. Aren't we all in this together? . . .

Overheard on a city bus, days after the attacks: "They will take them, like they did the Japanese, into camps. I think that's what they're going to do." In *Korematsu* v. *United States*, the infamous Supreme Court decision on Japanese internment, Justice Black wrote:

> It is said that we are dealing here with the case of imprisonment of a citizen in a concentration camp solely because of his ancestry, without evidence or inquiry concerning his loyalty and good disposition towards the United States. Our task would be simple, our duty clear, were this a case involving the imprisonment of a loyal citizen in a concentration camp because of racial prejudice. Regardless of the true nature of the assembly and relocation centers—and we deem it unjustifiable to call them concentration camps with all the ugly connotations that term implies—we are dealing specifically with nothing but an exclusion order.

In this exercise in rationalizing racism, Justice Black's backward logic is underscored by *his* taking offense at the term "concentration camp."

Of course, the Japanese American experience is not far from everyone's mind these days. Two months after the attack on Pearl Harbor, President Roosevelt signed Executive Order 9066, which led to the internment of over 110,000 Americans of Japanese descent. Without any need for evidence, anyone of Japanese ancestry, whether American-born or not, could be rounded up and placed in detention, all of course in the name of democracy and national security. What is less well-known is that in the 1980s, a multi-agency task force of the government, headed by the INS, had plans to round up citizens of seven Arab countries and Iran and place them in a camp in Oakdale, Louisiana, in the event of a war or action in the Middle East.

Will we see the return of the camps? I doubt it. How do you round up some seven or eight million people, geographically and economically dispersed throughout the society in ways people of Japanese descent were not in the 1940s? I suspect that this time we are not in for such measures, but we are already in the middle of something else. Over 1,000 people, most non-citizens, most Arab and Muslim, have been taken into custody under shadowy circumstances reminiscent of the *disappeared* of Argentina. Targeted for their looks, their opinions, or their associations, not one has yet to be indicted on any charge directly related to the attacks of September 11. Now, being Muslim means you are worthy of incarceration. INS administrative courts are the places where much of this happens, since non-citizens are the weakest segment of the population from a judicial point of view. Islam in this scenario becomes both racial and ideological.

In 1920, Attorney General A. Mitchell Palmer launched a nationwide assault on suspected communists and rounded up thousands without any judicial review (this event, an egregious abuse of authority, launched the ACLU). It too was directed mainly at immigrants to this country, and was covert and indiscriminate. This is what Palmer had to say about it: "How the Department of Justice discovered upwards of 60,000 of these organized agitators of the Trotsky doctrine in the United States, is the confidential information upon which the government is now sweeping the nation clean of such alien filth." John Ashcroft may be more circumspect in his language, but what we are facing now is a combination of both Yellow Peril and the Red Scare. Call it the Green Scare if you will, and recognize it is as a perilous path.

DISCUSSION QUESTIONS

1. How did the events of September 11, 2001, now called simply 9/11, redefine Arab Americans as a minority group?

2. How does their experience compare to that of Japanese Americans during World War II? How is it different?

5

The Importance of Collecting Data and Doing Social Scientific Research on Race

BY THE AMERICAN SOCIOLOGICAL ASSOCIATION

RACIAL CLASSIFICATIONS AS THE BASIS FOR SCIENTIFIC INQUIRY[1]

Race is a complex, sensitive, and controversial topic in scientific discourse and in public policy. Views on race and the racial classification system used to measure it have become polarized.[2] In popular discourse, racial groups are viewed as physically distinguishable populations that share a common geographically

From American Sociological Association, "The Importance of Collecting Data and Doing Social Scientific Research on Race, 2003." Reprinted by permission of the American Sociological Association.

[1]Editor's note: The complete reference list is available in the online version. See http://www.asanet.org/governance/racestmt.html

[2]The federal government defines race categories for statistical policy purposes, program administrative reporting, and civil rights compliance, and sets forth minimum categories for the collection and reporting of data on race. The current standards, adopted in October 1997, include five race categories: American Indian or Alaska Native; Asian; Black or African American; Native Hawaiian or Other Pacific Islander; and White. Respondents to federal data collection activities must be offered the option of selecting one or more racial designations. Hispanics or Latinos, whom current standards define as an ethnic group, can be of any race. However, before the government promulgated standard race categories in 1977, some U.S. censuses designated Hispanic groups as race categories (e.g., the 1930 census listed Mexicans as a separate race).

based ancestry. "Race" shapes the way that some people relate to each other, based on their belief that it reflects physical, intellectual, moral, or spiritual superiority or inferiority. However, biological research now suggests that the substantial overlap among any and all biological categories of race undermines the utility of the concept for scientific work in this field.

How, then, can it be the subject of valid scientific investigation at the social level? The answer is that social and economic life is organized, in part, around race as a social construct. When a concept is central to societal organization, examining how, when, and why people in that society use the concept is vital to understanding the organization and consequences of social relationships.

Sociological analysis of the family provides an analogue. We know that families take many forms; for example, they can be nuclear or extended, patrilineal or matrilineal. Some family categories correspond to biological categories; other do not. Moreover, boundaries of family membership vary, depending on a range of individual and institutional factors. Yet regardless of whether families correspond to biological definitions, social scientists study families and use membership in family categories in their study of other phenomena, such as well-being. Similarly, racial statuses, although not representing biological differences, are of sociological interest in their form, their changes, and their consequences.

THE SOCIAL CONCEPT OF RACE

Individuals and social institutions evaluate, rank, and ascribe behaviors to individuals on the basis of their presumed race. The concept of race in the United States—and the inevitable corresponding taxonomic system to categorize people by race—has changed, as economic, political, and historical contexts have changed. Sociologists are interested in explaining how and why social definitions of race persist and change. They also seek to explain the nature of power relationships between and among racial groups, and to understand more fully the nature of belief systems about race—the dimensions of how people use the concept and apply it in different circumstances.

SOCIAL REALITY AND RACIAL CLASSIFICATION

The way we define racial groups that comprise "the American mosaic" has also changed, most recently as immigrants from Asia, Latin America, and the Caribbean have entered the country in large numbers. One response to these demographic shifts has been the effort (sometimes contentious) to modify or add categories to the government's official statistical policy on race and ethnicity, which governs data collection in the census, other federal surveys, and administrative functions. Historically, changes in racial categories used for administrative purposes and self-identification have occurred within the context of a polarized biracialism of Black and White; other immigrants to the

United States, including those from Asia, Latin America, and the Caribbean, have been "racialized" or ranked in between these two categories.

Although racial categories are legitimate subjects of empirical sociological investigation, it is important to recognize the danger of contributing to the popular conception of race as biological. Yet refusing to employ racial categories for administrative purposes and for social research does not eliminate their use in daily life, both by individuals and within social and economic institutions. In France, information on race is seldom collected officially, but evidence of systematic racial discrimination remains. The 1988 Eurobarometer revealed that, of the 12 European countries included in the study, France was second (after Belgium) in both anti-immigrant prejudice and racial prejudice. Brazil's experience also is illustrative: The nation's then-ruling military junta barred the collection of racial data in the 1970 census, asserting that race was not a meaningful concept for social measurement. The resulting information void, coupled with government censorship, diminished public discussion of racial issues, but it did not substantially reduce racial inequalities. When racial data were collected again in the 1980 census, they revealed lower socio-economic status for those with darker skin.

THE CONSEQUENCES OF RACE AND RACE RELATIONS IN SOCIAL INSTITUTIONS

Although race is a social construct (in other words, a social invention that changes as political, economic, and historical contexts change), it has real consequences across a wide range of social and economic institutions. Those who favor ignoring race as an explicit administrative matter, in the hope that it will cease to exist as a social concept, ignore the weight of a vast body of sociological research that shows that racial hierarchies are embedded in the routine practices of social groups and institutions.

Primary areas of sociological investigation include the consequences of racial classification as:

- A sorting mechanism for mating, marriage and adoption.
- A stratifying practice for providing or denying access to resources.
- An organizing device for mobilization to maintain or challenge systems of racial stratification.
- A basis for scientifically investigating proximate causes.

Race as a Sorting Mechanism for Mating, Marriage, and Adoption

Historically, race has been a primary sorting mechanism for marriage (as well as friendship and dating). Until anti-miscegenation laws were outlawed in the United States in 1967, many states prohibited interracial marriage. Since then, intermarriage rates have more than doubled to 2.2 percent of all marriages, according to the latest census information. When Whites (the largest racial group in the United States) intermarry, they are most likely to marry Native Americans/American Indians and least likely to marry African Americans. Projections to the year 2010 suggest that intermarriage and, consequently, the universe of people identifying with two or more races is likely to increase, although most marriages still occur within socially designated racial groupings.

Race as a Stratifying Practice

Race serves as a basis for the distribution of social privileges and resources. Among the many arenas in which this occurs is education. On the one hand, education can be a mechanism for reducing differences across members of racial categories. On the other hand, through "tracking" and segregation, the primary and secondary educational system has played a major role in reproducing race and class inequalities. Tracking socializes and prepares students for different education and career paths. School districts continue to stratify by race and class through two-track systems (general and college prep/advanced) or systems in which all students take the same courses, but at different levels of ability. African Americans, Hispanics, American Indians, and students from low socioeconomic backgrounds, regardless of ability levels, are over-represented in lower level classes and in schools with fewer Advanced Placement classes, materials, and instructional resources.

Race as an Organizing Device for Mobilization to Maintain or Challenge Systems of Racial Stratification

Understanding how social movements develop in racially stratified societies requires scholarship on the use of race in strategies of mobilization. Racial stratification has clear beneficiaries and clear victims, and both have organized on racial terms to challenge or preserve systems of racial stratification. For example, the apartheid regime in South Africa used race to maintain supremacy and privilege for Whites in nearly all aspects of economic and political life for much of the 20th century. Blacks and others seeking to overthrow the system often were able to mobilize opposition by appealing to its victims, the Black population. The American civil rights movement was similarly successful in mobilizing resistance to segregation, but it also provoked some White citizens into organizing their own power base (for example, by forming White Citizen's Councils) to maintain power and privilege.

Race and Ethnicity as a Basis for the Scientific Investigation of Proximate Causes and Critical Interactions

Data on race often serve as an investigative key to discovering the fundamental causes of racially different outcomes and the "vicious cycle" of factors affecting these outcomes. Moreover, because race routinely interacts with other primary categories of social life, such as gender and social class, continued examination of these bases of fundamental social interaction and social cleavage is required. In the health arena, hypertension levels are much higher for African Americans than other groups. Sociological investigation suggests that discrimination and unequal allocation of society's resources might expose members of this racial group to higher levels of stress, a proximate cause of hypertension. Similarly, rates of prostate cancer are much higher for some groups of men than others. Likewise, breast cancer is higher for some groups of women than others. While the proximate causes may appear to be biological, research shows that environmental and socio-economic factors disproportionately place at greater risk members of socially subordinated racial and ethnic groups. For example, African Americans' and Hispanics' concentration in polluted and dangerous neighborhoods result in feelings of depression and powerlessness that, in turn, diminish the ability to improve these neighborhoods. Systematic investigation is necessary to uncover and distinguish what social forces, including race, contribute to disparate outcomes.

RESEARCH HIGHLIGHTS: RACE AND ETHNICITY AS FACTORS IN SOCIAL INSTITUTIONS

The following examples highlight significant research findings that illustrate the persistent role of race in primary social institutions in the United States, including the job market, neighborhoods, and the health care system. This scientific investigation would not have been possible without data on race.

Job Market

Sociological research shows that race is substantially related to workplace recruitment, hiring, firing, and promotions. Ostensibly neutral practices can advantage some racial groups and adversely affect others. For example, the majority of workers obtain their jobs through informal networks rather than through open recruitment and hiring practices. Business-as-usual recruitment and hiring practices include recruiting at predominantly White schools, advertising only in suburban newspapers, and employing relatives and friends of current workers. Young, White job seekers benefit from family connections, studies show. In contrast, a recent study revealed that word-of-mouth recruitment through family and friendship networks limited job opportunities for African Americans in the construction trades. Government downsizing provides another example of a "race neutral" practice with racially disparate consequences: Research shows that because African Americans have successfully established employment niches in the civil service, government workforce reductions displace disproportionate numbers of African American—and, increasingly, Hispanic—employees. These and other social processes, such as conscious and unconscious prejudices of those with power in the workplace, affecting the labor market largely explain the persistent two-to-one ratio of Black to White unemployment.

Neighborhood Segregation

For all of its racial diversity, the highly segregated residential racial composition is a defining characteristic of American cities and suburbs. Whites and African Americans tend to live in substantially homogenous communities, as do many Asians and Hispanics. The segregation rates of Blacks have declined slightly, while the rates of Asians and Hispanics have increased. Sociological research shows that the "hyper-segregation" between Blacks and Whites, for example, is a consequence of both public and private policies, as well as individual attitudes and group practices.

Sociological research has been key to understanding the interaction between these policies, attitudes, and practices. For example, according to attitude surveys, by the 1990s, a majority of Whites were willing to live next door to African Americans, but their comfort level fell as the proportion of African Americans in the neighborhood increased. Real estate and mortgage-industry practices also contribute to neighborhood segregation, as well as racially disparate homeownership rates (which, in turn, contribute to the enormous wealth gap between racial groups). Despite fair housing laws, audit studies show, industry practices continue to steer African American homebuyers away from White neighborhoods, deny African Americans information about available loans, and offer inferior property insurance.

Segregation profoundly affects quality of life. African American neighborhoods (even relatively affluent ones) are less likely than White neighborhoods to have high quality services, schools, transportation, medical care, a mix of retail establishments, and other amenities. Low capital investment, relative lack of political influence, and limited social networks contribute to these disparities.

Health

Research clearly documents significant, persistent differences in life expectancy, mortality, incidence of disease, and causes of death between racial groups. For example, African Americans have higher death rates than Whites for eight of the ten leading causes of death. While Asian–Pacific Islander babies have the lowest mortality rates of all broad racial categories, infant mortality for Native Hawaiians is nearly three times higher than for Japanese Americans. Genetics accounts for some health differences, but social and economic factors, uneven treatment, public health policy, and health and coping behaviors play a large role in these unequal health outcomes.

Socio-economic circumstances are the strongest predictors of both life span and freedom from disease and disability. Unequal life expectancy and mortality reflect racial disparities in income and incidence of poverty, education and, to some degree, marital status. Many studies have found that these characteristics and related environmental factors such as over-crowded housing, inaccessibility of medical care, poor sanitation, and pollution adversely impact life expectancy and both overall and cause-specific mortality for groups that have disproportionately high death rates.

Race differences in health insurance coverage largely reflect differences in key socio-economic characteristics. Hispanics are least likely to be employed in jobs that provide health insurance and relatively fewer Asian Americans are insured because they are more likely to be in small low-profit businesses that make it hard to pay for health insurance. Access to affordable medical care also affects health outcomes. Sociological research shows that highly segregated African American neighborhoods are less likely to have health care facilities such as hospitals and clinics, and have the highest ratio of patients to physicians. In addition, public policies such as privatization of medicine and lower Medicaid and Medicare funding have had unintended racial consequences; studies show a further reduction of medical services in African American neighborhoods as a result of these actions.

Even when health care services are available, members of different racial groups often do not receive comparable treatment. For example, African Americans are less likely to receive the most commonly performed diagnostic procedures, such as cardiovascular and orthopedic procedures. Institutional discrimination, including racial stereotyping by medical professionals, and systemic barriers, such as language difficulties for newer immigrants (the majority of whom are from Asia and Latin America), partly explain differential treatment patterns, stalling health improvements for some racial groups.

All of these factors interact to produce poorer health outcomes, indicating that racial stratification remains an important explanation for health disparities.

SUMMARY: THE IMPORTANCE OF SOCIOLOGICAL RESEARCH ON RACE

A central focus of sociological research is systematic attention to the causes and consequences of social inequalities. As long as Americans routinely sort each other into racial categories and act on the basis of those attributions, research on the role of race and race relations in the United States falls squarely within this scientific agenda. Racial profiling in law enforcement activities, "redlining" of predominantly minority neighborhoods in the mortgage and insurance industries, differential medical treatment, and tracking in schools, exemplify social practices that should be studied. Studying race as a social phenomenon makes for better science and more informed policy debate. As the United States becomes more diverse, the need for public agencies to continue to collect data on racial categories will become even more important. Sociologists are well qualified to study the impact of "race"—and all the ramifications of racial categorization—on people's lives and social institutions. The continuation of the collection and scholarly analysis of data serves both science and the public interest. For all of these reasons, the American Sociological Association supports collecting data and doing research on race.

DISCUSSION QUESTIONS

1. Why does the American Sociological Association think it is important to maintain race as a way of classifying humans?

2. What is some of the evidence of the significance of race in different social institutions?

InfoTrac College Edition:
Bonus Reading

http://infotrac.thomsonlearning.com

You can use the InfoTrac College Edition feature to find an additional reading pertinent to this section. You can also search on the author's name or keywords to locate the following reading:

Suarez-Balcazar, Yolanda, Lucia Orellana-Damacela, Nelson Portillo, Jean M. Rowan, and Chelsea Andrews-Guillen. 2003. "Experiences of Differential Treatment among College Students of Color." *Journal of Higher Education* 74 (July–August): 428–445.

In this article, the researchers examine how race matters in the experiences of college students who belong to racial-ethnic minority groups. What does their research find, and how does it illustrate the persistent reality of race in the United States? Were you to do a similar study on your campus, what would you expect to find? What would White students say? What would students of color say?

STUDENT EXERCISES

1. Observe your community and make note of its racial and ethnic makeup. What different groups do you find? Given the population of the United States, who is missing in the community? To find out, go to the U.S. Census Bureau home page (www.census.gov) and, using the American Fact Finder feature, select "Fact Sheet." Then fill in your own state and county and click on "Go." What does the resulting information reveal about the distribution of racial-ethnic groups in your area? How accurate were your own observations? Are these groups evenly distributed throughout the county or are they concentrated in particular areas?

2. For one week, keep a written log of every time the subject of race comes up in the conversations you hear around you. Make note of what people said and the tone in which they said it. You should also note, if possible, the age and race of the person making the comments. At the end of the week, review your log and answer the following questions:

    a. What evidence of racial attitudes did you find?

    b. How would you compare the attitudes of those belonging to different racial-ethnic groups?

    c. What do your observations reveal to you about the everyday reality of racism?

PART II

The Social Construction
of Race and Ethnicity

What does it mean to say that race is a social construction? The idea of race as a social construction is one of the most important things to learn in analyzing race in society. Most people tend to think of race in terms of skin color or other biological features that seemingly distinguish different groups, but is this a reasonable understanding of this concept? Race has been used historically to differentiate groups, but why not use eye color or height or some characteristic other than color to categorize people into so-called races? The answer lies in the historical treatment of different groups and the significance that society gives to features believed to mark different racial groups. Thus, to say that race is socially constructed is to say that what is important about race is not biological difference, but the different ways groups have been treated in society.

THE COMPLEXITY OF COMPLEXION

To begin with, race is not a matter of biology, although racist thinking claims this is so. Scientists in the Human Genome Project have concluded that there is no "race" gene. This project, which allows scientists to map people's genetic makeup, shows that there are far more similarities among people than there are differences, even when you take into account such things as eye color, skin color, and other physical characteristics. Scientists have concluded that genetic variation among human beings is indeed very small, and thus, that, as one of the biologists working in human genetics writes, "There is no biological basis for the separation of human beings into races" (Graves 2001: 1). What then is race?

The idea of race has developed within the context of the social institutions and practices in which groups socially defined as races have been exploited, controlled, and—in some cases—enslaved. Imagine this scenario: You have a social system in which some people are forced into slavery based on their presumed inferiority. The dominant group then has to create a belief system that supports this exploitation. Their solution is creating "race" as an idea they can use to justify the exploitation of groups perceived as different.

Sociologists define **race** as a group that is treated as distinct in society because of certain perceived characteristics that have been defined as signifying superiority or inferiority. Note that in this definition, perception, belief, and social treatment are the key elements defining race, not the actual characteristics of human groups. Furthermore, so-called "races" are created within a system of social dominance. *Race is thus a social construction, not an attribute of certain groups of people.*

Howard Taylor ("Defining Race") explores the complexity of defining race based on the multiple ways that race can be conceptualized. He helps us see that race is not a fixed personal attribute. Rather, it is fundamentally rooted in social definitions: how people see each other, how they define their own identity, and how they are situated within a social order—an order that has been structured along lines of inequality.

Abby Ferber ("Planting the Seed: The Invention of Race") takes us back to the establishment of race as a concept—a relatively recent development in modern history. She explains how race emerged through the work of quasi-scientists in the eighteenth century, as White Europeans sought to explain and rationalize the exploitation of African people. She shows how the process of developing systems of racial classification is "intertwined with the history of racism." This explanation is important for understanding how racist thinking has emerged and how it is tied to the exploitation of people of color worldwide. Her work also shows how racism arose co-terminously with the rise of science, and she will make you think twice about using the term Caucasian, commonly used to refer to White people, once you learn the term's racist origins.

Karen Brodkin ("How Did Jews Become White Folks?") illustrates the social construction of race in a different context. She writes about the long history of **anti-Semitism** (defined as the hatred of Jewish people) and tells us how at different times in world history anti-Semitic thinking constructed Jews, not as a religious ethnic group, but as a separate race. Thus, the Nazi regime of Germany in the 1930s and 1940s, constructed Jewish people as an inferior race—and systematically murdered them as a result. Brodkin's essay links anti-Semitism to other forms of racism in that anti-Semitism was supported by *scientific racism* (the use of quasi-science to support racism). As Brodkin shows, even in the United States, racism defined the only "real Americans" as native-born White people. American thinking, buttressed by exclusionary laws, defined Asian immigrants and many southern Europeans as outcasts, stripped them of rights and excluded them from many social institutions. On the other hand, as Brodkin shows, Jewish Americans have been upwardly mobile. Why? Because over time they became seen as "White" and conditions in the United States after World War II opened up opportunities to those defined as "White."

Brodkin's essay also brings out another important point: that ethnic groups can become racialized. **Ethnic groups** are those that share a common culture and that have a shared identity. Ethnicity can thus stem from religion, national origin, or other shared characteristics. Important to this definition of ethnic group is not just the shared culture, but the sense of group belonging. Thus, Jewish people are an ethnic group, as are Italian Americans, Cajuns, and Irish American Catholics. Ethnicity can exist even within a so-called racial group (such as Jamaicans, Haitians, or Cape Verdeans among African Americans). Some ethnic groups have historically been "racialized," as Brodkin's essay about Jews shows. This has typically occurred during periods of high rates of immigration when anti-immigrant sentiment swelled and groups like the Irish were defined by dominant groups as racially inferior. The fact that we do not now consider Jews or the Irish to be a "race" shows how powerful the social construction of race can be.

That concepts of race shift and change does not mean that race is not real. Race is real—but real in a social and historical sense. By recognizing the socially constructed character of race, you are not denying that it exists but are recognizing its social reality. As Joe Feagin argues in "Racist America," race is systemic. The term **systemic racism** refers to the "recurring and unequal relations between groups and individuals" (2000: 19) that are organized along lines of race. As Feagin argues, racism is not just *in*, but is *of* society. Race and racism, he shows, have been built into the institutions of society right from the founding of the nation.

As the result of systemic racism, dominant groups (namely, White elites) accumulate resources over time, providing them with what Feagin identifies as *unjust enrichment*. At the same time, those disadvantaged by a racist system experience *unjust impoverishment.* Within such a system, individuals—regardless of their own attitudes and beliefs—can benefit or be deprived of resources because of the systemic way that racism works. Thus, even the well-intentioned anti-racist White person can unintentionally benefit from what has been four centuries of racial hierarchy because racism, as we will see further in the following Part, extends beyond individual good and bad will.

Michael Omi and Howard Winant ("Racial Formation") introduce the concept of **racial formation** to refer to the social and historical process by which racial categories are created. Historically in the United States, one's racial membership was determined by law—though, interestingly enough, the meaning of race varied from state to state. In Louisiana, for example, you were defined as Black if you had one Black ancestor out of thirty-two; in Virginia, it was one in sixteen; in Alabama, you were Black by law if you had *any* Black ancestry. By just crossing state borders, a person would legally change his or her race!

Racial formation means that social structures, not biology, define race. Because a group's perceived racial membership has been the basis for group oppression, this understanding of race shows how systems of authority and governance construct concepts of race. There are, of course, sometimes observable differences between individuals, but it is what these differences have come to mean in history and society that matters.

In a society where certain groups are exploited—robbed of their lands, their labor appropriated for the benefit of others, their communities starved of resources—it does matter whether you are defined as Black, White, or some other so-called race. Historically, White, elite males determined the only groups who were counted as "White"—and, therefore, given full rights of citizenship. Who counts as a White person is a process of law and social judgments. Consider the case of Takao Ozawa, originally from Japan. Ozawa arrived in California in 1894, graduated from high school, and attended the University of California, Berkeley. After that, he worked for an American company, living with his family in the U.S. territory of Honolulu. Because he was not legally classified as White, he could not become a U.S. citizen, though he very much wanted and filed for citizenship in 1914. Ozawa argued that he was a "true American," a person of good character who neither drank, smoked, or gambled. His family went to an American church; his children went to American schools; he spoke only English at home and raised his children as Americans. In 1922 the U.S. Supreme Court denied his eligibility for citizenship based on the claim that he was not "Caucasian" (Takaki 1989; *Ozawa v. United States* 260 U.S. 178, 1922).

Why would our nation have such laws? Such laws emerge when social resources are distributed based on membership in groups being judged as superior or inferior, "alien," or "colored." During the history of racial formation in America many groups have been denied the rights of freedom and full citizenship, as many of the articles in this book show.

The authors in this part explore the social construction of race in diverse ways. Michael Omi and Howard Winant ("Racial Formation") establish the framework of racial formation as a way of locating race in the social and historical processes that create racial categories. They also develop the concept of *racial projects* to explain organized efforts by any group to distribute social and economic resources along racial lines. The authors note that resources are both material—such as money, housing, schooling, and so forth—as well as resources of representation—that is, how groups are defined and depicted within a society and culture. Omi and Winant's perspective shows us that racism is not just a matter of individual attitudes and prejudices, although those surely are important, as we will see in the next part.

REFERENCES

Graves, Joseph L., Jr. 2001. *The Emperor's New Clothes: Biological Theories of Race at the Millennium.* New Brunswick, NJ: Rutgers University Press.

Takaki, Ronald. 1989. *Strangers from a Different Shore: A History of Asian Americans.* New York: Penguin.

6

Defining Race

BY HOWARD F. TAYLOR

R ace is with us every minute of every day. Despite our constant protesta-
tions against it and its realities, it, like gender, permeates every fiber of our
very existence. Race is causally and intimately related to hundreds of
forces and has hundreds of consequences. Our race determines, beyond chance,
how long we will live. In general, racial minorities in America have a lower
life expectancy than Whites. American minorities have less access to medical
care; as a consequence, minorities are more burdened with chronic and life-
threatening illnesses. American minorities, particularly Blacks, Hispanics, and
American Indians, have considerably lower annual incomes relative to Whites,
even Whites who have the same level of education as their minority counter-
part. Thus Blacks, Hispanics, American Indians, and also Asians, have greater
odds of being poor than Whites. Blacks, Hispanics and American Indians are
on average promoted less often in the workplace than are Whites with the
same or even less education. All three racial minorities routinely suffer housing
discrimination and are less likely to be offered low-interest housing mortgages
in an urban area than are Whites who live in the same urban area. Surveys find
that members of all three minority groups feel more alienated from the institutions of society than do Whites of the same socioeconomic status. Finally,
Blacks and Hispanics are more likely to be arrested for crime, more likely to be
held without bail, more likely to be sentenced, and more likely to receive longer

From Howard F. Taylor, "Defining Race." Reprinted by permission of the author.

sentences than are Whites from the same part of the city or town, of the same or similar social class, and who have exactly the same record of prior arrests.

So: race matters. It matters a lot. What, then, is "race?" What are the definitions of race that are used in American society? How can one tell who is of which race? If the definitions of race in American society are inexact, as they are, then can one define race in some way so as to be able to research its effects? How can we actually measure one's race?

Race is *multiply defined* in this society; that is, there is no one single definition, but several definitions. All these definitions apply simultaneously, and no one definition takes precedence over another. These definitions do however have one thing in common: They are all creations of society. They have been put into place by humans and by their social interactions and their societal institutions. Let us have a look at these definitions.

RACE AS A BIOLOGICAL CATEGORY

Most people grew up thinking that the real definition of "race" is that it is a strictly biological category. This is only partly correct. We were taught to place people into "race" categories such as Caucasian (White), Negroid (Black; Negro; African American), Mongolian (Asian), and so on, on the basis of secondary physical characteristics such as skin color, hair texture, lip form, nose form, and so on. This definition and classification, put forth in the late nineteenth and early twentieth centuries by some physical anthropologists and now considered vastly outdated, has had considerable influence on how people think about the matter of race right up to the present time. It gave rise to the idea that racial classification was "scientific" since it was based on physical traits and upon thinking at the time in the field of physical anthropology. Even some sociology texts as recently as the 1940s defined race in this manner, and furthermore even defined race as a "subspecies" of humankind (Young, 1942)!

It is now generally agreed upon by researchers in both the physical as well as the social and behavioral sciences that only a small part of the definition of race in this society is based on these secondary physical characteristics. Such characteristics do play a part in how people themselves define race, to be sure, but only a part. This is because the human variability between what are regarded in society as "racial groups" occurs mostly *within* racial groups rather than between them. Thus, the skin color of African Americans varies from extremely light/white (even blond and blue-eyed) all the way to very dark brown. There are White people who have dark skin and very curly hair and full lips yet they are still White, and are so regarded in their immediate communities. Similarly, there are African Americans with white skin, thin lips, and straight hair. So skin color (and lip form and hair texture) is not a very good indicator of race.

The population geneticist Richard Lewontin (1996) notes that even for strictly physical traits involving body chemistry, blood type, and other strictly physical traits, almost all the variability on such traits is *within* racial categories; almost no variability in such traits exists between races. Lewontin estimates that the overlap in physical traits between any two groups designated as "races" in American society is more than 99 percent. That applies as well to genetics: Any two racial groups compared are 99 percent similar genetically. Clearly, then, in physical characteristics, the races are far more similar than they are different. It is in this sense that one hears that race is "not a true scientific concept." In the sense of thousands of physical traits and characteristics, this is certainly true.

Does this mean that race is no longer important, that race no longer matters? Certainly not. We have already noted that race is a fundamental and firmly ingrained part of human existence, not only in America, but in most other societies as well. How then is it that a concept with virtually no "scientific" validity (at least, in the physical science sense) has come to be so important and intrusive in human existence? The answer lies in the realization that *the definition of race in America, and in most other societies as well, is largely social.*

RACE AS A SOCIAL CONSTRUCTION

To say that race is a social construction means that it and its definition grow out of the process of human interaction. This means that race is what interacting humans define it to be. In this sense, how you are perceived in a community of peers in part defines your race: If your friends and associates see you as African American, then in that one respect you are indeed African American.

As another example of social construction, if race-like divisions among people in this society were made only on the basis of who had red hair and who did not, then people would come to think of races as being defined by hair color—a physical characteristic. Over time, this definition would come to be upheld by society's institutions—by the courts, by the educational system, by the federal government, and so on. It might well come to be thought of as a truly "scientific" classification, since it is based on a physical characteristic (hair color instead of skin color), and as everyone knows, redheads are different from everyone else—they have fiery tempers and argue a lot, don't they? Thus, *racial stereotypes* would soon be applied to redheads!

Social construction means that people learn, through socialization and interaction processes, to *attribute* certain characteristics to people who are classified into a racial category. These are just what racial stereotypes are: attributions that are for the most part not true and yet stubbornly persist over time. These stereotypes are social constructions. They are generally based on only a small truth (if on any truth at all) and are then thought of in society as applying to all members, and to "typical" members, of some racial category.

Stereotypes are generally negative. We have all heard the common stereotypes: Blacks are inherently musical, possess "natural" rhythm, are loud, and crime-prone; Asians are sneaky and overly conforming; Hispanics are naturally violence-prone and carry knives; American Indians are quiet, subservient, and underachievers. These traits are seen by society as inherent to any member of the particular group. In fact they are seen as *essential* to the group identified by the stereotype. Sociologists call this process *essentialization*: Such negative stereotypical traits are regarded by society as essential (inherent) to the character of any person identified by the stereotype. Negative stereotypes, thus negative essentializations, are applied far more to minorities than to Whites, and they thus help define what it means to be minority in society, even though Whites are sometimes ridiculed as having a few negatively stereotyped traits ("White boys can't jump," "blondes are dumb," and so forth).

RACE AS AN ETHNIC GROUP

A group is an ethnic group if its members are united by a common culture. Culture includes language, religion, tools, music, habits, socialization practices, and many other elements. Racial groups are ethnic groups. They generally share a common culture—not perfectly, of course, but to an extent great enough so as to be able to identify some set of cultural elements held in common. Thus African Americans are not only regarded as a racial group but they are also an ethnic group. They possess many elements of a common culture (music, art, linguistic similarities, a sense of "we" feeling and identity, and so on). There are even multiple ethnic groups among African Americans such as Cape Verdeans, Haitian Americans, Gullah Islanders, and so forth. Jews are also an ethnic group—a group sharing a common or nearly common religion (religious culture), but Jews are not regarded as a race.

Sometimes an ethnic group in a society comes to be thought of in that society as a race. The best example of this is Hitler's definition of Jews as a race in Nazi Germany in the 1930s and early 1940s—the "Jewish race." This led to heavy negative stereotyping of Jews in Germany and other countries and eventually to the killing of millions of Jews by means of starvation and gas chambers during the Holocaust. What had been an ethnic group came to be defined in Nazi society as a "race," though we now know that Jewish people are not a race.

When any group—whether an ethnic group or a social class—comes to be thought of as a race and then is actually defined by society as a race, this means the group has become *racialized*. Thus Hitler racialized Jews. By so doing, it was easier to then stereotype Jews and to regard them as a separate and inferior category of people. It was the process of racialization that allowed Hitler to do this and to convince many German citizens that Jews were bad and needed to be totally eliminated. He almost succeeded. This underscores the point that races tend to be defined in a society by *social* processes, less so than by physical characteristics. This shows again that race is a social construction.

RACE AS A SOCIAL CLASS
OR PRESTIGE RANK

In parts of Brazil, such as the province of Bahia, the higher one's social class status and wealth, the more likely one is to be formally listed in the country's census as "White" (Surratt and Inciardi 1998; Patterson 1982). This is true even if the person in question is dark-skinned and clearly of largely African ancestry. Thus while skin color is certainly important to defining who is of what race ("color") in Brazil, wealth and social standing are just as important. A favorite expression in Brazil is: "Money whitens" (*o dinheiro embranquence*)! People with mixed physical characteristics may be labeled "White" if they appear well-dressed and occupy a prestigious professional occupation. Similarly, a poor, light-skinned person may be labeled "Black" to indicate low social standing.

The use of color as a label for people in Brazil is far more complex than is the case in the United States. The simple distinctions of "Black," "White," "Yellow," and "Red" used in the United States are seen in Brazil as utterly ridiculous. There, a large number of labels are used to identify different colors of individuals—reddish brown; reddish dark brown; reddish brown with a hint of yellow; and so on. A fairly recent survey in Brazil revealed 143 color labels used for the population, including gray, pink, dirty white, and cinnamon (Ellison 1995). These race-like labels for individuals are used in conjunction with their wealth and social standing. This shows that how your color is perceived and thus labeled within a society can be significantly determined by your social class or prestige rank within that society.

RACE AS A "RACIAL FORMATION"
OF SOCIETY'S INSTITUTIONS

What a race is, and who is of what race, can be defined by one or more of society's institutions, such as the government, the educational institution, the state and federal legal system, and the criminal justice system. An insightful theory in this regard is Omi and Winant's (1994) *racial formation theory*. This theory argues that over time, society's powerful institutions define what race is and who is to be classified within what racial category.

Such has been the case in the United States. During slavery, and after the Emancipation Proclamation of 1863, Black people were defined in the law as "Negro" if they had only a small fraction of Black ancestry ("Black blood"), even so small an amount as one great-great-great grandparent who was Black. This is the origin of the *one-drop rule*, namely, one was "Negro" if he or she had but one drop of Negro blood. The amount of ancestry ("blood") required varied somewhat from state to state, but it was a small amount. The negative stigma of Blackness needed to be only minuscule for one to be totally labeled.

(In some states, one Black great-great-great-great grandparent was sufficient to render one a "total" Negro!) The point is that by law it was the government, via the law, that determined how race was to be defined. The individual was not allowed to define her or his own race; it was up to the government, and the government carried the strong sanction of the law. This is what is meant by racial formation. The "one-drop rule" exists even to this day in the public mind.

A clear illustration of how the government determines race is how one is designated as American Indian (Native American Indian) in the United States. According to the U.S. government, one cannot simply decide to call oneself "Indian." Through what is called the federal acknowledgement process, the U.S. government declares one to be an American Indian, but one must go through an elaborate bureaucratic process, involving paper forms and legal documents, and then be certified as a member of some Indian tribe. The U.S. government maintains a list of Indian tribes that it considers to be "legitimate," and the individual must be able to prove membership in one of those pre-designated tribes. Moreover, the tribes themselves must be certified as legitimate—and experts estimate that about 569 American Indian groups have been so designated. This means that the government, through this type of racial formation, has more say in whether or not American Indian people are American Indians than do the people themselves! Obviously, not all Native American Indians agree with this procedure.

RACE AS SELF-DEFINED

Finally, a person's race may be defined in part simply by what that person calls herself or himself. When asked "What race are you?" (or even simply "What are you"?), what then does the person say? A person with one Black parent and one White parent may well say "I am Black" (an implicit use of the one-drop rule); or, they may say "I am White" (which may conflict with how they appear to others); or, they may say "I am bi-racial" or "I am of mixed race." Or, still further, the person may refuse to be classified by race, arguing that race is a false classification in the first place. (If such a person has light skin and as a result chooses to live as a White person—severing all ties with Black relatives and also the Black community—that person by tradition could be accused by the Black community of "passing," that is, passing for White. Designating someone as "passing" is less common now than it was a decade or two ago, but it is still done.) On some present-day college campuses, a light-skinned Black person who appears to be passing for White and who denies being Black is (only somewhat) jokingly called an "incogNegro" (or simply, "incog")—a pun on "incognito!"

A person is, in principle, free to designate himself or herself in any way they might wish. The problem is that all the other five definitions that we have given also, simultaneously, come into play. What if the person with one Black parent and one White parent has brown skin and *looks* to everyone else like a Black person, that is, the person is dark enough of skin in order to be called "Black" among her or his peers? This would show intervention of both the biological as well as the social constructionist definitions, and perhaps the racial formation definition as well. (Golf star Tiger Woods is of African American and Asian parentage and has racially designated himself—though somewhat jokingly—as Cablinasian. What would you call him?)

The point is that there is more to race than one's personal definition even as applied to one's own self. It may sound odd to quarrel with a person about what they choose to call themselves racially. But such is the reality of race in the United States—and most other countries as well: It is not totally up to the person alone!

We are reminded in this context of the story of a dark-skinned Black man in the southern U.S. who in the early 1950s boarded a bus and proceeded to sit in the front row. The laws and norms of segregation in much of the South [at that time] prevented Black people from sitting up front on a bus. Consequently, the bus driver turned to the man and said: "Boy, you can't sit there!" The man replied, "Well, why not?" "Because," said the bus driver, "you are a Negro, and Negroes cannot sit up here in the front of the bus!" "Oh, well in that case," said the man, "I can stay right here. I have resigned from the Negro race!"

One is left to contemplate what race would be like in this country if one could easily "resign" from a race!

CONCLUSION

"Race" in the United States is not defined by one single definition, but simultaneously by several definitions. Six definitions of race were explored here. One's race is defined by a combination of the following: by one's physical appearance, such as by skin color (the biological definition); by social construction (any definition arising out of the process of human interaction, such as how those around you define you); as an ethnic group (Hitler defined the Jews as a race); as a social class rank (as in Brazil); as racial formation (such as the U.S. government's definition of who is American Indian); and finally, by one's own self-definition. No one definition is dominant over another in U.S. society. Each definition shows the significance of society in defining race.

REFERENCES

Ellison, K. 1995. "Brazil's Blacks, Building Pride, Invoke the Legend of Rebel Warrior Zumbi." *Miami Herald* (November 19): 1A.

Lewontin, Richard. 1966. *Human Diversity*. New York: W. H. Freeman.

Omi, Michael, and Howard Winant. 1994. *Racial Formation in the United States*, 2nd ed. New York: Routledge.

Patterson, Orlando. 1982. *Slavery and Social Death: A Comparative Study*. Cambridge: Harvard University Press.

Surratt, Hilary L., and James A. Inciardi. 1998. "Unraveling the Concept of Race in Brazil: Issues for the Rio de Janeiro Cooperative Agreement Site." *Journal of Psychoactive Drugs* 30 (July– September): 255–260.

Young, Kimball. 1942. *Sociology: A Study of Society and Culture*. New York: American Book Co.

DISCUSSION QUESTIONS

1. What are the different ways race can be defined in society? How does this challenge the understanding of race as simply a fixed, biological category?

2. How do social beliefs shape our understanding of race?

3. How does Taylor's essay indicate that race matters, even if it is a social construction?

7

Planting the Seed

The Invention of Race

BY ABBY L. FERBER

My students are always surprised to learn that race is a relatively recent invention. In their minds, race and racial antagonisms have taken on a universal character; they have always existed, and probably always will, in some form or another. Yet this fatalism belies the reality—that race is indeed a modern concept and, as such, does not have to be a life sentence.

Winthrop Jordan has suggested that ideas of racial inferiority, specifically that blacks were savage and primitive, played an essential role in rationalizing slavery.[1] There was no conception of race as a physical category until the eighteenth century.[2] There was, however, a strong association between blackness and evil, sin, and death, long grounded in European thought. The term "race" is believed to have originated in the Middle Ages in the romance languages, first used to refer to the breeding of animals. Race did not appear in the English language until the sixteenth century and was used as a technical term to define human groups in the seventeenth century. By the end of the eighteenth century, as emphasis upon the observation and classification of human differences grew, "race" became the most commonly employed concept for differentiating human groups according to Northern European standards. Audrey Smedley argues that because "race" has its roots in the breeding of

From Abby L. Ferber, *White Man Falling: Race, Gender, and White Supremacy* (Lanham, MD: Rowman & Littlefield, 1998), pp. 27–43. Reprinted by permission of Rowman & Littlefield Publishers, Inc.

animal stock, unlike other terms used to categorize humans, it came to imply an innate or inbred quality, believed to be permanent and unchanging.[3]

Until the nineteenth century, the Bible was consulted and depended upon for explanations of human variation, and two schools of thought emerged. The first asserted that there was a single creation of humanity, monogenesis, while the second asserted that various human groups were created separately, polygenesis. Polygenesis and ideas about racial inferiority, however, gained few believers, even in the late 1700s when the slave trade was under attack, because few were willing to support doctrines that conflicted with the Bible.[4]

While European Americans remained dedicated to a biblical view of race, the rise of scientific racism in the middle of the eighteenth century shaped debate about the nature and origins of races.[5] The Enlightenment emphasized the scientific practices of observing, collecting evidence, measuring bodies, and developing classificatory schemata. In the early stages of science, the most prevalent activity was the collection, examination, and arrangement of data into categories. Carolus Linnaeus, a prominent naturalist in the eighteenth century, developed the first authoritative racial division of humans in his *Natural System*, published in 1735.[6] Considered the founder of scientific taxonomy, he attempted to classify all living things, plant and animal, positioning humans within the matrix of the natural world. As Cornel West demonstrates, from the very beginning, racial classification has always involved hierarchy and the linkage of physical features with character and cultural traits.[7] For example, in the descriptions of his racial classifications, Linnaeus defines Europeans as "gentle, acute, inventive . . . governed by customs," while Africans are "crafty, indolent, negligent . . . governed by caprice."[8] Like most scientists of his time, however, Linnaeus considered all humans part of the same species, the product of a single creation.

Linnaeus was followed by Georges Louis Leclerc, Comte de Buffon, who is credited with introducing the term "race" into the scientific lexicon. Buffon also believed in monogenesis and in his 1749 publication *Natural History*, suggested that human variations were the result of differences in environment and climate. Whiteness, of course, was assumed to be the real color of humanity. Buffon suggested that blacks became dark-skinned because of the hot tropical sun and that if they moved to Europe, their skin would eventually lighten over time. Buffon cited interfertility as proof that human races were not separate species, establishing this as the criterion for distinguishing a species.

Buffon and Johann Friedrich Blumenbach are considered early founders of modern anthropology. Blumenbach advanced his own systematic racial classification in his 1775 study *On the Natural Varieties of Mankind*, designating five human races: Caucasian, Mongolian, Ethiopian, American, and Malay. While he still considered races to be the product of one creation, he ranked them on a scale according to their distance from the "civilized" Europeans.[9] He introduced the term "Caucasian," chosen because he believed that the Caucasus region in Russia produced the world's most beautiful women. This assertion typifies the widespread reliance upon aesthetic judgments in ranking races. . . .

The science of racial classifications relied upon ideals of Greek beauty, as well as culture, as a standard by which to measure races. Race became central to the definition of Western culture, which became synonymous with "civilization."[10] . . .

The history of racial categorizations is intertwined with the history of racism. Science sought to justify a priori racist assumptions and consequently rationalized and greatly expanded the arsenal of racist ideology. Since the eighteenth century, racist beliefs have been built upon scientific racial categorizations and the linking of social and cultural traits to supposed genetic racial differences. While some social critics have suggested that contemporary racism has replaced biology with a concept of culture, the [1994] publication of *The Bell Curve*[11] attests to the staying power of these genetic notions of race. Today, as in the past, racism weaves together notions of biology and culture, and culture is assumed to be determined by some racial essence.

Science defined race as a concept believed to be hereditary and unalterable. The authority of science contributed to the quick and widespread acceptance of these ideas and prevented their interrogation. Equally important, the study of race and the production of racist theory also helped establish scientific authority and aided discipline building. While the history of the scientific concept of race argues that race is an inherent essence, it reveals, on the contrary, that race is a social construct. Young points out that "the different Victorian scientific accounts of race each in their turn quickly became deeply problematic; but what was much more consistent, more powerful and long-lived, was the cultural construction of race."[12]

Because race is not grounded in genetics or nature, the project of defining races always involves drawing and maintaining boundaries between those races. This was no easy task. It is important to pay attention to the construction of those borders: how was it decided, in actual policy, who was considered white and who was considered black? What about those who did not easily fit into either of those categories? What were the dangers of mixing? How could these dangers be avoided? These issues preoccupied policy makers, popular culture, and the public at large. . . .

Throughout the second half of the nineteenth century, discussion of race and racial purity grew increasingly popular in both academic and mainstream circles as Americans developed distinctive beliefs and theories about race for the first time. As scientific beliefs about race were increasingly accepted by the general public, support for the one-drop rule became increasingly universal. Popular opinion grew to support the belief that no matter how white one appeared, if one had a single drop of black blood, no matter how distant, one was black. . . .

Throughout the history of racial classification in the West, miscegenation and interracial sexuality have occupied a place of central importance. The science of racial differences has always displayed a preoccupation with the risks of interracial sexuality. Popular and legal discourses on race have been preoccupied with maintaining racial boundaries, frequently with great violence. This [essay] suggests that racial classification, the maintenance of racial boundaries, and racism are inexorably linked. The construction of biological races and the belief in maintaining the hierarchy and separation of races has led to widespread fears of integration and interracial sexuality. . . .

The history of racial classification, and beliefs about race and interracial sexuality, can be characterized as inherently white supremacist. White supremacy has been the law and prevailing worldview throughout U.S. history, and the ideology of what is today labeled the white supremacist movement is firmly rooted in this tradition. Accounts that label the contemporary white supremacist movement as fringe and extremist often have the consequence of rendering this history invisible. Understanding this history, however, is essential to understanding and combating both contemporary white supremacist and mainstream racism.

REFERENCES

1. Jordan, Winthrop. 1969. *White over black*. Chapel Hill: University of North Carolina Press.

2. Banton, Michael, and Jonathan Harwood. 1975. *The race concept*. New York: Praeger; Mencke, John G. 1979. *Mulattoes and race mixture: American attitudes and images, 1865–1918*. Ann Arbor, Mich.: University Microfilms Research Press.

3. Smedley, Audrey. 1993. *Race in North America: Origin and evolution of a worldview*. Boulder, Colo.: Westview Press.

4. Banton and Harwood 1975, 19.

5. Banton and Harwood 1975, 24.

6. West, Cornel. 1982. *Prophesy deliverance! An Afro-American revolutionary Christianity*. Philadelphia: Westminster Press.

7. West 1982.

8. West 1982, 56.

9. Smedley 1993, 166.

10. Young, Robert J. C. 1995. *Colonial desire: Hybridity in theory, culture and race*. New York: Routledge.

11. Hernstein, Richard J., and Charles Murray. 1994. *The bell curve: Intelligence and class structure in American life*. New York: Free Press.

12. Young. 1995, 94.

DISCUSSION QUESTIONS

1. What does Ferber mean when she writes that "the history of racial categorizations is intertwined with the history of racism"?

2. What role have science and religion played in the social construction of racism?

8

How Did Jews Become
White Folks?

BY KAREN BRODKIN

The American nation was founded and developed by the Nordic race, but if a few more million members of the Alpine, Mediterranean and Semitic races are poured among us, the result must inevitably be a hybrid race of people as worthless and futile as the good-for-nothing mongrels of Central America and Southeastern Europe.
—Kenneth Roberts, "Why Europe Leaves Home"

The late nineteenth century and early decades of the twentieth saw a steady stream of warnings by scientists, policymakers, and the popu-
. . . lar press that "mongrelization" of the Nordic or Anglo-Saxon race—the real Americans—by inferior European races (as well as by inferior non-European ones) was destroying the fabric of the nation.

I continue to be surprised when I read books that indicate that America once regarded its immigrant European workers as something other than white, as biologically different. My parents are not surprised; they expect anti-Semitism to be part of the fabric of daily life, much as I expect racism to be part of it. They came of age in the Jewish world of the 1920s and 1930s, at the peak of anti-Semitism in America. They are rightly proud of their upward mobility and think of themselves as pulling themselves up by their own boot-straps. I grew up during the 1950s in the Euro-ethnic New York suburb of Valley Stream, where Jews were simply one kind of white folks and where ethnicity meant little more to my generation than food and family heritage. Part of my ethnic heritage was the belief that Jews were smart and that our success was due to our own efforts and abilities, reinforced by a culture that valued sticking together, hard work, education, and deferred gratification.

I am willing to affirm all those abilities and ideals and their contribution to Jews' upward mobility, but I also argue that they were still far from sufficient to account for Jewish success. I say this because the belief in a Jewish version of Horatio Alger has become a point of entry for some mainstream Jewish organizations to adopt a racist attitude against African Americans especially and to oppose affirmative action for people of color. Instead I want to suggest that Jewish success is a product not only of ability but also of the removal of powerful social barriers to its realization.

It is certainly true that the United States has a history of anti-Semitism and of beliefs that Jews are members of an inferior race. But Jews were hardly alone. American anti-Semitism was part of a broader pattern of late-nineteenth-century racism against all southern and eastern European immigrants, as well as against Asian immigrants, not to mention African Americans, Native Americans, and Mexicans. These views justified all sorts of discriminatory treatment, including closing the doors, between 1882 and 1927, to immigration from Europe and Asia. This picture changed radically after World War II. Suddenly, the same folks who had promoted nativism and xenophobia were eager to believe that the Euro-origin people whom they had deported, reviled as members of inferior races, and prevented from immigrating only a few years earlier, were now model middle-class white suburban citizens.

It was not educational epiphany that made those in power change their hearts, their minds, and our race. Instead, it was the biggest and best affirmative action program in the history of our nation, and it was for Euromales. That is not how it was billed, but it is the way it worked out in practice. I tell this story to show the institutional nature of racism and the centrality of state policies to creating and changing races. Here, those policies reconfigured the category of whiteness to include European immigrants. There are similarities and differences in the ways each of the European immigrant groups became "whitened." I tell the story in a way that links anti-Semitism to other varieties of anti-European racism because this highlights what Jews shared with other Euro-immigrants.

EURORACES

The U.S. "discovery" that Europe was divided into inferior and superior races began with the racialization of the Irish in the mid-nineteenth century and flowered in response to the great waves of immigration from southern and eastern Europe that began in the late nineteenth century. Before that time, European immigrants—including Jews—had been largely assimilated into the white population. However, the 23 million European immigrants who came to work in U.S. cities in the waves of migration after 1880 were too many and too concentrated to absorb. Since immigrants and their children made up more than 70 percent of the population of most of the country's largest cities, by the 1890s urban American had taken on a distinctly southern and eastern European immigrant flavor. Like the Irish in Boston and New York, their urban concentrations in dilapidated neighborhoods put them cheek by jowl next to the rising elites and the middle class with whom they shared public space and to whom their working-class ethnic communities were particularly visible.

The Red Scare of 1919 clearly linked anti-immigrant with anti-working-class sentiment—to the extent that the Seattle general strike by largely native-born workers was blamed on foreign agitators. The Red Scare was fueled by an economic depression, a massive postwar wave of strikes, the Russian Revolution, and another influx of postwar immigration. Strikers in the steel and garment industries in New York and New England were mainly new immigrants. . . .

Not surprisingly, the belief in European races took root most deeply among the wealthy, U.S.-born Protestant elite, who feared a hostile and seemingly inassimilable working class. By the end of the nineteenth century, Senator Henry Cabot Lodge pressed Congress to cut off immigration to the United States; Theodore Roosevelt raised the alarm of "race suicide" and took Anglo-Saxon women to task for allowing "native" stock to be outbred by inferior immigrants. In the early twentieth century, these fears gained a great deal of social legitimacy thanks to the efforts of an influential network of aristocrats and scientists who developed theories of engenics—breeding for a "better" humanity—and scientific racism.

Key to these efforts was Madison Grant's influential *The Passing of the Great Race*, published in 1916. Grant popularized notions developed by William Z. Ripley and Daniel Brinton that there existed three or four major European races, ranging from the superior Nordics of northwestern Europe to the inferior southern and eastern races of the Alpines, Mediterraneans, and worst of all, Jews, who seemed to be everywhere in his native New York City. Grant's nightmare was race-mixing among Europeans. For him, "the cross between any of the three European races and a Jew is a Jew." He didn't have good things to say about Alpine or Mediterranean "races" either. For Grant, race and class were interwoven: the upper class was racially pure Nordic; the lower classes came from the lower races.

Far from being on the fringe, Grant's views were well within the popular mainstream. Here is the *New York Times* describing the Jewish Lower East Side of a century ago:

> The neighborhood where these people live is absolutely impassable for wheeled vehicles other than their pushcarts. If a truck driver tries to get through where their pushcarts are standing they apply to him all kinds of vile and indecent epithets. The driver is fortunate if he gets out of the

street without being hit with a stone or having a putrid fish or piece of meat thrown in his face. This neighborhood, peopled almost entirely by the people who claim to have been driven from Poland and Russia, is the eyesore of New York and perhaps the filthiest place on the western continent. It is impossible for a Christian to live there because he will be driven out, either by blows or the dirt and stench. Cleanliness is an unknown quantity to these people. They cannot be lifted up to a higher plane because they do not want to be. If the cholera should ever get among these people, they would scatter its germs as a sower does grain.[1]

Such views were well within the mainstream of the early twentieth-century scientific community. . . .

By the 1920s, scientific racism sanctified the notion that real Americans were white and that real whites came from northwest Europe. Racism by white workers in the West fueled laws excluding and expelling the Chinese in 1882. Widespread racism led to closing the immigration door to virtually all Asians and most Europeans between 1924 and 1927, and to deportation of Mexicans during the Great Depression.

Racism in general, and anti-Semitism in particular, flourished in higher education. Jews were the first of the Euro-immigrant groups to enter college in significant numbers, so it was not surprising that they faced the brunt of discrimination there. The Protestant elite complained that Jews were unwashed, uncouth, unrefined, loud, and pushy. Harvard University President A. Lawrence Lowell, who was also a vice president of the Immigration Restriction League, was open about his opposition to Jews at Harvard. The Seven Sister schools had a reputation for "flagrant discrimination." M. Carey Thomas, Bryn Mawr president, may have been some kind of feminist, but she was also an admirer of scientific racism and an advocate of immigration restriction. . . .

Jews are justifiably proud of the academic skills that gained them access to the most elite schools of the nation despite the prejudices of their gatekeepers. However, it is well to remember that they had no serious competition from their Protestant classmates. This is because college was not about academic pursuits. It was about social connection—through its clubs, sports and other activities, as well as in the friendships one was expected to forge with other children of elites. From this, the real purpose of the college experience, Jews remained largely excluded.

This elite social mission had begun to come under fire and was challenged by a newer professional training mission at about the time Jews began entering college. Pressures for change were beginning to transform the curriculum and to reorient college from a gentleman's bastion to a training ground for the middle-class professionals needed by an industrial economy. . . . Occupational training was precisely what had drawn Jews to college. In a setting where disparagement of intellectual pursuits and the gentleman C were badges of distinction, it certainly wasn't hard for Jews to excel. Jews took seriously what their affluent Protestant classmates disparaged, and, from the perspective of nativist elites, took unfair advantage of a loophole to get where they were not wanted.

Patterns set by these elite schools to close those "loopholes" influenced the standards of other schools, made anti-Semitism acceptable, and "made the aura of exclusivity a desirable commodity for the college-seeking clientele."[2] Fear that colleges "might soon be overrun by Jews" were publicly expressed at a 1918 meeting of the Association of New England Deans. In 1919 Columbia University took steps to decrease the number of its Jewish students by a set of practices that soon came to be widely adopted. They developed a psychological test based on the World War I army intelligence tests to measure "innate ability—and middle-class home environment"; and they redesigned the admission application to ask for religion, father's name and birth place, a photo, and personal interview. Other techniques for excluding Jews, like a fixed class size, a chapel requirement, and preference for children of alumni, were less obvious. . . .

Columbia's quota against Jews was well known in my parents' community. My father is very proud of having beaten it and been admitted to Columbia Dental School on the basis of his skill at carving a soap ball. Although he became a teacher instead because the tuition was too high, he took me to the dentist every week of my childhood and prolonged the agony by discussing the finer points of tooth-filling and dental care. My father also almost failed the speech test required for his teaching license because he didn't speak "standard," i.e., nonimmigrant, nonaccented English. For my parents and most of their friends. English was the language they had learned when they went to school, since their home and neighborhood language was Yiddish. They saw the speech test as designed to keep all ethnics, not just Jews, out of teaching.

There is an ironic twist to this story. My mother always urged me to speak well, like her friend Ruth Saronson, who was a speech teacher. Ruth remained my model for perfect diction until I went away to college. When I talked to her on one of my visits home, I heard the New York accent of my version of "standard English," compared to the Boston academic version.

My parents believe that Jewish success, like their own, was due to hard work and a high value placed on education. They attended Brooklyn College during the Depression. My mother worked days and went to school at night; my father went during the day. Both their families encouraged them. More accurately, their families expected it. Everyone they knew was in the same boat, and their world was made up of Jews who were advancing just as they were. The picture for New York—where most Jews lived—seems to back them up. In 1920, Jews made up 80 percent of the students at New York's City College, 90 percent of Hunter College, and before World War I, 40 percent of private Columbia University. By 1934, Jews made up almost 24 percent of all law students nationally and 56 percent of those in New York City. Still, more Jews became public school teachers, like my parents and their friends, than doctors or lawyers. . . .

How we interpret Jewish social mobility in this milieu depends on whom we compare them to. Compared with other immigrants, Jews were upwardly mobile. But compared with nonimmigrant whites, that mobility was very limited and circumscribed. The existence of anti-immigrant, racist, and anti-Semitic barriers kept the Jewish middle class confined to a small number of occupations. Jews were excluded from mainstream corporate management and corporately employed professions, except in the garment and movie industries, in which they were pioneers. Jews were almost totally excluded from university faculties (the few who made it had powerful patrons). Eastern European Jews were concentrated in small businesses, and in professions where they served a largely Jewish clientele. . . .

Although Jews, as the Euro-ethnic vanguard in college, became well established in public school teaching—as well as visible in law, medicine, pharmacy, and librarianship before the postwar boom—these professions should be understood in the context of their times. In the 1930s they lacked the corporate context they have today, and Jews in these professions were certainly not corporation-based. Most lawyers, doctors, dentists, and pharmacists were solo practitioners, depended upon other Jews for their clientele, and were considerably less affluent than their counterparts today.

Compared to Jewish progress after World War II, Jews' pre-war mobility was also very limited. It was the children of Jewish businessmen, but not those of Jewish workers, who flocked to college. Indeed, in 1905 New York, the children of Jewish workers had as little schooling as the children of other immigrant workers. My family was quite the model in this respect. My grandparents did not go to college, but they did have a modicum of small business success. My father's family owned a pharmacy. Although my mother's father was a skilled garment worker, her mother's family was large and always had one or another grocery or deli in which my grandmother participated. It was the relatively privileged children of upwardly mobile Jewish immigrants like my grandparents who began to push on the doors to higher education even before my parents were born.

Especially in New York City—which and almost one and a quarter million Jews by 1910 and retained the highest concentration of the nation's 4 million Jews in 1924—Jews built a small-business-based middle class and began to develop a second-generation professional class in the interwar years. Still, despite the high percentages of Jews in eastern colleges, most Jews were not middle class, and fewer than 3 percent were professionals—compared to somewhere between two-thirds and three-quarters in the postwar generation.

My parents' generation believed that Jews overcame anti-Semitic barriers because Jews are special. My answer is that the Jews who were upwardly mobile were special among Jews (and were also well placed to write the story). My generation might well respond to our parents' story of pulling themselves up by their own bootstraps with "But think what you might have been without the racism and with some affirmative action!" And that is precisely what the post–World War II boom, that decline of systematic, public, anti-Euro racism and anti-Semitism, and governmental affirmative action extended to white males let us see.

WHITENING EURO-ETHNICS

By the time I was an adolescent, Jews were just as white as the next white person. Until I was eight, I was a Jew in a world of Jews. Everyone on Avenue Z in Sheepshead Bay was Jewish. I spent my days playing and going to school on three blocks of Avenue Z, and visiting my grandparents in the nearby Jewish neighborhoods of Brighton Beach and Coney Island. There were plenty of Italians in my neighborhood, but they lived around the corner. They were a kind of Jew, but on the margins of my social horizons. Portuguese were even more distant, at the end of the bus ride, at Sheepshead Bay. The *shul*, or temple, was on Avenue Z, and I begged my father to take me like all the other fathers took their kids, but religion wasn't part of my family's Judaism. Just how Jewish my neighborhood was hit me in first grade, when I was one of two kids to go to school on Rosh Hashanah. My teacher was shocked—she was Jewish too—and I was embarrassed to tears when she sent me home. I was never again sent to school on Jewish holidays. We left that world in 1949 when we moved to Valley Stream, Long Island, which was Protestant and Republican and even had farms until Irish, Italian, and Jewish ex-urbanites like us gave it a more suburban and Democratic flavor.

Neither religion not ethnicity separated us at school or in the neighborhood. Except temporarily. During my elementary school years, I remember a fair number of dirt-bomb (a good suburban weapon) wars on the block. Periodically, one of the Catholic boys would accuse me or my brother of killing his god, to which we'd reply, "Did not," and start lobbing dirt bombs. Sometimes he'd get his friends from Catholic school and I'd get mine from public school kids on the block, some of whom were Catholic. Hostilities didn't last for more than a couple of hours and punctuated an otherwise friendly relationship. They ended by our junior high years, when other things became more important. Jews, Catholics and Protestants, Italians, Irish, Poles, "English" (I don't remember hearing WASP as a kid), were mixed up on the block and in school. We thought of ourselves as middle class and very enlightened because our ethnic backgrounds seemed so irrelevant to high school culture. We didn't see race (we thought), and racism was not part of our peer consciousness. Nor were the immigrant or working-class histories of our families.

As with most chicken-and-egg problems, it is hard to know which came first. Did Jews and other Euro-ethnics become white because they became middle-class? That is, did money whiten? Or did being incorporated into an expanded version of whiteness open up the economic doors to middle-class status? Clearly, both tendencies were at work.

Some of the changes set in motion during the war against fascism led to a more inclusive version of whiteness. Anti-Semitism and anti-European racism lost respectability. The 1940 Census no longer distinguished native whites of native parentage from those, like my parents, of immigrant parentage, so Euro-immigrants and their children were more securely white by submersion in an expanded notion of whiteness.

Theories of nurture and culture replaced theories of nature and biology. Instead of dirty and dangerous races that would destroy American democracy, immigrants became ethnic groups whose children had successfully assimilated into the mainstream and risen to the middle class. In this new myth, Euro-ethnic suburbs like mine became the measure of American democracy's victory over racism. . . . Jewish mobility became a new Horatio Alger story. In time and with hard work, every ethnic group would get a piece of the pie, and the United States would be a nation with equal opportunity for all its people to become part of a prosperous middle-class majority. And it seemed that Euro-ethnic immigrants and their children were delighted to join middle America.

REFERENCES

1. *The New York Times*, July 30, 1893; cited in Allon Schoener, 1967. *Portal to America: The Lower East Side 1870-1925*. New York: Holt, Rinehart, and Winston, pp. 57-58.

2. Synott, Marcia Graham. 1986. "Anti-Semitism and American Universities: Did Quotas Follow the Jews?" In *Anti-Semitism in American History*, edited by David A. Gerber. Urbana, IL: University of Illinois Press, p. 250.

DISCUSSION QUESTIONS

1. How does Brodkin see anti-Semitism as linked to other forms of racism? How is this revealed by historical events?

2. What does Brodkin mean by arguing the "Jews have *become* white folks?" How does her own family history reveal this process?

9

Systemic Racism

A Comprehensive Perspective

BY JOE R. FEAGIN

We the people of the United States, in order to form a more perfect union, establish
justice, insure domestic tranquility, provide for the common defense, promote the
general welfare, and secure the blessings of liberty to ourselves and our posterity,
do ordain and establish this Constitution for the United States of America.
—Preamble, U.S. Constitution

Culturally the Negro represents a paradox: Though he is an organic part of the
nation, he is excluded by the entire tide and direction of American culture. . . .
Therefore if, within the confines of its present culture, the nation ever seeks to purge
itself of its color hate, it will find itself at war with itself, convulsed by a spasm of
emotional and moral confusion. —Richard Wright, Black Boy

The year is 1787, the place Philadelphia. Fifty-five men are meeting in
summer's heat to write a constitution for what will be called the "first
democratic nation." These founders create a document so radical in
breaking from monarchy and feudal institutions that it will be condemned and
attacked in numerous European countries. These radicals are men of European
origin, and most are well-off by the standards of their day. Significantly, at least
40 percent have been or are slave owners, and a significant proportion of the
others profit to some degree as merchants, shippers, lawyers, and bankers from
the trade in slaves, commerce in slave-produced agricultural products, or sup-
plying provisions to slaveholders and slavetraders.[1] Moreover, the man who
pressed hard for this convention and now chairs it, George Washington, is one
of the richest men in the colonies because of the hundreds of black men,
women, and children he has held in bondage. Washington and his colleagues
create the first democratic nation, yet one for whites only. In the preamble to

From *Racist America: Roots, Current Realities, and Future Reparations* by Joe R. Feagin
(New York: Routledge, 2001), pp. 9–24. Copyright © 2001. Reprinted by permission of
Routledge/Taylor & Francis Books, Inc.

their bold document, the founders cite prominently "We the People," but this phrase does not encompass the fifth of the population that is enslaved.

Laying a Racist Foundation

Many historical analysts have portrayed slavery as a side matter at the 1787 Constitutional Convention. Slavery was central however, as a leading participant, James Madison, made clear in his important notes on the convention's debates. Madison accented how the convention was scissored across a north/south, slave/not-slave divide among the states.[2] The southern and northern regions were gradually diverging in their politicoeconomic frameworks. Slavery had once been of some, albeit greatly varying, importance in all states, but the northern states were moving away from chattel slavery as a part of their local economies, and some were seeing a growing abolitionist sentiment. Even so, many northern merchants, shippers, and consumers still depended on products produced by southern slave plantations, and many merchants sold goods to the plantations. . . .

The trade in, and enslavement of, people of African descent was an important and divisive issue for the convention. Most of these prominent, generally well-educated men accepted the view that people of African descent could be the chattel property of others—and not human beings with citizens' rights. At the heart of the Constitution was protection of the property and wealth of the affluent bourgeoisie in the new nation, including property in those enslaved. There was near unanimity on the idea, as delegate Gouverneur Morris (New York) put it, that property is the "main object of Society."[3] For the founders, freedom meant the protection of unequal accumulation of property, particularly property that could produce a profit in the emerging capitalist system. Certain political, economic, and racial interests were conjoined. This was not just a political gathering with the purpose of creating a major new bourgeois-democratic government; it was also a meeting to protect the racial and economic interests of men with substantial property and wealth in the colonies. . . .

The new nation formed by European Americans in the late eighteenth century was openly and officially viewed as a *white* republic. These founders sought to build a racially based republic in the face of monarchical opposition and against those people on the North American continent whom they defined as inferior and as problems. . . .

The U.S. Constitutional Convention, the first such in the democratic history of the modern world, laid a strong base for the new societal "house" called the United States. Yet from the beginning this house's foundation was fundamentally flawed. While most Americans have thought of this document and the sociopolitical structure it created as keeping the nation together, in fact this structure was created to maintain separation and oppression at the time and for the foreseeable future. The framers reinforced and legitimated a system of racist oppression that they thought would ensure that whites, especially white men of means, would rule for centuries to come. . . .

Understanding the centrality and harsh realities of slavery leads to some tough questions for today: How are we to regard the "founding fathers"? They are our founding fathers, yet many were oppressors who made their living by killing, brutalizing, and exploiting other human beings. The combination of white freedom and black enslavement seems radically contradictory. However, William Wiecek notes that " the paradox dissolves when we recall that American slavery was racial. White freedom was entirely compatible with black enslavement. African Americans were, as the framers of Virginia's first Constitution determined, simply not part of the Lockean body politic."[4] Indeed, the work of those enslaved brought the wealth and leisure that whites, especially those in the ruling class, could use to pursue their own liberty and freedom. . . .

Into the mid-nineteenth century, the majority of whites—in the elites and among ordinary folk—either participated directly in slavery or in the trade around slavery, or did not object to those who did so. The antihuman savagery called *slavery* was considered *normal* in what was then seen as a white republic. This point must be understood well if one is to probe deeply into the origins, maintenance, and persistence of racist patterns and institutions in North America. . . .

Frederick Douglass was one of the first U.S analysts to develop a conceptual approach accenting *institutionalized* racism across many sectors of society. In 1881, speaking about the ubiquitous impact of racist prejudice and discrimination, he argued that "[i]n nearly every department of American life [black Americans] are confronted by this insidious influence. It fills the air. It meets them at the workshop and factory, when they apply for work. It meets them at the church, at the hotel, at the ballot-box, and worst of all, it meets them in the jury-box. . . . [the black American] has ceased to be a *slave of an individual*, but has in some sense become *the slave of society*."[5]. . .

Historically the social-science study of racial oppression has been identified by such terms as "intergroup relations" or "race relations." These somewhat ambiguous and euphemistic phrases are accented by many analysts who prefer to view an array of racial groups as more or less responsible for the U.S. "race problem." Such terminology, however, can allow the spotlight to be taken off the whites who have created and maintained the system of racism. Moreover, many white analysts have written about the "race problem" as "in" U.S. society. Yet, race relations—or, more accurately, *racist relations*—are not *in*, but rather *of* this society. In the American case systemic racism began with European colonists enriching themselves substantially at the expense of indigenous peoples and the Africans they imported for enslavement. This brutally executed enrichment was part of the new society's foundation, not something tacked onto and otherwise healthy and egalitarian system. . . .

Unjust enrichment is an old legal term associated with relationships between individuals. One legal dictionary defines the concept as "circumstances which gives rise to the obligation of restitution, that is, the receiving and retention of property, money, or benefits which in justice and equity belong to another."[6] This legal concept encompasses not only the receiving of benefits that justly belong to another but also the obligation to make restitution for that injustice. This idea can be extended beyond individual relationships envisioned in the traditional legal argument to the unjust theft of labor or resources by one group, such as white Americans, from another group, such as black Americans. I suggest here the parallel idea of *unjust impoverishment* to describe the conditions of those who suffer oppression. . . . For over fourteen generations the exploitation of African Americans has redistributed income and wealth earned by them to generations of white Americans, leaving the former relatively impoverished as a group and the latter relatively privileged and affluent as a group.

Racial Classes with Vested Group Interests

Understanding how this undeserved impoverishment and enrichment gets transmitted, reproduced, and institutionalized over generations of white and black Americans is an important step in developing an adequate conceptual framework. Black labor was used unjustly for building up the wealth of this white-dominated nation from the 1600s to at least the 1960s—the slavery and legal segregation periods. . . .

The racial-class system, initially created by the white ruling class, provided benefits to most white Americans. From the seventeenth century onward, the farms and plantations run with enslaved laborers brought significant income and wealth to many white Americans, and not just to their particular owners. These enterprises multiplied economic development for white Americans well outside the farms' and plantations' immediate geographical areas. . . . Ordinary whites for the most part bought into the identity of whiteness, thereby binding themselves to the white racial class.

Slavery's impact extended well beyond the economy. Each institutional arena in the new nation was controlled by whites and was closely linked to other major arenas. As we have seen, the new Constitution and its "democratic" political system were grounded in the racist thinking and practices of white men, many of whom had links to slavery. Those who dominated the economic system crafted the political system. Likewise the religious, legal, educational, and media systems were interlinked with the slavery economy and polity. Woven through each institutional area was a broad racist ideology— a set of principles and views—centered on rationalizing white-on-black domination and creating positive views of whiteness. . . .

Systemic racism is not just about the construction of racial images, attitudes, and identities. It is even more centrally about the creation, development, and maintenance of white privilege, economic wealth, and sociopolitical power over nearly four centuries. It is about hierarchical interaction. The past and present worlds of racism include not only racist relations at work but also the racist relations that black Americans and other Americans of color encounter in trying to secure adequate housing, consumer goods, and public accommodations for themselves and their families. Racism reaches deeply into family lives, shaping who has personal relations with whom, and who gets married to whom. Racism shapes which groups have the best health, get the best medical care, and live the longest lives. . . .

UNDESERVED IMPOVERISHMENT
AND UNDERSERVED ENRICHMENT

The Wealth of a White Nation

Not long ago, Bob Dole, onetime Senate Majority Leader and presidential candidate, spoke in a television interview of "displaced" white men who must compete with black workers because of affirmative action. He said that he was not sure that "people in America" (presumably he meant whites) should now be paying a price for racial discrimination that occurred "before they were born." He was fairly candid about the past, saying, "We did discriminate. We did suppress people. It was wrong. Slavery was wrong." Yet Dole added that he was not sure any compensation for this damage was now due. Similarly, Representative Henry Hyde (R–Illinois), who served as chair of the House Judiciary Committee, has commented that the idea of collective guilt for slavery in the distant past "is an idea whose time has gone. I never owned a slave. I never oppressed anybody. I don't know that I should have to pay for someone who did generations before I was born."[7] This questioning of the relevance of the racist past for the present is commonplace among white Americans.

Sources of White Wealth: A Vignette

Consider four young children coming into the American colonies in the late seventeenth century. An African brother and sister are ripped from their homes and imported in chains into Virginia, the largest slaveholding colony. Their African names being ignored, they are renamed "negro John" and "negro Mary" (no last name) by the white family that purchased them from a slave ship. This white (Smith) family has young twins, William and Priscilla. Their first and last names are those given to them by their parents, and they never wear chains. The enslaved children are seen as "black" by the Smith family, while the twins will live as "white."

What do these children and their descendants have to look forward to? Their experiences will be very different as a result of the system of racist oppression. William's and Priscilla's lives may be hard because of the physical environment, but they and their descendants will likely build lives with an array of personal choices and the passing down of significant social resources. As a girl and later as a woman, Priscilla will not have the same privileges as William, but her life is more likely to be economically supported and protected than Mary's. Indeed, John and Mary face a stark, often violent existence, with most of their lives determined by the whims of the slaveholder, who has stolen not only their labor but their lives. They will never see their families or home societies again. They can be radically separated at any time, a separation much less likely for William and Priscilla. From their and other slaves' labor some wealth will be generated for the Smith family and passed on to later generations. Unlike the white twins, John and Mary will not be allowed to read or write and will be forced to replace their African language with English. Where they eat and sleep will be largely determined by whites. As they grow older, major decisions about their personal and family relationships will be made by whites. Mary will face repeated sexual threats, coercion, and rape at the hands of male overseers and slaveholders, perhaps including William. Moreover, if John even looks at Priscilla the wrong way, he is likely to be punished severely.

If John and Mary are later allowed to have spouses and children, they will face a much greater infant mortality rate than whites. And their surviving children may well be taken from them, so that they and later generations may have great difficulty in keeping the full memory of their ancestors, a problem not faced by William and Priscilla. If John or Mary resist their oppression, they are likely to be whipped, put in chains, or have an iron bit put in their mouths. If John is rebellious or runs away too much, he may face castration. John and Mary will have to struggle very hard to keep their families together because the slaveholders can destroy them at any moment. Still, together with other black Americans, they build a culture of resistance carried from generation to generation in oral traditions. Moreover, for many more generations John's and Mary's descendants will suffer similarly severe conditions as the property of white families. Few if any of their descendants will see freedom until the 1860s.

The end of slavery does not end the large-scale oppression faced by John's and Mary's descendants. For four more generations after 1865 the near-slavery called legal or de facto segregation will confront them, but of course will not affect the descendants of William and Priscilla Smith. The later black generations will also be unable to build up resources and wealth; they will have their lives substantially determined by the white enforcers of comprehensive segregation. Where they can get a job, where they can live, whether and where they can go to school, and how they can travel will still be significantly determined by whites. Some may face brutal beatings or lynching by whites, especially if they resist oppression. They will have inherited no wealth from many generations of enslaved ancestors, and they are unlikely to garner resources themselves to pass along to later generations. From the late 1600s to the 1960s, John and Mary and their descendants have been at an extreme economic, political, and social disadvantage compared to William and Priscilla Smith and their descendants. The lives of these black Americans have been shortened, their opportunities severely limited, their inherited resources all but nonexistent, and their families pressured by generations of well-organized racial oppression. . . .

From this vignette we can begin to see how racism is economically and systematically constructed. Unjust impoverishment for John and Mary and undeserved enrichment for William and Priscilla become bequeathed inheritances for many later generations. Undeserved impoverishment and enrichment are at the heart of colonial land theft and the brutal slavery system. Over time this ill-gotten gain has been used and invested by white colonizers and their descendants to construct a prosperous white-dominated nation. Today there is a general denial in the white population that black Americans have contributed much to American (or Western) development and civilization. This denial is part of contemporary white racist misunderstandings of the reality of the history of the West. However, the facts are clear: The slavery system provided much stimulus for economic development and generated critical surplus capital for the new nation.... Without the enslaved labor of millions of black Americans, there might well not be a prosperous United States today.

REFERENCES

1. National Archives and Records Administration, *Framers of the Constitution* (Washington, DC: National Archives Trust Fund Board, 1986.)

2. Max Farrand, ed., *Records of the Federal Convention of 1787*, vol. 1. (New Haven, CT: Yale University, 1911) p. 486.

3. Herbert Aptheker. *Early Years of the Republic: From the End of the Revolution to the First Administration of Washington, 1783–1793.* (New York: International Publishers, 1976) p. 570.

4. William M. Wiecek. "The Origins of the Law of Slavery in British North America." *Cardozo Law Review* 17. (May 1996) p. 1791.

5. Frederick Douglass, "The Color Line," *North American Review.* June 1881, as excerpted in *Jones et ux. v. Alfred H. Mayer Co.*, 392 U.S. 409, 446–447; emphasis added.

6. James A. Ballentine. *Ballentine's Law Dictionary*, 3rd ed., ed. William S. Anderson, (San Francisco: Bancroft-Whiteney, 1969) p. 1320.

7. Dole comments on "Meet the Press," NBC Television, February 5, 1995. Hyde is quoted in Kevin Merida, "Did Freedom Alone Pay a Nation's Debt." *The Washington Post,* November 23, 1999. p. C1.

DISCUSSION QUESTIONS

1. How does Feagin define "systemic racism?" How does this differ from the common understanding of racism as a matter of individual prejudice?

2. What does Feagin mean by the concepts of *unjust enrichment* and *undeserved impoverishment*? How have they emerged over time?

10

Racial Formation

MICHAEL OMI
AND HOWARD WINANT

I n 1982–83, Susie Guillory Phipps unsuccessfully sued the Louisiana Bureau of Vital Records to change her racial classification from black to white. The descendant of an 18th-century white planter and a black slave, Phipps was designated "black" in her birth certificate in accordance with a 1970 state law which declared anyone with at least 1/32nd "Negro blood" to be black.

The Phipps case raised intiguing questions about the concept of race, its meaning in contemporary society, and its use (and abuse) in public policy. Assistant Attorney General Ron Davis defended the law by pointing out that some type of racial classification was necessary to comply with federal record-keeping requirements and to facilitate programs for the prevention of genetic diseases. Phipps's attorney, Brian Begue, argued that the assignment of racial categories on birth certificates was unconstitutional and that the 1/32nd designation was inaccurate. He called on a retired Tulane University professor who cited research indicating that most Louisiana whites have at least 1/20th "Negro" ancestry.

In the end, Phipps lost. The court upheld the state's right to classify and quantify racial identity. . . .

Phipps's problematic racial identity, and her effort to resolve it through state action, is in many ways a parable of America's unsolved racial dilemma. It illustrates the difficulties of defining race and assigning individuals or groups to racial categories. It shows how the racial legacies of the past—slavery and bigotry—continue to shape the present. It reveals both the deep involvement of the state in the organization and interpretation of race, and the inadequacy of state institutions to carry out these functions. It demonstrates how deeply Americans both as individuals and as a civilization are shaped, and indeed haunted, by race.

Having lived her whole life thinking that she was white, Phipps suddenly discovers that by legal definition she is not. In U.S. society, such an event is indeed catastrophic. But if she is not white, of what race is she? The *state* claims that she is black, based on its rules of classification . . . and another state agency, the court, upholds this judgment. But despite these classificatory standards which have imposed an either-or logic on racial identity, Phipps will not in fact "change color." Unlike what would have happened during slavery times if one's claim to whiteness was successfully challenged, we can assume that despite the outcome of her legal challenge, Phipps will remain in most of the social relationships she had occupied before the trial. Her socialization, her familial and friendship networks, her cultural orientation, will not change. She will simply have to wrestle with her newly acquired "hybridized" condition. She will have to confront the "Other" within.

The designation of racial categories and the determination of racial identity is no simple task. For centuries, this question has precipitated intense debates and conflicts, particularly in the U.S.—disputes over natural and legal rights, over the distribution of resources, and indeed, over who shall live and who shall die.

A crucial dimension of the Phipps case is that it illustrates the inadequacy of claims that race is a mere matter of variations in human physiognomy, that it is simply a matter of skin color. But if race cannot be understood in this manner, how *can* it be understood? We cannot fully hope to address this topic—no less than the meaning of race, its role in society, and the forces which shape it—in one [article], nor indeed in one book. Our goal in this [article], however, is far from modest: we wish to offer at least the outlines of a theory of race and racism.

WHAT IS RACE?

There is a continuous temptation to think of race as an *essence*, as something fixed, concrete, and objective. And there is also an opposite temptation: to imagine race as a mere *illusion*, a purely ideological construct which some ideal nonracist social order would eliminate. It is necessary to challenge both these positions, to disrupt and reframe the rigid and bipolar manner in which they are posed and debated, and to transcend the presumably irreconcilable relationship between them.

The effort must be made to understand race as an unstable and "decentered" complex of social meanings constantly being transformed by political struggle. With this in mind, let us propose a definition: *race is a concept which signifies and symbolizes social conflicts and interests by referring to different types of human bodies*. Although the concept of race invokes biologically based human characteristics (so-called "phenotypes"), selection of these particular human features for purposes of racial signification is always and necessarily a social and histori-cal process. In contrast to the other major distinction of this type, that of gen-der, there is no biological basis for distinguishing among human groups along the lines of race. . . . Indeed, the categories employed to differentiate among human groups along racial lines reveal themselves, upon serious examination, to be at best imprecise, and at worst completely arbitrary.

If the concept of race is so nebulous, can we not dispense with it? Can we not "do without" race: at least in the "enlightened" present? This question has been posed often, and with greater frequency in recent years. . . . An affirma-tive answer would of course present obvious practical difficulties: it is rather difficult to jettison widely held beliefs, beliefs which moreover are central to everyone's identity and understanding of the social world. So the attempt to banish the concept as an archaism is at best counterintuitive. But a deeper dif-ficulty, we believe, is inherent in the very formulation of this schema, in its way of posing race as a *problem*, a misconception left over from the past, and suitable now only for the dustbin of history.

A more effective starting point is the recognition that despite its uncertain-ties and contradictions, the concept of race continues to play a fundamental role in structuring and representing the social world. The task for theory is to explain this situation. It is to avoid both the utopian framework which sees race as an illusion we can somehow "get beyond," and also the essentialist for-mulation which sees race as something objective and fixed, a biological datum. Thus we should think of race as an element of social structure rather than as an irregularity within it; we should see race as a dimension of human repre-sentation rather than an illusion. These perspectives inform the theoretical approach we call racial formation.

Racial Formation

We define *racial formation* as the sociohistorical process by which racial cate-gories are created, inhabited, transformed, and destroyed. Our attempt to elab-orate a theory of racial formation will proceed in two steps. First, we argue that racial formation is a process of historically situated *projects* in which human bodies and social structures are represented and organized. Next we link racial formation to the evolution of hegemony, the way in which society is orga-nized and ruled. Such an approach, we believe, can facilitate understanding of a whole range of contemporary controversies and dilemmas involving race, including the nature of racism, the relationship of race to other forms of dif-ferences, inequalities, and oppression such as sexism and nationalism, and the dilemmas of racial identity today.

From a racial formation perspective, race is a matter of both social structure and cultural representation. Too often, the attempt is made to understand race simply or primarily in terms of only one of these two analytical dimensions. . . . For example, efforts to explain racial inequality as a purely social structural phenomenon are unable to account for the origins, patterning, and transfor-mation of racial difference.

Conversely, many examinations of racial difference—understood as a matter of cultural attributes à la ethnicity theory, or as a society-wide signification system, à la some poststructuralist accounts—cannot comprehend such structural phenomena as racial stratification in the labor market or patterns of residential segregation.

An alternative approach is to think of racial formation processes as occurring through a linkage between structure and representation. Racial *projects* do the ideological "work" of making these links. *A racial project is simultaneously an interpretation, representation, or explanation of racial dynamics, and an effort to recognize and redistribute resources along particular racial lines. Racial projects connect what race means* in a particular discursive practice and the ways in which both social structures and everyday experiences are racially *organized*, based upon that meaning. Let us consider this proposition, first in terms of large-scale or macro-level social processes, and then in terms of other dimensions of the racial formation process.

Racial Formation as a Macro-Level Social Process   *To interpret the meaning of race is to frame it social structurally.* Consider for example, this statement by Charles Murray on welfare reform:

> My proposal for dealing with the racial issue in social welfare is to repeal every bit of legislation and reverse every court decision that in any way requires, recommends, or awards differential treatment according to race, and thereby put us back onto the track that we left in 1965. We may argue about the appropriate limits of government intervention in trying to enforce the ideal, but at least it should be possible to identify the ideal: Race is not a morally admissible reason for treating one person differently from another. Period. . . .

Here there is a partial but significant analysis of the meaning of race: it is not a morally valid basis upon which to treat people "differently from one another." We may notice someone's race, but we cannot act upon that awareness. We must act in a "color-blind" fashion. This analysis of the meaning of race is immediately linked to a specific conception of the role of race in the social structure: it can play no part in government action, save in "the enforcement of the ideal." No state policy can legitimately require, recommend, or award different status according to race. This example can be classified as a particular type of racial project in the present-day U.S.—a "neoconservative" one.

Conversely, *to recognize the racial dimension in social structure is to interpret the meaning of race.* Consider the following statement by the late Supreme Court Justice Thurgood Marshall on minority "set-aside" programs:

> A profound difference separates governmental actions that themselves are racist, and governmental actions that seek to remedy the effects of prior racism or to prevent neutral government activity from perpetuating the effects of such racism. . . .

Here the focus is on the racial dimensions of *social structure*—in this case of state activity and policy. The argument is that state actions in the past and present have treated people in very different ways according to their race, and thus the government cannot retreat from its policy responsibilities in this area. It cannot suddenly declare itself "color-blind" without in fact perpetuating the same type of differential, racist treatment.... Thus, race continues to signify difference and structure inequality. Here, racialized social structure is immediately linked to an interpretation of the meaning of race. This example too can be classified as a particular type of racial project in the present-day U.S.—a "liberal" one.

To be sure, such political labels as "neoconservative" or "liberal" cannot fully capture the complexity of racial projects, for these are always multiply determined, politically contested, and deeply shaped by their historical context. Thus encapsulated within the neoconservative example cited here are certain egalitarian commitments which derive from a previous historical context in which they played a very different role, and which are rearticulated in neoconservative racial discourse precisely to oppose a more open-ended, more capacious conception of the meaning of equality. Similarly, in the liberal example, Justice Marshall recognizes that the contemporary state, which was formerly the architect of segregation and the chief enforcer of racial difference, has a tendency to reproduce those patterns of inequality in a new guise. Thus he admonishes it (in dissent, significantly) to fulfill its responsibilities to uphold a robust conception of equality. These particular instances, then, demonstrate how racial projects are always concretely framed, and thus are always contested and unstable. The social structures they uphold or attack, and the representations of race they articulate, are never invented out of the air, but exist in a definite historical context, having descended from previous conflicts. This contestation appears to be permanent in respect to race.

These two examples of contemporary racial projects are drawn from mainstream political debate; they may be characterized as center-right and center-left expressions of contemporary racial politics.... We can, however, expand the discussion of racial formation processes far beyond these familiar examples. In fact, we can identify racial projects in at least three other analytical dimensions: first, the political spectrum can be broadened to include radical projects, on both the left and right, as well as along other political axes. Second, analysis of racial projects can take place not only at the macro-level of racial policy-making, state activity, and collective action, but also at the micro-level of everyday experience. Third, the concept of racial projects can be applied across historical time, to identify racial formation dynamics in the past.

DISCUSSION QUESTIONS

1. What do Omi and Winant mean by *racial formation*? What role does the law play in such a process? How is this shown in the history of the United States?

2. What difference does it make to conceptualize race as a property of social structures versus as a property (or attribute) of individuals?

InfoTrac College Edition: Bonus Reading

http://infotrac.thomsonlearning.com

You can use the InfoTrac College Edition feature to find an additional reading pertinent to this section. You can also search on the author's name or keywords to locate the following reading:

> Garroutte, Eva Marie. 2001. "The Racial Formation of American Indians: Negotiating Legitimate Identities within Tribal and Federal Law." *The American Indian Quarterly* 25 (Spring): 224–241.

Garroutte uses the concept of racial formation to show how American Indians are defined as citizens in a given group. She also shows that this is a social process that is different in federal and tribal law. How does Garroutte's analysis illustrate the concept of the social construction of race and the role played by law (both federal and tribal) in this process. How does the experience of American Indians in this regard compare to and contrast with other racial-ethnic groups (such as African Americans, Latinos/Latinas, Asian Americans)?

STUDENT EXERCISES

1. Using some random method of assignment, your instructor will divide your class (or some other grouping) into two groups, one of which is designated the Blues and the other the Greens. Over a period of a week, the Greens should serve the Blues in any way the Blues ask—such as carrying their books, running errands for them, delivering meals to places they designate, or any other job that the Blues design. (Since this is a course assignment, you should be reasonable in your demands).

   As the week progresses, observe how the Blues act among themselves and in front of the Greens. Also observe how the Greens act among themselves and in front of the Blues. What attitudes do the two groups develop toward each other and toward themselves? How do they talk publicly about the other group? Do classmates begin to generalize about the assumed characteristics of the two different groups?

   What does the experiment reveal about the social construction of race?

2. Have members of your class describe the ethnic background of family members. You can describe such things as when and how your family arrived in the United States, ethnic traditions that your families may observe, whether ethnic pride is a part of your family experience. After hearing from classmates belonging to different ethnic groups, list what you learned about ethnicity from listening to these different experiences. Is ethnicity more significant for some groups than others? Is ethnicity more important to some generations within a family than others? Why? What does this teach you about the social construction of ethnicity in society?

PART III

INTRODUCTION

Representations of Race and Group Beliefs: Prejudice, and Racism

When you think about beliefs about race, the term *prejudice* most likely comes to mind. Prejudice is an attitude that tends to denigrate individuals and groups who are perceived to be somehow different and undesirable. The social scientific definition of prejudice dates back to the 1950s and the work of psychologist Gordon Allport. Allport defined **prejudice** as "a hostile attitude directed toward a person or group simply because the person is presumed to be a member of that group and is perceived as having the negative characteristics associated with the group" (Allport 1954: 7).

Prejudice can be directed at many groups. One can be prejudiced against women or gays or athletes or foreigners—anyone who is perceived as a member of an "out-group," that is, one different from one's own. Although prejudice can be positive (as when you think all women are nurturing), it is generally a negative attitude, involving hostile or derogatory feelings as well as false generalizations about people in the out-groups. Prejudice can also be expressed by any group—whether a dominant or subordinate group; thus, racial minorities may be prejudiced against other racial minorities or against the dominant group, just as more powerful people may be prejudiced against less powerful people. In other words, prejudice is a prejudgment, and is the basis for much racial intolerance.

Prejudice rests on social **stereotypes**—that is, oversimplified beliefs about members of a particular social group (Andersen and Taylor 2004). Stereotypes categorize people based on false generalizations along a narrow range of presumed characteristics, such as the belief that all Jewish people are greedy or that all blondes are dumb. Although stereotypes are perpetuated in many ways in society (e.g., in families, where parents teach children about other groups), one of the most influential is through popular culture—music, magazines, films, and television, among others. Thus, because men of color are portrayed in the media as criminals, this is the most common way that they are stereotyped. Or, Asian American women may be stereotyped as sexy and beguiling—an image repeatedly shown in magazines, videos, and other popular forms.

Because stereotypes are not just neutral depictions of people, they can be enormously harmful in that they influence how people define others and themselves. Research has found, for example, that the exposure of young Native American children to cartoonish depictions of Native Americans as school mascots actually depresses the children's self-esteem, that is, how they value themselves (Fryberg 2003). Stereotypes control how people come to define each other and, as such, are "controlling images" (Collins 1990) in U.S. culture.

The prejudice that grows out of social stereotypes is an *attitude.* This is distinct from discrimination, which is a behavior. **Discrimination** is the negative and unequal *treatment* of members of a social group based on their perceived membership in a particular group (Andersen and Taylor 2004: 320). Although the term *discrimination* sometimes has a positive connotation (as in "she has a discriminating attitude"), such behavior generally categorically excludes people from their civil rights.

Generally, prejudice and discrimination are thought of as related—prejudice causes discrimination—but things are not that simple. Over fifty years ago, sociologist Robert Merton (1949) developed a four-square typology, showing different ways that prejudice and discrimination are and are not related. Look at the following:

	PREJUDICE:	*POSITIVE* (+)	*NEGATIVE* (−)
Discrimination:	*Positive(+)*	Case 1: The bigot (++)	Case 2: Non-prejudiced discriminator (−+)
	Negative (−)	Case 3: Prejudiced non- discriminator (+−)	Case 4: The all-weather liberal (−−)

In Case 1, someone may be both prejudiced and discriminate; the classic bigot. Both prejudice and discrimination are overt, intentional, and hostile. In Case 4, someone may be free of prejudice and not discriminate (the person Merton called the "all-weather liberal"). It is in Cases 2 and 3 that we see that prejudice and discrimination may not have a causal relationship. In Case 2, one may not be prejudiced, but still discriminate, such as a homeowner who holds no racial prejudice but will only buy a home in a White people's neighborhood "to protect their property value." Such a person may say they hold no prejudice but most look out for their own interests. In Case 3, someone may be prejudiced, but *not* discriminate—for example, when the law prohibits discrimination. A landlord may, for example, rent to a Black tenant despite holding prejudice. The point is that both prejudice and discrimination occur in a larger context—that of society as a whole. This societal context as much predicts the occurrence of discrimination as do people's individual attitudes.

Prejudice is not just a free-floating attitude, however. It is linked to group positions in society (Blumer 1958; Bobo and Hutchings 1996). It is a learned behavior and it can be "unlearned." It can be intentional or unintentional. Public opinion surveys show that although most people do not now express overt prejudice, they do blame minorities themselves for their socioeconomic standing and they are resistant to policy changes that would actually bring about more racial equality. Lawrence Bobo refers to this trend in "The Color Line, The Dilemma, and the Dream" as laissez-faire racism. In **laissez-faire** ("hands off") **racism**, antiracist practices are abandoned and overt bigotry is minimal. Bobo shows how most people support equal rights in principle, but are unwilling to support practices and policies that would move our society in that direction. The problem is institutional racism, not prejudice per se.

Racism is a principle of social domination in which a group that is seen as inferior or different because of presumed biological or cultural characteristics is oppressed, controlled, and exploited—socially, economically, culturally, politically, psychologically—by a dominant group (Wilson 1973). Note the key elements of this definition. First, racism is a *principle of domination*—that is, embedded in this definition is the thought that racism involves a group or groups' subordinate position within a system of racial inequality.

Second the emphasis is on "presumed." As we learned in Part II, race is not "real" in the biological or cultural sense, but it has real meaning in society and through history. Race takes on meaning in the context of power relations (Feagin 2000)—thus how people are perceived within a system of hierarchy, and power is the key to understanding racism. Who gets the power to define different groups and what are the means by which they do it? Law? Media?

Schools? All of the above? Those who shape how people are represented have enormous power to shape people's consciousness about race.

Third, racism involves domination on a number of fronts—social, economic, political, cultural, and psychological. Although its economic effects are more easily seen, racism also involves the suppression of some groups' cultures and can shape the psychology of both dominant and subordinate groups. Most people think of racism only at the individual level, but it is present in society (the social front), and, as such, shapes people's minds, their interactions with others, and the opportunity structures in which groups find themselves.

Institutional racism is the complex and cumulated pattern of racial advantage and disadvantage built into the structure of a society (see also Feagin's discussion of systemic racism in the Part II). Institutional racism, reflected in the prejudice and discrimination seen in a society, comprises more than an attitude or behavior. It is a system of power and privilege that systematically advantages some groups over others. Thus you might say that prejudice is lodged in people's minds, but racism is lodged in society.

As many of the authors in Part III show, many people who benefit from institutional racism are often blind to the systemic advantage that it gives them. Thus, just being a White person may open to people some opportunities that might not otherwise be as readily available to others—independent of a White person's own attitudes and behavior.

The invisibility of racial privilege to dominant groups is reflected in the growth of so-called **color-blind racism**—the belief that race should be ignored and that race-conscious practices and policies only foster more racism. When dominant groups think that racism is no longer an issue, despite its ongoing reality, they are not likely to engage in practices or support policies that challenge racism (Brown et al. 2003; Bonilla-Silva 2003). To be color-blind in a society in which race still structures people's relationships, identities, and opportunities is to be blind to the continuing realities of race.

Charles Gallagher examines color-blind privilege in his essay, "Color-Blind Privilege: The Social and Political Functions of Erasing the Color Line in Post-Race America." He points out that we live with the appearance of a multiracial society in which we "all get along." In this context, race has actually become a commodity—something that White people can buy and display, while at the same time not challenging the privilege that underlies a system of racial stratification. Products are mass-marketed using multiracial images and presumably sell across color lines, but such images legitimate color-blindness. Gallagher's research shows that while many White people believe themselves to be color-blind, once you scratch the surface of this belief, they are quick to defend the status quo.

In "Learning to be White through the Movies," Hernán Vera and Andrew M. Gordon show another dimension of racial representations in everyday life—popular culture. People generally go to the movies to just be entertained, but movies, like other visual and printed media, communicate powerful images of race, images that influence how we see ourselves and others. Vera and Gordon examine films critically, asking who do we see in films and how are they portrayed? They find that films implicitly celebrate and promote White people, while at the same time obscuring White privilege by making it seem natural or "just normal." The next time you see a movie or a television show, or read a popular magazine, or watch the news, go beyond the surface. Rather than just taking what you see at face value, try to examine how White people and people of color are portrayed. Think about how these portrayals shape people's understandings of race in society.

Racial minority groups often develop an improvisational culture that flows from limited resources but much creativity. Often the culture that emerges is not only expressive about the lived experience of oppressed groups and their cultures, but it also provides a critical perspective on the dominant culture. Tricia Rose ("Hidden Politics") discusses how rap music, especially when associated with young, urban, African American men, is perceived as a threat to society and thus is heavily policed. Her analysis shows that, even in oppressive circumstances, people develop oppositional cultures—that is, they do not just passively acquiesce to how dominant groups see them. Rap and hip-hop are good examples, although, ironically, they have now become the popular music of White youth as well.

Finally, Charles Springwood and C. Richard King ("Playing Indian": Why Native American Mascots Must End") take on the use of Native American icons in popular culture. These icons, mostly in the form of caricatures, are everywhere—as part of team sports names, on products in the grocery store, in school textbooks. These images distort understanding of the realities of Native American life and stereotype Indians as warriors, on the one hand, savages, cartoonish on the other. Note, too, how gender is displayed in such images. Although some claim that such portrayals are all in fun, you might ask: Would it seem so light-hearted if their own group were depicted in such narrow, repetitive, and joking ways?

Racial inequality in the United States has resulted in enormous misunderstanding of and misinformation about different groups. And, this is true for both dominant and subordinate groups. Because many people, especially White people, do not interact regularly with people outside their own group, understanding across racial-ethnic lines is blunted. As you read the articles in this section, think about the racial representations you see around you and how they might be transformed or challenged.

REFERENCES

Allport, Gordon. 1954. *The Nature of Prejudice*. Reading, MA: Addison-Wesley.

Andersen, Margaret L., and Howard F. Taylor. 2004. *Sociology: Understanding a Diverse Society*. Belmont, CA: Wadsworth.

Blumer, Herbert. 1958. "Race Prejudice as a Sense of Group Position." *Pacific Sociological Review* 1 (Spring): 3–7.

Bobo, Lawrence, and Vincent L. Hutchings. 1996. "Perceptions of Racial Group Competition: Extending Blumer's Theory of Group Position to a Multiracial Social Context." *American Sociological Review* 25 (December): 951–972.

Bonilla-Silva, Eduardo. 2003. *Racism without Racists: Colorblind Racism and the Persistence of Racial Inequality in the United States*. Lanham, MD: Rowman & Littlefield.

Brown, Michael, Martin Carnoy, Elliott Currie, Troy Duster, David Oppenheimer, Marjorie M. Schultz, and David Wellman. 2003. *Whitewashing Race: The Myth of a Color-Blind Society*. New York: Oxford University Press.

Collins, Patricia Hill. 1990. *Black Feminist Thought: Knowledge, Consciousness, and the Politics of Empowerment*. Boston: Unwin Hyman.

Feagin, Joe. 2000. *Racist America: Roots, Realities, and Reparations*. New York: Routledge.

Fryberg, Stephanie. 2003. "Really, You Don't Look Like an American Indian: Social Representations and Social Group Identities." Ph.D. dissertation, Department of Psychology, Stanford University.

Merton, Robert. 1949. "Discrimination and the American Creed." *Discrimination and the National Welfare*, ed. Robert W. MacIver. New York: Harper and Brothers. pp. 99–126.

Wilson, William Julius. 1973. *Power, Racism, and Privilege: Race Relations in Theoretical and Sociohistorical Perspectives*. New York: Macmillan.

11

The Color Line, the Dilemma, and the Dream

Race Relations in America at the Close of the Twentieth Century

BY LAWRENCE D. BOBO

At the dawning of the twentieth century W. E. B. Du Bois forecast that the defining problem of the twentieth century would be "the color line."[1] His analysis was penetrating. He wrote at a time when most African Americans lived as disenfranchised, purposely miseducated, and brutally oppressed second-class citizens. He wrote at a time when popular conceptions of African Americans were overtly racist, even among the well-educated white elite. As we stand near the dawning of the twenty-first century, the problem to which Du Bois so presciently drew our attention appears no closer to a fundamental resolution in the United States or in much of the rest of the world than it did a century ago. The color line endures. . . .

The review of information of changing racial attitudes discussed below will support three conclusions. . . . First, the available data suggest that the United States has experienced a genuine and tremendous positive transformation in racial attitudes. A once predominant ideology of Jim Crow racism has, over the past five decades, steadily receded from view. Nonetheless, . . . the collapse of

From *Civil Rights and Social Wrongs: Black-White Relations Since World War II*, ed. by John Higham (University Park, PA: Pennsylvania State University Press, 1999), pp. 33–55. Reprinted by permission.

Jim Crow racism has not been followed by a full embrace of African Americans as true co-equals with whites, deserving of a complete measure of all the fruits of membership in the polity. Instead, a new configuration of negative racial attitudes has recently crystallized. This new cultural pattern of understanding the actual and normative position of blacks in American society is appropriately labeled "laissez-faire racism."

Second, racial discrimination remains a barrier to blacks' full economic, political, and social participation in American institutions. The problem of racial discrimination today is less extreme, absolute, and all-encompassing. Hence, I reject claims that only superficial change in the position of blacks has taken place. However, direct discrimination in jobs, in housing, and in myriad forms of interpersonal interaction continue to face African Americans almost irrespective of the social class background and achievements of the individual black person. In short, the significance of race in social life goes on.

This is an extremely nettlesome issue. Black and white Americans, the surveys show, could not be further apart in their thinking about the problem of discrimination. Blacks perceive it, experience it, and feel it acutely. They become frustrated when whites do not see the problem in the same way. Many whites see tremendous positive gains, cannot understand the steady litany of complaints, and have grown resentful and impatient. Miscommunication and mounting resentments accumulate on both sides as a result.

Third, the problem of social breakdown occurring in poor urban communities has a strong racial overlay and increasing political potence as a device to mobilize voters at the local, state, and national levels. The linkage in the minds of many white Americans between black culture and the problem of family dissolution, welfare dependency, crime, failing schools, and drug use may be setting the stage for a new period of deep retrenchment in civil rights and social welfare provision. All too often, a major subtext of campaigns about reducing welfare and fighting crime is a narrative about generally retaining white status privilege over blacks, and specifically about controlling and punishing poor black communities. This thinly veiled racial subtext of American politics is not lost on the black community. It feeds a growing suspicion and distrust among African Americans that white-dominated institutions may be moving toward overt hostility to the aspirations of African American communities. . . .

THE DECLINE OF JIM CROW RACISM

The available survey data suggest that antiblack attitudes associated with Jim Crow racism were once widely accepted. The Jim Crow social order called for a society based on deliberate segregation by race. It gave positive sanction to antiblack discrimination in economics, education, and politics. It prohibited race-mixing, especially in the form of miscegenation or racial intermarriage. It involved an etiquette of interaction designed to reinforce the inferior status imposed on African Americans. All of this was expressly premised on the notion that blacks were the innate intellectual, cultural, and temperamental inferiors to whites.

Survey-based questions dealing with racial principles essentially asked about the degree of popular endorsement of these tenets of Jim Crow racism. The evidence from national sample surveys of the white population show that at one time the Jim Crow ideology was widely accepted, especially among residents of southern states but outside the South as well. Over the ensuing five decades since the first baseline surveys were conducted in the early 1940s, however, support for Jim Crow has steadily declined. It has been replaced by popular support for integration and equal treatment as principles that should guide relations between blacks and whites.

For example, whereas a solid majority, 68 percent, of white Americans in 1942 favored racially segregated schools, only 7 percent took such a position as early as 1985. Similarly, 55 percent of whites surveyed in 1944 believed whites should receive preference over blacks in access to jobs, as compared with only 3 percent who offered such an opinion as early as 1972. Indeed, so few people are willing to endorse a discriminatory response to either question that both have been dropped from ongoing social surveys. On both of these issues, once-pivotal features of the Jim Crow racist ideology—majority endorsements of the principles of segregation and discrimination—have given way to overwhelming support for the principles of integration and equal treatment.

This pattern of movement away from support for Jim Crow toward apparent support for racial egalitarianism holds with equal force for those questions dealing with issues of residential integration, access to public transportation and public accommodations, choice among qualified candidates for political office, and even racial intermarriage. It is important to note that the high absolute levels of support seen for the principles of school integration and equal access to jobs (both better than 90 percent nationwide) is not achieved for all racial principle–type questions. Despite improvement from extraordinarily low levels of support in the 1950s and 1960s, survey data continue to show substantial levels of white discomfort with the prospect of interracial dating and marriage.

Opinions among whites have never been uniform or monolithic. Both historical and sociological research have pointed to lines of cleavage and debate in the thinking of whites about the place of African Americans. The survey-based literature has shown that views on issues of racial principle vary greatly by region (at least along the traditional South versus non-South divide), level of education, age or generation, and other ideological factors. As might be expected, opinions in the South more lopsidedly favored segregation and discrimination at the time baseline surveys were conducted than was true outside the South. Patterns of change, save for a period of unusually rapid change in the South, have usually been parallel.

Level of education matters for racial attitudes. The highly educated are also typically found to express greater support for principles of racial equality and integration. Indeed, one can envision separating the white population into a multitiered reaction to issues of racial justice based on the interaction of level of education and region. At the more progressive and liberal end, one finds college-educated whites who live outside the South. At the bottom, one finds Southern whites with the least amount of schooling.

The degree of expressed support for racial integration and equality is also responsive to a person's age. Younger people are usually more racially tolerant than older people, where issues of principle are concerned. A small but meaningful part of the apparent generational differences in racial attitudes is attributable to the increasing levels of education obtained by younger cohorts. As the average level of education has risen, so too has the level of support for racial equality and integration. . . .

The transformation of attitudes regarding the rules that should guide interaction between blacks and whites in public and impersonal spheres of life has been large, steady, and sweeping. Individuals living outside the South, the highly educated, and younger-age cohorts led the way to these changes. However, positive changes usually occurred across regions, age-groups, and education levels. At least with regard to racial principles, this change was so sweeping that Schuman, Steeh, and myself characterized it as a fundamental transformation of social norms with regard to race. Analysts of in-depth interview material reached similar conclusions. Bob Blauner's discussions with a small group of blacks and whites living in the San Francisco Bay area led him to conclude:

> The belief in a right to dignity and fair treatment is now so widespread and deeply rooted, so self-evident, that people of all colors would vigorously resist any effort to reinstate formalized discrimination. This consensus may be the most profound legacy of black militancy, one that has brought a truly radical transformation in relations between the races (1989).[2]

Some read this change as having far-reaching implications for transcending the color line and overcoming the American Dilemma. Based on their assessment of the available trend data, one group of survey researchers concluded: "Without ignoring real signs of enduring racism, it is still fair to conclude that America has been successfully struggling to resolve its Dilemma and that equality has been gaining ascendancy over racism."[3]

THE EMERGENCE OF
LAISSEZ-FAIRE RACISM

The sanguine picture of change, however, changes substantially when attention shifts from principle to policy. Where issues of implementing the social changes needed to bring about greater integration of communities, schools, and workplaces are concerned, we find evidence of an important qualification on the extent of change in racial attitudes. Likewise, where issues of enforcing antidiscrimination laws and taking steps to improve the economic standing of African Americans are concerned, the survey data again point to significant bounds on the scope of positive change in racial attitudes.

At one level, it is not surprising that there are sharp differences in level of support between racial principles and policy implementation. Principles, when viewed in isolation, need not conflict with other principles, interests, or needs. At another, more concrete level, however, choices must often be made and priorities set. Particularly in the domain of racial attitudes and relations, there are large gaps between the principles most white Americans advocate and the practical steps they are willing to undertake to pursue those ends. For example, a 1964 survey showed that 64 percent of whites nationwide supported the principle of integrated schooling, but that only 38 percent believed that the federal government had a role to play in bringing about greater integration. The gap had actually grown larger by 1986. At that time, 93 percent of whites supported the principle of integrated schooling, but only 26 percent endorsed government efforts to increase the level of school integration. Analysis suggests that little of this disjuncture can be accounted for by distrust of government.

Similar patterns emerge in the areas of jobs and housing. Support for the principle of equal access to jobs stood at 97 percent in 1972, while support for federal efforts to prevent discrimination in jobs had only reached 39 percent. A 1976 national survey showed that 88 percent of whites supported the idea that blacks have a right to live wherever they can afford, that only 35 percent would vote in favor of a law prohibiting homeowners from discriminating on the basis of race when selling a home. While the sort of near-exact pairing of principle and implementation items that Schuman, Steeh, and myself were able to compare for the 1942-85 period is no longer possible with the available data, it seems likely that this disjuncture continues.

Implementation items not only typically exhibit lower absolute levels of support than principle items do, and less evidence of positive change, they also are less responsive to a number of other factors. There are smaller differences by age, education level, and region on questions of implementation. The lack of strong age-group differences, education effects, and regional differences implies that policy change in the area of race is not likely to witness the sort of great positive transformation seen for broad issues of principle.

Furthermore, the black-white divide on implementation questions is also sharp. Most of the implementation questions Schuman, Steeh, and myself analyzed showed majority—sometimes overwhelming majority—black support for government action to bring about integration, to fight discrimination, or to improve the economic conditions of the black community. Responses of whites were often a mirror opposite, with a clear majority opposing such a role for government.

The black-white division of opinion is often sharpest on questions of affirmative action. Bobo and Smith reported that 69 percent of white respondents in a major 1990 national survey opposed affirmative action in higher education for blacks, while only 26 percent of blacks opposed it. An equally large gap between black and white views emerged for the question of affirmative action in employment. Whereas 82 percent of whites opposed giving preference to blacks in hiring and promotion, only 37 percent of blacks adopted a similar position.

Bobo and Smith also attempted to determine whether the black-white gap in opinions could be explained by differences in social class background, political ideology and values, and racial attitudes. Their analyses of opinions on several race-specific social policies showed that virtually none of the black-white gap in opinion could be explained by social class blackground factors, that a small part reflected differences in political ideology and basic beliefs about economic inequality, and that a somewhat larger fraction was attributable to different racial outlooks. However, a substantial difference in views persisted, despite controls for a wide range of factors. This suggests, as myself and James Kluegel argue, that there is a set of collective interests that divide blacks and whites on policy questions of race.

Part of the reason for the opposition of whites to policy changes favorable to blacks may be found in the persistence of antiblack stereotypes. A 1990 national survey found evidence of widespread negative stereotypes of blacks, Hispanics, and Asian Americans. The study showed that whites tend to perceive blacks as more likely than whites to be unpatriotic, violence-prone, unintelligent, lazy, and to prefer to live off of welfare rather than being self-supporting. Fully 78 percent of whites in this national survey adopted this position. Such high levels of negative stereotyping were found because of a measurement procedure that did not force people to make categorical judgments, but rather allowed the expression of differing magnitudes of group difference. Whereas an overall 78 percent of whites saw blacks as more likely than white people to prefer to live off welfare, only a small number understood this to be a stark difference. Most whites expressed only a small difference between the races in tendencies on this and the other trait dimensions examined. The evidence of persistent negative stereotyping is not at odds with evidence of positive change in racial principles. These data merely qualify the scope and meaning of those positive changes.

Negative stereotypes help explain the gap between principle and implementation described above and shed light on current racial tensions. The more negative stereotypes a person holds about blacks, the less likely he or she is to support affirmative action policies. In addition, those with the highest negative stereotypes are strongly predisposed to maintain social distance between themselves and blacks. In sum, while Jim Crow racism has fallen into disrepute, images of blacks appear to have remained sufficiently negative to prompt many whites to reject affirmative action and to resist close association with blacks.

Another factor contributing to black-white polarization on the policy solutions to racial inequality is a substantial difference in their thinking about the problem of racial discrimination. The available survey data suggest that most blacks see racial discrimination as a more prevalent problem than do most whites, as having a stronger institutional base than do most whites, and as carrying responsibility for the generally disadvantaged position blacks occupy in American society than do most whites. The black-white gap in thinking about discrimination may be the core factor underlying modern misunderstanding and miscommunication across the color line.

In their 1992 study of the Detroit housing market, Farley and colleagues asked a general question about how much discrimination blacks faced in finding housing wherever they want. They found that 85 percent of blacks perceived "a lot" or at least "some" discrimination, compared with 80 percent of whites. While this difference is not large, they also found that blacks perceived discrimination as having changed little or gotten worse, whereas whites saw discrimination as declining. In addition, they wrote: "Blacks see much more institutionalized discrimination than do whiter. . . . While 86 percent of blacks believe that blacks miss out on good housing because real estate agents discriminate, only 61 percent of whites believe this. When it comes to banks and lenders, 89 percent of blacks see discrimination, in contrast, to 56 percent of whites."[4]

Thus, although many whites acknowledge the existence of discrimination, especially in the housing market, they are much less likely than blacks to think of discrimination as having a systematic, institutionalized social basis. . . .

According to Bobo and Smith, laissez-faire racism is a new form of racist ideology that has emerged in the post–Jim Crow, post–civil rights era. Although the full argument about laissez-faire racism cannot be developed here, its emergence reflects crucial changes in the economy and polity that at once undermined the structural basis for the Jim Crow social order of the American South and yet left in place the patterns of residential segregation, economic inequality, racialized identities, and antiblack outlooks that existed on a national basis. This new pattern of belief involves staunch rejection of an active role for government in undoing racial segregation and inequality, an acceptance of negative stereotypes of African Americans, a denial of discrimination as a current societal problem, and attribution of primary responsibility for black disadvantage to blacks themselves. . . .

Does the color line still divide us? Does the dilemma still persist? Is the dream ever to be realized? There is little question that the United States is still sharply divided by race, that the struggle to fulfill national democratic ideals goes on, and that the dream remains far, very far, from realization. As pessimistic as these answers sound, they do not, however, amount to an acceptance of the idea that no meaningful change in American racism has or can take place. The "racial chasm" and "permanence of racism" to which [some authors]. . . have written are seriously overstated.

We have witnessed the substantial disappearance of one epochal form of racist ideology. Jim Crow racism once dominated the views white Americans had of black people. This ideology is no longer widely accepted or publicly espoused by any significant number of people. While segregationist notions, including explicitly biological racist ideas, have not completely vanished, they show no sign of making a quick or easy return to popular acceptance.

Nonetheless, a new epochal form of racism has emerged. Bobo and Smith label it "laissez-faire racism." It is a cultural pattern of belief that connects opposition to substantial policy change and activism with regard to improving the status of blacks with negative stereotyping of African Americans, a tendency to deny the potency of modern discrimination, and a view of blacks as largely responsible for their own disadvantaged circumstances. The emergence of this new form of American racism can be traced, in the first instance, to the historical erosion of the structural basis for the Jim Crow economic, political, and social institutions. In the second instance, it can be traced to the important but partial victories of the civil rights movement. The 1954 *Brown v. Board of Education* decision, the Civil Rights Act of 1964, and the Voting Rights Act of 1965 secured the basic citizenship rights of African Americans. As fundamental as these gains were, these accomplishments did not directly undo racial residential segregation or the individual and institutional actors that sustain it; these accomplishments did not directly undo tremendous disparities in earnings, prospects for employment, and wealth-holding that constituted the core structural bases of racial economic inequality; and these accomplishments did not wipe away racial identities and a long cultural heritage of disparaging views of African Americans. Thus, a new antiblack ideology has crystallized, an ideology that is appropriate to a historical epoch in which we have a formally race-neutral state and economy but a still racially divided social order and quality of life experience.

Laissez-faire racism has crystallized despite mounting evidence of discrimination against blacks with regard to housing, jobs, and access to many public spaces. The political climate makes the challenge of improving race relations that much greater. The very center of American politics is now infused with thinly veiled racial appeals that call for rolling back the welfare state, erecting more prisons, lengthening prison sentences, making it easier to convict the accused, and bringing greater certainty and swiftness to the execution of death sentences. While not based solely in efforts to harm the black community and its interests, or simply in the cynical pursuit of votes by opportunistic politicians, the current political climate has a powerful and undeniably racial subtext. The product of these trends is a potentially widening gulf of perception, of understanding, of feelings, and of interests across the color line.

Some, . . . look at these complex problems, see no real change, and see little hope for progress in the future, but that prognosis is difficult to accept. Racial categories and identities are social constructs. They are not given in nature, and they are subject to enormous variation over historical time and space. To be sure, the black-white divide in the United States is deeply entrenched structurally, culturally, and psychologically. But simply as a matter of logic and historical experience, the color line is not unmodifiable.

If historical figures like Thurgood Marshall, Fannie Lou Hamer, Rosa Parks, and Martin Luther King Jr. had simply accepted Jim Crow racism rather than challenging it head on, we almost surely would not have had a *Brown* decision, the Civil Rights Act, or the Voting Rights Act. We would not have witnessed the emergence of a black middle class that is at least modestly better residentially integrated and as large and accomplished as we have now. We would just as surely not have as many appointed and elected black public officials.

We stand at a moment of great ambiguity, uncertainty, and potentially momentous change in race relations. Lack of clarity about the future, however, does not warrant pessimistic certainty. The present is a time of deeply contradictory trends, not one of unequivocal backlash and polarization. Positive changes in racial attitudes and relations do not simply happen. They are made, in both intended and unintended ways. Laissez-faire racism, modern discrimination, and the current subtext of race in American politics are harder to confront directly than the obvious racism of the Jim Crow era. But the only path to transcending the color line is a continuous struggle to resolve the . . . dilemma of race and to steadfastly pursue Dr. King's dream.

REFERENCES

1. W. E. B. Du Bois, *The Souls of Black Folk* (Greenwich, Conn.: Fawcett, 1903).

2. Bob Blauner, *Black Lives, White Lives: Three Decades of Race Relations in America* (Berkeley and Los Angeles: University of California Press, 1989), 317.

3. Richard G. Niemi, John Mueller, and Tom W. Smith, *Trends in Public Opinion: A Compendium of Survey Data* (New York: Greenwood Press, 1989), 168.

4. Farley, Reynolds, Charlotte Steeh, Tara Jackson, Maria Kryan, and Heith Reeves. 1993. "Continued Racial Residential Segregation in Detroit; (Chocolate City, Vanilla Suburb) Revisited." *Journal of Housing Research 4:* 1–38.

DISCUSSION QUESTIONS

1. What does Bobo mean by *laissez faire racism*? How does it contrast with *Jim Crow racism*?

2. What evidence does Bobo present to support his argument that there is a gap in the principles and policies that most White Americans are willing to support with regard to race in the Untied States?

12

Color-Blind Privilege

The Social and Political Functions of Erasing the Color Line in Post Race America

BY CHARLES A. GALLAGHER

The young white male sporting a FUBU (African-American owned apparel company "For Us By Us") shirt and his white friend with the tightly set, perfectly braided cornrows blended seamlessly into the festivities at an all white bar mitzvah celebration. A black model dressed in yachting attire peddles a New England, yuppie boating look in Nautica advertisements. It is quite unremarkable to observe white, Asian or African-Americans with dyed purple, blond or red hair. White, black and Asian students decorate their bodies with tattoos of Chinese characters and symbols. In cities and suburbs young adults across the color line wear hip-hop clothing and listen to white rapper Eminem and black rapper 50-cent. It went almost unnoticed when a north Georgia branch of the NAACP installed a white biology professor as its president. Subversive musical talents like Jimi Hendrix, Bob Marley and The Who are now used to sell Apple Computers, designer shoes and SUVs. Du-Rag kits, complete with bandana headscarf and elastic headband, are on sale for $2.95 at hip-hop clothing stores and family centered theme parks like Six Flags. Salsa has replaced ketchup as the best selling condiment in the United Sates. Companies as diverse as Polo, McDonalds, Tommy Hilfiger, Walt Disney World, Master Card, Skechers sneakers, IBM, Giorgio Armani and Neosporin antibiotic ointment have each crafted advertisements that show an integrated, multiracial cast of characters interacting and consuming their products in post-race, color-blind world.

From *Race, Gender & Class*, 10, 2003, pp. 22–37. Reprinted by permission.

Americans are constantly bombarded by depictions of race relations in the media which suggest that discriminatory racial barriers have been dismantled. Social and cultural indicators suggest that America is on the verge, or has already become, a truly color-blind nation. National polling data indicate that a majority of whites now believe discrimination against racial minorities no longer exists. A majority of whites believe that blacks have "as good a chance as whites" in procuring housing and employment or achieving middle class status while a 1995 survey of white adults found that a majority of whites (58%) believed that African Americans were "better off" finding jobs than whites (Gallup, 1997; Shipler, 1998). Much of white America now see a level playing field, while a majority of black Americans sees a field which is still quite uneven. . . . The color-blind or race neutral perspective holds that in an environment where institutional racism and discrimination have been replaced by equal opportunity, one's qualifications, not one's color or ethnicity, should be the mechanism by which upward mobility is achieved. Color as a cultural style may be expressed and consumed through music, dress, or vernacular but race as a system which confers privileges and shapes life chances is viewed as an atavistic and inaccurate accounting of U.S. race relations.

Not surprisingly, this view of society blind to color is not equally shared. Whites and blacks differ significantly, however, on their support for affirmative action, the perceived fairness of the criminal justice system, the ability to acquire the "American Dream," and the extent to which whites have bene-fited from past discrimination (Moore, 1995; Moore & Saad, 1995; Kaiser, 1995). This article examines the social and political functions colorblindness serves for whites in the United States. Drawing on interviews and focus groups with whites from around the country I argue that colorblind depictions of U.S. race relations serves to maintain white privilege by negating racial inequality. Embracing a colorblind perspective reinforces whites' belief that being white or black or brown has no bearing on an individual's or a group's relative place in the socio-economic hierarchy.

DATA AND METHOD

I use data from seventeen focus groups and thirty-individual interviews with whites from around the country. Thirteen of the seventeen focus groups were conducted in a college or university setting, five in a liberal arts college in the Rocky Mountains and the remaining eight at a large urban university in the Northeast. Respondents in these focus groups were selected randomly from the student population. Each focus group averaged six respondents . . . equally divided between males and females. An overwhelming majority of these respondents were between the ages of eighteen and twenty-two years of age. The remaining four focus groups took place in two rural counties in Georgia and were obtained through contacts from educational and social service providers in each county. One county was almost entirely white (99.54%) and in the other county whites constituted a racial minority. These four focus groups allowed me to tap rural attitudes about race relations in environments where whites had little or consistent contact with racial minorities. . . .

COLORBLINDNESS AS NORMATIVE IDEOLOGY

The perception among a majority of white Americans that the socio-economic playing field is now level, along with whites' belief that they have purged themselves of overt racist attitudes and behaviors, has made colorblindness the dominant lens through which whites understand contemporary race relations. Colorblindness allows whites to believe that segregation and discrimination are no longer an issue because it is now illegal for individuals to be denied access to housing, public accommodations or jobs because of their race. Indeed, lawsuits alleging institutional racism against companies like Texaco, Denny's, Coke, and Cracker Barrel validate what many whites know at a visceral level is true; firms which deviate from the color blind norms embedded in classic liberalism will be punished. As a political ideology, the commodification and mass marketing of products that signify color but are intended for consumption across the color line further legitimate colorblindness. Almost every household in the United States has a television that, according to the U.S. Census, is on for seven hours every day (Nielsen 1997). Individuals from any racial background can wear hip-hop clothing, listen to rap music (both purchased at Wal-Mart) and root for their favorite, majority black, professional sports team. Within the context of racial symbols that are bought and sold in the market, colorblindness means that one's race has no bearing on who can purchase a Jaguar, live in an exclusive neighborhood, attend private schools or own a Rolex.

The passive interaction whites have with people of color through the media creates the impression that little, if any, socio-economic difference exists between the races. . . .

Highly visible and successful racial minorities like [former] Secretary of State Colin Powell and . . . [Secratary of State] Condelleeza Rice are further proof to white America that the state's efforts to enforce and promote racial equality have been accomplished.

The new color-blind ideology does not, however, ignore race; it acknowledges race while disregarding racial hierarchy by taking racially coded styles and products and reducing these symbols to commodities or experiences that whites and racial minorities can purchase and share. It is through such acts of shared consumption that race becomes nothing more than an innocuous cultural signifier. Large corporations have made American culture more homogenous through the ubiquitousness of fast food, television, and shopping malls but this trend has also created the illusion that we are all the same through consumption. Most adults eat at national fast food chains like McDonalds, shop at mall anchor stores like Sears and J.C. Penney's and watch major league sports, situation comedies or television drama. Defining race only as cultural symbols that are for sale allows whites to experience and view race as nothing more than a benign cultural marker that has been stripped of all forms of institutional, discriminatory or coercive power. The post-race, color-blind perspective allows whites to imagine that depictions of racial minorities working in high status jobs and consuming the same products, or at least appearing in commercials for products whites desire or consume, is the same as living in a society where color is no longer used to allocate resources or shape group outcomes. By constructing a picture of society where racial harmony is the norm, the color-blind perspective functions to make white privilege invisible while removing from public discussion the need to maintain any social programs that are race-based.

How then, is colorblindness linked to privilege? Starting with the deeply held belief that America is now a meritocracy, whites are able to imagine that the socio-economic success they enjoy relative to racial minorities is a function of individual hard work, determination, thrift and investments in education. The color-blind perspective removes from personal thought and public discussion any taint or suggestion of white supremacy or white guilt while legitimating the existing social, political and economic arrangements which privilege whites. This perspective insinuates that class and culture, and not institutional racism, are responsible for social inequality. Colorblindness allows whites to define themselves as politically and racially tolerant as they proclaim their adherence to a belief system that does not see or judge individuals by the "color of their skin." This perspective ignores, as Ruth Frankenberg puts it, how whiteness is a "location of structural advantage societies structured in racial dominance" (2001 p. 76). . . . Colorblindness hides white privilege behind a mask of assumed meritocracy while rendering invisible the institutional arrangements that perpetuate racial inequality. The veneer of equality implied in colorblindness allows whites to present their place in the racialized social structure as one that was earned.

OPPORTUNITY HAS NO COLOR

Given this norm of colorblindness it was not surprising that respondents in this study believed that using race to promote group interests was a form of (reverse) racism. . . .

Believing and acting as if America is now color-blind allows whites to imagine a society where institutional racism no longer exists and racial barriers to upward mobility have been removed. The use of group identity to challenge the existing racial order by making demands for the amelioration of racial inequities is viewed as racist because such claims violate the belief that we are a nation that recognizes the rights of individuals not rights demanded by groups. . . .

The logic inherent in the colorblind approach is circular; since race no longer shapes life chances in a color-blind world there is no need to take race into account when discussing differences in outcomes between racial groups.

This approach erases America's racial hierarchy by implying that social, economic and political power and mobility is equally shared among all racial groups. Ignoring the extent or ways in which race shapes life chances validates whites' social location in the existing racial hierarchy while legitimating the political and economic arrangements that perpetuate and reproduce racial inequality and privilege.

REFERENCES

Frankenberg, R. (2001). The mirage of an unmarked whiteness. In B.B. Rasmussen, E. Klineberg, I.J. Nexica & M. Wray (eds.) *The making and unmaking of whiteness.* Durham: Duke University Press.

Gallup Organization. (1997). Black/white relations in the U.S. June 10, pp. 1–5.

Kaiser Foundation. (1995). *The four Americas: Government and social policy through the eyes of America's multi-racial and multi-ethnic society.* Menlo Park, CA: Kaiser Family Foundation.

Moore, D. (1995). "Americans" most important sources of information: Local news." *The Gallup Poll Monthly,* September, pp. 2–5.

Moore, D. & Saad, L. (1995). No immediate signs that Simpson trial intensified racial animosity. *The Gallup Poll Monthly,* October, pp. 2–5.

Nielsen, A.C. (1997). *Information please almanac* (Boston: Houghton Mifflin).

Shipler, D. (1998). *A country of strangers: Blacks and whites in America.* New York: Vintage Books.

DISCUSSION QUESTIONS

1. How does Gallagher see color-blind racism as resulting from White people's privilege? How does privilege influence what White people can understand about racism?

2. In what ways does color-blind racism support the traditional American ideal that any individual can succeed if they only try hard enough?

13

Learning to Be White through the Movies

BY HERNÁN VERA
AND ANDREW M. GORDON

Why does Gone with the Wind *touch such deep chords inside me? Maybe because it put those chords there in the first place. This is the movie that taught me and three generations how to be Southerners. It doesn't move us because we are Southern; we are Southern because we have taken this movie to heart.*
—*Susan Stewart*[1]

I was the only Negro in the theater, and when Butterfly McQueen went into her act, I felt like crawling under the rug. —*Malcolm X*[2]

W e need to study movies because ordinarily we do not want to think about the influence that they have on us and on our society. We tend to dismiss the cinema as mere entertainment; yet it has profound effects, shaping our thinking and our behavior. As Susan Stewart proposed at the beginning of this chapter, movies can teach us who we are: what our identity is and what it should be. "Radio, television, film and the other products of the culture industries," Douglas Kellner argues, "provide the models of what it means to be male or female, successful or a failure, powerful or powerless. Media culture also provides the materials out of which many people construct their sense of class, of ethnicity and race, of nationality, of sexuality, of "us" and "them." Movies manufacture the way we see, think of, feel, and act towards others."[3]

We need to study movies not only because of what they tell us about the world we live in but also, and most importantly, because movies are a crucial part of that world. In the simulations of the moving pictures we learn who has

From Hernán Vera and Andrew M. Gordon, *Screen Saviors: Hollywood Fictions of Whiteness* (Lanham, MD: Rowman & Littlefield, 2003), pp. 8–13. Reprinted by permission of Rowman & Littlefield Publishers, Inc.

the power and who is powerless, who is good and who is evil. "Media spectacles," Kellner writes, "dramatize and legitimate the power of the forces that be and demonstrate to the powerless that if they fail to conform, they risk incarceration or death."[4]

We live in a cinematic society, one that presents and represents itself through movie and television screens. By 1930 the movies had become a weekly pastime for a majority of Americans. After 1950, with the advent of television, watching moving pictures became a daily activity, even an addiction, in the United States and other countries in the industrialized world. One report projected that for the year 2001, the average American spent 1577 hours in front of the TV set, 13 hours in movie theaters, and 55 hours watching prerecorded videos at home.[5] This represents 28 percent of our waking time. It is also five times the number of hours the average American spent in 2001 reading books, newspapers, and magazines.

In the same way that literate societies are dramatically different from illiterate societies, the social organization of cinematic societies is dramatically different from that of noncinematic ones. Without taking into account the impact that the moving pictures in television and cinema screens have on the people of a country, we could no more understand contemporary society than we could understand it without realizing the impact of literacy. The daily rhythms of our lives, what we know and what we ignore, are set by the rhythm of and the information contained in the screens of cinema and television. Countries without a film industry can be considered colonies for foreign filmmakers. Within countries, one can, of course, speak of diversely cinematized segments of the population because the time, energy, and money spent on media consumption vary greatly by age, class, religion, income, race, geography, and other such sociodemographic variables.

The Hollywood film industry does not portray all the segments of society and the world populations equally, with the same frequency, accuracy, or with the same respect. Consider that Latinos, who according to the 2000 U.S. Census constitute one of the largest U.S. minority groups, have seldom been represented as the protagonist of Hollywood films. The *Video Hound Golden Movie Retriever Index*, for example, lists only 17 films under the category of "Hispanic America," roughly half of which are Hollywood main releases. In contrast, the same index lists 69 films in the category "Ireland," 151 under "Judaism," 45 under "British Royalty," and even 119 under "Zombies"![6]

Allan G. Johnson notes that of the films that have won the Academy Award in the category Best Picture from 1965 through 1999, "none set in the United States places people of color at the center of the story without their having to share it with white characters of equal importance" (e.g., *In the Heat of the Night* [1967] or *Driving Miss Daisy* [1989]). "Anglo, heterosexual males, even though they are less than twenty percent of the U.S. population," he proposes, "represent ninety percent of the characters in the most important movies ever made."[7] Until recently, most minority characters in Hollywood movies have usually been caricatured and portrayed with disrespect.

Much of what we know about people we consider to be "others" we learn through the movies. The moving pictures allow access to private spaces, scenes that would normally be out of the reach of our eyes. Through the film media we learn what life supposedly is like or used to be like and what it is in distant lands and in private places.

Films also represent us, the spectators, who find enjoyment and solace in them. The streets we walk; the landmarks in our cities we go by; the appliances, furniture, and gadgets we use every day; the cars and buses we ride; and the music we listen to all appear in the movies. The social roles we play—as children, parents, workers, and lovers—are also recognizable in films. The words and the jargon the characters use in the movies are part of the language we speak. In this sense, we, the audience, watch ourselves. Much of the attraction and power of film, its ability to make us laugh or cry and to teach us about the world and about ourselves, rests on our being, simultaneously, spectators and subjects being gazed on.

The cinematic viewing experience, in our opinion, is one of recognizing and mis-recognizing ourselves in the moving pictures. Watching a movie is the experience of sharing—or sometimes, of resisting—the way of seeing, the ideology, and the values of the filmmakers, their gaze, and their imagination. Through their technology and their language, films implement ways of looking at class, gender, and race differences. Filmmakers can make us see these differences, but they can also hide them from our sight by creating pleasing fictions. This way of seeing carries the individual and social biases of the filmmakers but also the biases and standpoints of the culture of the people for which films are produced, the culture to which the film belongs. . . .

DIALECTICS OF RACE IN FILM

. . . We regard film in a dialectical fashion, considering them in two opposed ways. First, we consider them as a means to celebrate whiteness, to teach what it is like to be white and to enjoy the privilege of being white. Second, we consider movies as social therapeutic devices to help us cope with the unjust racial divide by denying or obscuring white privilege and the practices on which it depends. . . . We believe that unless we capture the tension and contradictions between these two intentions and the central need of Hollywood to entertain and to be profitable, we would miss critical elements of the role films play in the production and reproduction of racism in the United States and around the world.

Consider that in Susan Stewart's earlier remark about *Gone with the Wind,* although she recognizes that the film taught her how to be "Southern," she fails to recognize that it only taught her how to be a *white* Southerner: white becomes so normative and universal a category that she does not even need to mention it. She does not notice the ways the film forces African Americans into the background and occludes their story. Recall Malcolm X's humiliation, when he was the sole black patron in a white theater, at seeing how blacks were portrayed in *Gone with the Wind*. Lorraine Hansberry, the black playwright, confirms Malcolm X's response when she writes that *Gone with the Wind* did not teach blacks to be blacks. The fact that in the United States white "goes without saying" in statements such as Susan Stewart's is an important trait of what . . . we call "the white self." . . .

The concept of self, of the white self—the portrayal of which we will be examining—is used by scholars to designate who and what we are. The self is the human person, the place in which all experience—our memories, our pain and pleasure, our emotions—is organized. By self, we mean the sense of being a person, the experience of existing as an individual contained in the space of a body over time. As universal as this notion might be, it is highly culture specific. In the United States, the fundamental entries on our birth certificates are name, birth date, gender, and race. These constitute our legal sense of self. Race is also crucial to our psychological sense of self. Without it we would be fatally disoriented, like Joe Christmas in Faulkner's *Light in August*, who goes mad and is destroyed because he never knows whether he is white or black.

WHITENESS

The key element to understanding racial thinking in the United State and in much of the world today is white supremacy. The modern concept of race and the notion of whiteness were invented during the period of European colonization of the Americas and Africa. The stock of knowledge we call racism has been developed in the past five hundred years precisely to establish the superiority of whites and to contribute a veneer of legitimacy to colonial domination, exploitation, or extermination of people of color, both domestically and internationally, by whites.

One difficulty in studying "whiteness" is that, until recently, it was an empty or invisible category, not perceived as a distinctive racial identity. Richard Dyer writes, "As long as race is something only applied to non-white peoples, as long as white people are not racially seen and named, they/we function as a human norm. Other people are raced, we are just people."[8] Thus, most white Americans either do not think of their "whiteness" or think of it as neutral. The power of whiteness rests in its apparent universality and invisibility, in the way it has gone unexamined. Nevertheless, the images of film, especially of films in which whites interact with persons of another color, offer a way to study white self-representation across the twentieth century. As has often been said, "Whites don't have a color until a person of color enters the room."

Until recently, sociologists and culture critics concentrated on prejudice, that is, on the distorted images that we construct of others we perceive as different. For example, Bogle (1997), Cripps (1997), Snead (1994), and Guerrero (1993), among others, have studied the prejudicial images of African Americans in American films.[9] We want to shift the focus to the representation of the white self-concept.

Whiteness as we understand it today in the United States is a construct, a public fiction that has evolved throughout American history in response to changing political and economic needs and conditions. Whiteness has always been a shifting category used to police class and sexual privilege. At the beginning, only "free white persons" could become American citizens. White privilege depended on the exclusion of "others," but the definition of who was non-white constantly changed. Thus, previous historic categories such as Celt, Slav, Alpine, Hebrew, Iberic, Anglo-Saxon, and Nordic have been incorporated into the contemporary concept of "white" or "Caucasian." "Caucasians are made and not born."[10] We argue that the notion of whiteness has become so integral to the American identity that it is embedded in the national unconscious.

RACE

In practice, the term "race" designates one or more biological traits (e.g., skin color) from which a sociopolitical hierarchy is derived and the assumption that some races are superior and therefore deserve to be more powerful than others. In spite of the concentrated efforts by scientists over the past one hundred years, the concept of "race" has become progressively more elusive, to the extent that today we can say that race is an illusion, a fiction that no longer leads to a meaningful classification of humans in the biological or social sciences. It is not an objective or fixed category. Race, according to Omi and Winant, is "an unstable and 'decentered' complex of social meanings constantly being transformed by political struggle." Although the concept of race may be a fiction, we cannot simply jettison it because it "continues to be central to everyone's identity and understanding of the social world."[11]

Today the vast majority of humans across the globe still think, feel, and act as if "race" were real, as if it pointed to true, useful differences among people. Furthermore, no one alive today has lived in a world in which race did not matter. In the United States, race matters in the chance each of us had of being born alive and healthy. Race matters in the neighborhoods where we grow up, the quality of the education we obtain, the persons we choose as friends, spouses, or lovers, the careers we pursue, the health and opportunities of the children we are going to have, and the churches we attend. Race matters in the length of our life span. Finally, race matters even in the cemetery where we lie after death.

At the beginning of the twenty-first century, the memories of the horrors in which race was the operative concept are still fresh. Among others, the horror of racial segregation and lynching in the United States, of the Nazi Holocaust in Europe, of apartheid in South Africa, and of the "ethnic cleansing" in Bosnia and Kosovo cannot be ignored. One can also not deny that members of oppressed groups find identity, self-expression, and solidarity in racial and ethnic categories. Historically and today, race and the violent or subtle practices we call racism shape both the structures of our societies and the daily rhythms of our lives.

We cannot begin to explain the contradiction between the scientific use-lessness of the concept of race and the real consequences the application of racial categories bring about. At the interpersonal level, the biological trait or set of traits thought to reveal "race" are used as assumptions about other physi-cal, intellectual, emotional, or spiritual traits of persons with those characteris-tics. In the United States, for example, those who are not considered white are often automatically assumed to be smelly, "greasy," less intelligent, lazy, dirty, not in control of their emotions, unreliable, and so on. The category of race, however fictional, is taken for real and is real in its consequences.

Today, white supremacy still dominates America. Consider that the Constitution of the United States of 1789, the fundamental document of the first democratic society, accepted the slavery of Africans and African Americans within its borders and gave Congress the authority to suppress slave insurrec-tions. For tax distribution purposes, a slave was counted as three-fifths of a per-son. In 1861, both houses of the U.S Congress passed a bill that would have made slavery a permanent feature of the American legal system. Today, decades after the Civil Rights revolution of the 1960s, the enforcement of civil rights laws is very weak, at best. Film production is one of the resources through which power is wielded by the classes that benefit from the racial status quo.

REFERENCES

1. Stewart, Susan. 2000. "Lessons Learned: The Enduring Truths of *Gone with the Wind*," *TV Guide*, December 23–29, p. 26.

2. Malcolm X, with Alex Haley. 1965. *The Autobiography of Malcolm X*. New York: Grove, p. 42.

3. Kellner, Douglas. 1995. *Media Culture: Cultural Studies, Identity, and Politics between the Modern and the Postmodern*. New York: Routledge, p. 1.

4. Ibid, p. 2.

5. Veronis, Suhler & Associates. 1999. "Table 920. Media Usage and Consumer Spending: 1992 to 2002: Communications Industry Report." New York.

6. Craddock, Jim, ed. 2001. *Video Hound's Golden Movie Retriever 2001: The Complete Guide to Movies on Videocassette, DVD, and Laserdisc*. Detroit, Mich.: Visible Ink.

7. Johnson, Alan G. 2001. *Privilege, Power and Difference*. Mountain View, Calif.: Mayfield, p. 108.

8. Dyer, Richard. 1997. *White*. New York: Routledge, p. 1.

9. Bogle, Donald. 1997. *Toms, Coons, Mulattoes, Mammies, and Bucks*. New York: Continuum; Cripps, Thomas. 1997. *Slow Fade to Black: The Negro in American Film, 1900–1942*. New York: Oxford University Press; Guerrero, Ed. 1993. *Framing Blackness: The African American Image in Film*. Philadelphia: Temple University Press; Snead, James A., Colin MacCabe, and Cornel West. 1994. *White Screens, Black Images: Hollywood from the Dark Side*. New York: Routledge.

10. Jacobson, Mathew Frye. 1998. *Whiteness of a Different Color: European Immigrants and the Alchemy of Race*. Cambridge, Mass.: Harvard University Press, p. 4. See also Roediger, David. 1991. *The Wages of Whiteness: Race and the Making of the American Working Class*. London: Verso; Allen, Theodore. 1994. *The Invention of the White Race*. London: Verso; Ignatiev, Noel. 1995. *How the Irish Became White*. New York: Routledge; Brodkin, Karen. 1998. *How Jews Became White Folks & What that Says About Race in America*. New Brunswick, N.J.: Rutgers University Press.

11. Omi, Michael, and Howard Winant. 1994. *Racial Formation in the United States: From the 1960s to the 1990s*. 2d ed. New York: Routledge, p. 55.

DISCUSSION QUESTIONS

1. Think of the last three movies that you saw (and, if you can, view them again). How are White people depicted in these movies? How are other groups depicted? How do your observations relate to Vera and Gordon's argument that movies are one of the lenses through which we come to understand race?

2. What other forms of popular culture do you think contribute to the "celebration of White privilege" that Vera and Gordon identify? Do you agree with Vera and Gordon that movies (and other forms of popular culture) have the ability to define who is powerful and who is powerless?

14

Hidden Politics

Discursive and Institutional Policing of Rap Music

BY TRICIA ROSE

T he way rap and rap-related violence are discussed in the popular media is fundamentally linked to the larger social discourse on the spatial control of black people. Formal policies that explicitly circumscribe housing, school, and job options for black people have been outlawed; however, informal, yet trenchant forms of institutional discrimination still exist in full force. Underwriting these de facto forms of social containment is the understanding that black people are a threat to social order. Inside of this, black urban teenagers are the most profound symbolic referent for internal threats to social order. Not surprisingly, then, young African Americans are in fundamentally antagonistic relationships to the institutions that most prominently frame and constrain their lives. The public school system, the police, and the popular media perceive and construct young African Americans as a dangerous internal element in urban America; an element that if allowed to roam freely, will threaten the social order; an element that must be policed. Since rap music is understood as the predominant symbolic voice of black urban males, it heightens this sense of threat and reinforces dominant white middle-class objections to urban black youths who do not aspire to (but are haunted by) white middle-class standards.

My experiences and observations while attending several large-venue rap concerts in major urban centers serve as disturbingly obvious cases of how

From Tricia Rose, *Black Noise: Rap Music and Black Culture in Contemporary America* (Middletown, CT: Wesleyan University Press, 1994). Copyright © 1994 by Tricia Rose. Reprinted by permission of Wesleyan University Press.

black urban youth are stigmatized, vilified, and approached with hostility and suspicion by authority figures. I offer a description of my confrontation and related observations not simply to prove that such racially and class-motivated hostility exists but, instead, to use it as a case from which to tease out how the public space policing of black youth and rap music feeds into and interacts with other media, municipal, and corporate policies that determine who can publicly gather and how.

Thousands of young black people milled around waiting to get into the large arena. The big rap summer tour was in town, and it was a prime night to see and be seen. The "pre-show show" was in full effect. Folks were dressed in the latest fly-gear: bicycle shorts, high-top sneakers, chunk jewelry, baggie pants, and polka-dotted tops. Hair style was a fashion show in itself: high-top fade designs, dreads, corkscrews, and braids with gold and purple sparkles. Crews of young women were checking out the brothers; posses of brothers were scooping out the sisters, each comparing styles among themselves. Some wide-eyed pre-teenyboppers were soaking in the teenage energy, thrilled to be out with the older kids.

As the lines for entering the arena began to form, dozens of mostly white private security guards hired by the arena management (many of whom are off-duty cops making extra money), dressed in red polyester V-neck sweaters and gray work pants, began corralling the crowd through security checkpoints. The free-floating spirit began to sour, and in its place began to crystallize a sense of hostility mixed with humiliation. Men and women were lined up separately in preparation for the weapon search. Each of the concertgoers would go through a body patdown, pocketbook, knapsack, and soul search. Co-ed groups dispersed, people moved toward their respective search lines. The search process was conducted in such a way that each person being searched was separated from the rest of the line. Those searched could not function as a group, and subtle interactions between the guard and person being searched could not be easily observed. As the concertgoers approached the guards, I noticed a distinct change in posture and attitude. From a distance, it seemed that the men were being treated with more hostility than the women in line. In the men's area, there was an almost palpable sense of hostility on behalf of the guards as well as the male patrons. Laughing and joking among men and women, which had been loud and buoyant up until this point, turned into virtual silence.

As I approached the female security guards, my own anxiety increased. What if they found something I was not allowed to bring inside? What was prohibited, anyway? I stopped and thought: All I have in my small purse is my wallet, eyeglasses, keys, and a notepad—nothing "dangerous." The security woman patted me down, scanned my body with an electronic scanner while she anxiously kept an eye on the other black women in line to make sure that no one slipped past her. She opened my purse and fumbled through it pulling out a nail file. She stared at me provocatively, as if to say "why did you bring this in here?" I didn't answer her right away and hoped that she would drop it back into my purse and let me go through. She continued to stare at me, sizing me up to see if I was "there to cause trouble." By now, my attitude had turned foul; my childlike enthusiasm to see my favorite rappers had all but fizzled out. I didn't know the file was in my purse, but the guard's accusatory posture rendered such excuses moot. I finally replied tensely, "It's a nail file, what's the problem?" She handed it back to me, satisfied, I suppose, that I was not intending to use it as a weapon, and I went in to the arena. As I passed her, I thought to myself, "This arena is a public place, and I am entitled to come here and bring a nail file if I want to." But these words rang empty in my head; the language of entitlement couldn't erase my sense of alienation. I felt harassed and unwanted. This arena wasn't mine, it was hostile, alien territory. The unspoken message hung in the air: "You're not wanted here, let's get this over with and send you all back to where you came from."

I recount this incident for two reasons. First, a hostile tenor, if not actual verbal abuse, is a regular part of rap fan contact with arena security and police. This is not an isolated or rare example, incidents similar to it continue to take place at many rap concerts. Rap concertgoers were barely tolerated and regarded with heightened suspicion. Second, arena security forces, a critical facet in the political economy of rap and its related sociologically based crime discourse, contribute to the high level of anxiety and antagonism that confront young African Americans. Their military posture is a surface manifestation of a complex network of ideological and economic processes that "justify" the policing of rap music, black youths, and black people in general. Although my immediate sense of indignation in response to public humiliation may be related to a sense of entitlement that comes from my status as a cultural critic, thus separating me from many of the concertgoers, my status as a young African American woman is a critical factor in the way I was *treated* in this instance, as well as many others.

Rap artists articulate a range of reactions to the scope of institutional policing faced by many young African Americans. However, the lyrics that address the police directly—what Ice Cube has called "revenge fantasies"—have caused the most extreme and unconstitutional reaction from law enforcement officials in metropolitan concert arena venues. . . .

Rap music is by no means the only form of expression under attack. Popular white forms of expression, especially heavy metal, have recently been the target of increased sanctions and assaults by politically and economically powerful organizations, such as the Parent's Music Resource Center, The American Family Association, and Focus on the Family. These organizations are not fringe groups, they are supported by major corporations, national-level politicians, school associations, and local police and municipal officials.

However, there are critical differences between the attacks made against black youth expression and white youth expression. The terms of the assault on rap music, for example, are part of a long-standing sociologically based discourse that considers black influences a cultural threat to American society. Consequently, rappers, their fans, and black youths in general are constructed as coconspirators in the spread of black cultural influence. For the antirock organizations, heavy metal is a "threat to the fiber of American society," but the fans (e.g., "our children") are *victims* of its influence. Unlike heavy metal's victims, rap fans are the youngest representatives of a black presence whose cultural difference is perceived as an internal threat to America's cultural development. *They* victimize *us*. These differences in the ideological nature of the sanctions against rap and heavy metal are of critical importance, because they illuminate the ways in which racial discourses deeply inform public transcripts and social control efforts. This racial discourse is so profound that when Ice-T's speed metal band (*not rap group*) Body Count was forced to remove "Cop Killer" from its debut album because of attacks from politicians, these attacks consistently referred to it as a rap song (even though it in no way can be mistaken for rap) to build a negative head of steam in the public. As Ice-T describes it, "There is absolutely no way to listen to the song 'Cop Killer' and call it a rap record. It's so far from rap. But, politically, they know by saying the word *rap* they can get a lot of people who think, 'Rap-black-rap-black-ghetto,' and don't like it. You say the word *rock*, people say, 'Oh, but I like Jefferson Airplane, I like Fleetwood Mac—that's rock.' They don't want to use the word rock & roll to describe this song."[1] . . .

The social construction of "violence," that is, when and how particular acts are defined as violent, is part of a larger process of labeling social phenomena. Rap-related violence is one facet of the contemporary "urban crisis" that consists of a "rampant drug culture" and "wilding gangs" of black and Hispanic youths. When the *Daily News* headline reads, "L.I. Rap-Slayers Sought" or a *Newsweek* story is dubbed "The Rap Attitude," these labels are important, because they assign a particular meaning to an event and locate that event in a larger context. Labels are critical to the process of interpretation, because they provide a context and frame for social behavior. As Stuart Hall et al. point out in *Policing the Crisis*, once a label is assigned, "the use of the label is likely to mobilize this whole referential context, with all its associated meaning and connotations."[2] The question then, is not "is there really violence at rap concerts," but how are these crimes contextualized, labeled? . . . Whose interests do these interpretive strategies serve? What are the repercussions?

Venue owners have the final word on booking decisions, but they are not the only group of institutional gatekeepers. The other major powerbroker, the insurance industry, can refuse to insure an act approved by venue management. In order for any tour to gain access to a venue, the band or group hires a booking agent who negotiates the act's fee. The booking agent hires a concert promoter who "purchases" the band and then presents the band to both the insurance company and the venue managers. If an insurance company will not insure the act, because they decide it represents an unprofitable risk, then the venue owner will not book the act. Furthermore, the insurance company and the venue owner reserve the right to charge whatever insurance or permit fees they deem reasonable on a case-by-case basis. So, for example, Three Rivers Stadium in Pittsburgh, Pennsylvania, tripled its normal $20,000 permit fee for the Grateful Dead. The insurance companies who still insure rap concerts have raised their minimum coverage from about $500,000 to between $4 and $5 million worth of coverage per show.[3] Several major arenas make it almost impossible to book a rap show, and others have refused outright to book rap acts at all.

These responses to rap music bear a striking resemblance to the New York City cabaret laws instituted in the 1920s is response to jazz music. A wide range of licensing and zoning laws, many of which remained in effect until the late 1980s, restricted the places where jazz could be played and how it could be played. These laws were attached to moral anxieties regarding black cultural effects and were in part intended to protect white patrons from jazz's "immoral influences." They defined and contained the kind of jazz that could be played by restricting the use of certain instruments (especially drums and horns) and established elaborate licensing policies that favored more established and mainstream jazz club owners and prevented a number of prominent musicians with minor criminal records from obtaining cabaret cards.

During an interview with "Richard" from a major talent agency that books many prominent rap acts, I asked him if booking agents had responded to venue bans on rap music by leveling charges of racial discrimination against venue owners. His answer clearly illustrates the significance of the institutional power at stake:

> These facilities are privately owned, they can do anything they want. You say to them: "You won't let us in because you're discriminating against black kids." They say to you, "Fuck you, who cares. Do whatever you got to do, but you're not coming in here. You, I don't need you, I don't want you. Don't come, don't bother me. I will book hockey, ice shows, basketball, country music and graduations. I will do all kinds of things 360 days out of the year. But I don't need you. I don't need fighting, shootings and stabbings." Why do they care? They have their image to maintain.[4]

Richard's imaginary conversation with a venue owner is a pointed description of the scope of power these owners have over access to large public urban spaces and the racially exclusionary silent policy that governs booking policies. . . .

Because rap has an especially strong urban metropolitan following, freezing it out of these major metropolitan arenas has a dramatic impact on rappers' ability to reach their fan base in live performance. Public Enemy, Queen Latifah, and other rap groups use live performance settings to address current social issues, media miscoverage, and other problems that especially concern black America. For example, during a December 1988 concert in Providence, R.I., Chuck D from Public Enemy explained that the Boston arena refused to book the show and read from a *Boston Herald* article that depicted rap fans as a problematic element and that gave its approval of the banning of the show. To make up for this rejection, Chuck D called out to the "Roxbury crowd in the house," to make them feel at home in Providence. Each time Chuck mentioned Roxbury, sections of the arena erupted in especially exuberant shouts and screams. Because black youths are constructed as a permanent threat to social order, large public gatherings will always be viewed as dangerous events. The larger arenas possess greater potential for mass access and unsanctioned behavior. And black youths, who are highly conscious of their alienated and marginalized lives, will continue to be hostile toward those institutions and environments that reaffirm this aspect of their reality.

The presence of a predominantly black audience in a 15,000 capacity arena, communicating with major black cultural icons whose music, lyrics, and attitude illuminate and affirm black fears and grievances, provokes a fear of the consolidation of black rage. Venue owner and insurance company anxiety over broken chairs, insurance claims, or fatalities are not important in and of themselves, they are important because they symbolize a loss of control that might involve challenges to the current social configuration. They suggest the possibility that black rage can be directed at the people and institutions that support the containment and oppression of black people. As West Coast rapper Ice Cube points out in *The Nigga Ya Love to Hate*, "Just think if niggas decided to retaliate?"[5]

The coded familiarity of the rhythms and hooks that rap samples from other black music, especially funk and soul music, carries with it the power of black collective memory. These sounds are cultural markers, and responses to them are not involuntary at all but in fact densely and actively intertextual; they immediately conjure collective black experience, past and present. . . .

Rap music is fundamentally linked to larger social constructions of black culture as an internal threat to dominant American culture and social order. Rap's capacity as a form of testimony, as an articulation of a young black urban critical voice of social protest, has profound potential as a basis for a language of liberation. Contestation over the meaning and significance of rap music and its ability to occupy public space and retain expressive freedom constitutes a central aspect of contemporary black cultural politics.

During the centuries-long period of Western slavery, there were elaborate rules and laws designed to control slave populations. Constraining the mobility of slaves, especially at night and in groups, was of special concern; slave masters reasoned that revolts could be organized by blacks who moved too freely and without surveillance. Slave masters were rightfully confident that blacks had good reason to escape, revolt, and retaliate. Contemporary laws and practices curtailing and constraining black mobility in urban America function in much the same way and for similar reasons. Large groups of African Americans, especially teenagers, represent a threat to the social order of oppression. Albeit more sophisticated and more difficult to trace, contemporary policing of African Americans resonates with the legacy of slavery.

Rap's poetic voice is deeply political in content and spirit, but rap's hidden struggle, the struggle over access to public space, community resources, and the interpretation of black expression, constitutes rap's hidden politics.

REFERENCES

1. Light, Alan. "Ice-T." *Rolling Stone*, 20 August 1992, pp. 32, 60.

2. Stuart Hall et al., *Policing the Crisis* (London: Macmillan, 1977), p. 19.

3. Interview with "Richard," a talent agency representative from a major agency that represents dozens of major rap groups, October 1990.

4. Rose interview with "Richard." I have decided not to reveal the identity of this talent agency representative, because it serves no particular purpose here and may have a detrimental effect on his employment.

5. Ice Cube, "The Nigga Ya Love to Hate," *AmeriKKKa's Most Wanted* (Priority Records, 1990).

DISCUSSION QUESTIONS

1. Why does Rose think that rap music is so heavily "policed?" According to Rose, how does rap challenge the dominant culture?

2. What images of different groups are portrayed in the rap music being listened to today? As rap has become more popular among White audiences, do you think its critical perspective has been toned down?

3. How are women (and particularly women of color) depicted in contemporary rap music? Are there forms of popular culture that challenge the status quo without belittling and stereotyping women?

15

"Playing Indian"

Why Native American Mascots Must End

BY CHARLES FRUEHLING SPRINGWOOD AND C. RICHARD KING

Amerian Indian icons have long been controversial, but 80 colleges still use them, according to the National Coalition on Racism in Sports and Media. Recently, the struggles over such mascots have intensified, as fans and foes across the country have become increasingly outspoken.

At the University of Illinois at Urbana-Champaign, for example, more than 800 faculty members . . . signed a petition against retaining Chief Illiniwek as the university's mascot. Students at Indiana University of Pennsylvania have criticized the athletics teams' name, the Indians. The University of North Dakota has experienced rising hostilities on campus against its Fighting Sioux. Meanwhile, other students, faculty members, and administrators have vehemently defended those mascots.

Why, nearly 30 years after Dartmouth College and Stanford University retired their American Indian mascots, do similar mascots persist at many other institutions? And why do they evoke such passionate allegiance and strident criticism?

American Indian mascots are important as symbols because they are intimately linked to deeply embedded values and worldviews. To supporters, they honor indigenous people, embody institutional tradition, foster shared identity, and intensify the pleasures of college athletics. To those who oppose them, however, the mascots give life to racial stereotypes as well as revivify historical

From *Chronicle of Higher Education*, Dec. 9, 2001. Reprinted by permission of the authors.

patterns of appropriation and oppression. They often foster discomfort, pain, and even terror among many American Indian people.

The December 1999 cover of *The Orange and Blue Observer*, a conservative student newspaper at Urbana-Champaign, graphically depicts the multilayered and value-laden images that American Indian mascots evoke. Beneath the publication's masthead, a white gunslinger gazes at the viewer knowingly while pointing a drawn pistol at an Indian dancer in full regalia. A caption in large letters spells out the meaning of the scene: "Manifest Destiny: Go! Fight! Win!" Although arguably extreme, the cover, when placed alongside what occurs at college athletic events—fans dressing in paint and feathers, half-time mascot dances, crowds cheering "the Sioux suck"—reminds us that race relations, power, and violence are inescapable aspects of mascots.

We began to study these mascots while we were graduate students in anthropology at the University of Illinois in the early 1990s. American Indian students and their allies were endeavoring to retire Chief Illiniwek back then, as well, and the campus was the scene of intense debates. Witnessing such events inspired us to move beyond the competing arguments and try to understand the social forces and historical conditions that give life to American Indian mascots—as well as to the passionate support of, and opposition to, them. We wanted to understand the origins of mascots: how and why they have changed over time: how arguments about mascots fit into a broader racial context: and what they might tell us about the changing shape of society.

Over the past decade, we have developed case studies on the role that mascots have played at the halftime ceremonies of the University of Illinois. Marquette University, Florida State University, and various other higher-education institutions. Recently, we published an anthology, *Team Spirits: The Native American Mascots Controversy*, in which both American Indian and European American academics explored "Indian-ness," "whiteness," and American Indian activism. They also suggested strategies for change—in a variety of contexts that included Syracuse University and Central Michigan University, the Los Angeles public schools, and the Washington Redskins. Our scholarship and that of others have confirmed our belief that mascots matter, and that higher-education institutions must retire these hurtful symbols.

The tradition of using the signs and symbols of American Indian tribes to identify an athletic team is part of a much broader European American habit of "playing Indian," a metaphor that Philip Joseph Deloria explores in his book of that title (Yale University Press, 1998). In his historical analysis, Deloria enumerates how white people have appropriated American Indian cultures and symbols in order to continually refashion North American identities. Mimicking the indigenous, colonized "other" through imaginary play—as well as in literature, in television, and throughout other media—has stereotyped American Indian people as bellicose, wild, brave, pristine, and even animalistic.

Educators in particular should realize that such images, by flattening conceptions of American Indians into mythological terms, obscure the complex histories and misrepresent the identities of indigenous people. Moreover, they literally erase from public memory the regnant terror that so clearly marked the encounter between indigenous Americans and the colonists from Europe.

That higher-education institutions continue to support such icons and ensure their presence at athletics games and other campus events—even in the face of protest by the very people who are ostensibly memorialized by them—suggests not only an insensitivity to another race and culture, but also an urge for domination. Power in colonial and postcolonial regimes has often been manifested as the power to name, to appropriate, to represent, and to speak—and to use such powers over others. American Indian mascots are expressive practices of precisely those forms of power.

Consider, for example, the use of dance to feature American Indian mascots. Frequently, the mascot, adorned in feathers and paint, stages a highly caricatured "Indian dance" in the middle of the field or court during halftime. At Urbana-Champaign, Chief Illiniwek sports an Oglala war bonnet to inspire the team: at Florida State University, Chief Osceola rides across the football field, feathered spear held aloft.

Throughout U.S. history, dance has been a controversial form of expression. Puritans considered it sinful; when performed by indigenous people, the federal government feared it as a transgressive, wild, and potentially dangerous form of expression. As a result, for much of the latter half of the 19th century, government agents, with the support of conservative clergy, attempted to outlaw native dance and ritual. In 1883, for example, the Department of the Interior established rules for Courts of Indian Offenses. Henry Teller, the secretary of the department, anticipated the purpose of such tribunals in a letter that he wrote to the Bureau of Indian Affairs stating that they would end the "heathenish practices" that hindered the assimilation of American Indian people. As recently as the 1920s, representatives of the federal government criticized American Indian dance, fearing the "immoral" meanings animated by such performances.

The majority of Indian mascots were invented in the first three decades of the 20th century, on the heels of such formal attempts to proscribe native dance and religion, and in the wake of the massive forced relocation that marked the 19th-century American Indian experience. European Americans so detested and feared native dance and culture that they criminalized those "pagan" practices. Yet at the same time they exhibited a passionate desire for certain Indian practices and characteristics—evidenced in part by the proliferation of American Indian mascots.

Although unintentional perhaps. the mascots' overtones of racial stereotype and political oppression have routinely transformed intercollegiate-athletic events into tinderboxes. Some 10 years ago at Urbana-Champaign, several Fighting Illini boosters responded to American Indian students who were protesting Chief Illiniwek by erecting a sign that read "Kill the Indians, Save the Chief." And, in the wake of the North Dakota controversy, faculty members who challenged the Fighting Sioux name have reported to us that supporters of the institution's symbol have repeatedly threatened those who oppose it.

Although many supporters of such mascots have argued that they promote respect and understanding of American Indian people, such symbols and the spectacles associated with them are often used in insensitive and demeaning ways that further shape how many people perceive and engage American Indians. Boosters of teams employing American Indians have enshrined largely romanticized stereotypes—noble warriors—to represent themselves. Meanwhile, those who support competitive teams routinely have invoked images of the frontier, Manifest Destiny, ignoble savages, and buffoonish natives to capture the sprit of impending athletics contests and their participants. In our studies, we find countless instances of such mockery on the covers of athletics programs, as motifs for homecoming floats, in fan cheers, and in press coverage.

For example, in 1999, *The Knoxville News-Sentinel* published a cartoon in a special section commemorating the appearance of the University of Tennessee at the Fiesta Bowl. At the center of the cartoon, a train driven by a team member in a coonskin cap plows into a buffoonish caricature of a generic Indian, representing the team's opponent. the Florida State Seminoles. As he flies through the air, the Seminole exclaims. "Paleface speak with forked tongue! This land is ours as long as grass grows and river flows. Oof!"

The Tennessee player retorts. "I got news, pal. This is a desert. And we're painting it orange!" Below them, parodying the genocide associated with the conquest of North America Smokey, a canine mascot of the University of Tennessee, and a busty Tennessee fan speed down Interstate 10, dubbed "The New and Improved Trail of Tears." What effect can such a cartoon have on people whose ancestors were victims of the actual Trail of Tears?

The tradition of the Florida State Seminoles bears its share of responsibility for inviting that brand of ostensibly playful, yet clearly demeaning, discourse. For, at FSU, the image of the American Indian as warlike and violent is promoted without hesitation. Indeed, the Seminoles' football coach, Bobby Bowden, is known to scribble "Scalp 'em" underneath his autograph.

Such images and performances not only deter cross-cultural understanding and handicap social relations, they also harm individuals because they deform indigenous traditions, question identities, and subject both American Indians and European Americans to threatening experiences. For example, according to a *Tampa Tribune* article, a Florida resident and Kiowa tribe member, Joe Quetone, took his son to a Florida State football game during the mid-1990s. As students ran through the stands carrying tomahawks and sporting war paint, loincloths, and feathers, Quetone and his son overheard a man sitting nearby turn to a little boy and say, "Those are real Indians down there. You'd better be good, or they'll come up and scalp you!"

Environmental historian Richard White has suggested that "[White Americans] are pious toward Indian peoples, but we don't take them seriously; we don't credit them with the capacity to make changes. Whites readily grant certain nonwhites a 'spiritual' or 'traditional' knowledge that is timeless. It is not something gained through work or labor; it is not contingent knowledge in a contingent world." The omnipresence of American Indian mascots serves only to advance the inability to accept American Indians as indeed contingent, complicated, diverse, and genuine Americans.

Ultimately, American Indian mascots cannot be separated from their origins in colonial conditions of exploitation. Because the problem with such mascots is one of context, they can never be anything more than a white man's Indian.

Based on our research and observations, we cannot imagine a middle ground for colleges with Indian mascots to take—one that respects indigenous people, upholds the ideals of higher education, or promotes cross-cultural understanding. For instance, requiring students to take courses focusing on American Indian heritage, as some have suggested, reveals a troubling vision of the fit between curriculum, historic inequities, and social reform. Would we excuse colleges with active women's studies curricula if their policies and practices created a hostile environment for women?

Others have argued that colleges with American Indian mascots can actively manage them, promoting positive images and restricting negative uses. Many institutions have already exerted greater control over the symbols through design and licensing agreements. But they can't control the actions of boosters at their institutions or competitors at others. For example, the University of North Dakota would probably not prefer fans at North Dakota State University to make placards and T-shirts proclaiming that the "Sioux suck." Such events across the nation remind us that mascots are useful and meaningful because of their openness and flexibility—the way that they allow individuals without institutional consent or endorsement to make interpretations of self and society.

American Indian mascots directly contradict the ideals that most higher-education institutions seek—those of transcending racial and cultural boundaries and encouraging respectful relations among all people who live and work on their campuses. Colleges and universities bear a moral responsibility to relegate the unreal and unseemly parade of "team spirits" to history.

DISCUSSION QUESTIONS

1. What arguments do different groups make in favor of eliminating the use of Native American mascots? What arguments are there to keep them?

2. What would happen if similar caricatures of other groups were used as school mascots?

InfoTrac College Edition: Bonus Reading

http://infotrac.thomsonlearning.com

You can use the InfoTrac College Edition feature to find an additional reading pertinent to this section. You can also search on the author's name or keywords to locate the following reading:

Coltrane, Scott and Melinda Messineo. 2000. "The Perpetuation of Subtle Prejudice: Race and Gender Imagery in 1990s Television Advertising." *Sex Roles: A Journal of Research* (March): 363–389.

In their article, Coltrane and Messineo show the strong impact that popular culture has on the reproducing of race (and gender) stereotypes. Particularly in a racially segregated society in which groups often have limited closed contact with other groups, mass-media images can profoundly influence what people think about others. What specific images do Coltrane and Messineo find in television advertising? Suppose you were to do similar research on some part of popular culture familiar to you. What do you think you'd find in terms of race and gender images in this media form?

STUDENT EXERCISES

1. Identify a particular form of media that interests you—film, television, magazines, or books for example) and design a research plan that will examine some aspect of the images you find of a racial-ethnic groups. Narrow your topic so it won't be overly general. For example, if you choose films, pick only those nominated for Best Film in a given year, or if you choose television, look only at prime-time situation comedies. Alternatively, you could examine images of women of color in top fashion magazines, or watch Saturday morning children's cartoons to see how people of color are portrayed. Once you have narrowed your topic, design a systematic way to catalog your observations, such as, count the number of times people of color are represented in the medium you select, list the type of characters portrayed by Asian men, or compare the portrayal of White men and men of color in women's fashion magazines.

    What do your observations tell you about the representation of race in the form you chose? If you were to design your project to study such images as seen now and in the past, what might you expect to find? What impact do you think the images you found have on the beliefs of different racial-ethnic groups?

2. The readings in Part III identify *color-blind racism* as a new form of racism in which dominant group members (and some subordinate group members) think that race no longer matters and that to recognize race is to be

racist. Some of the authors claim that when you delve under the surface of these beliefs, you will find that people still harbor stereotypical ideas about racial minorities. Design a series of interview questions, perhaps modeled on some of the research reported, and then interview a small sample of people. Do you find evidence of color-blind racism among those you interviewed? How does the race and ethnicity of those you interviewed influence your findings?

PART IV

Race, Relationships, and Identity

This book shows how race and ethnicity are part of social structure. As such, you might be beginning to think they are "out there"—but they are also in us and in our relationships with other people and groups. As Peter Berger (1963) once wrote about the sociological perspective: Yes, people live in society but society also lives in people. Similarly, in the United States people live within a system of race and ethnic relations, but race and ethnic relations also live within us. How society has organized race and ethnicity is reflected in our identities and in our relationships with others. And, as society becomes more racially and ethnically diverse, so do people's identities become more diverse, and the possibilities for multiracial identities and relationships increase.

Identity means the self-definition of a person or group, but it is not free-floating. Identity is anchored in a social context: We define ourselves in relationship to the social structures that surround us. Moreover, identities are multidimensional and thus include many of the social spaces we occupy. At any given time, some identities may be more salient than others—age as you grow older, sexuality if you are questioning your sexual orientation, race as you confront the realities of a racially stratified society, and so forth.

Racial identity is learned early in life, although those in the dominant group may take it for granted. People of color likely learn explicit lessons about racial identity early on, as parents prepare them for living in a society in which their racial status will make them vulnerable to harm. Forming a positive racial identity—for both dominant and subordinate groups—means having to grapple with the realities of race. For people of color, this can mean surrounding themselves with others of their group (even though they may then be blamed for "self-segregating" by whites who do not understand or appreciate the support this affiliation can provide (see Tatum 1997).

The formation of racial identity is especially complex when multiple races are involved. The children of biracial couples may define themselves as being of two or more races. Thus, a child born to a White parent and a Black parent may identify as Black, but appear White to others, and then identify herself or himself as "biracial." As the society becomes more diverse, multiracial identities are becoming increasingly common. Racial and ethnic identities can also be complex because we have so many immigrants from nations in which race and ethnicity may be "constructed" differently than in the United States. Such complex identities hold out the possibility that the rigid thinking about race that has prevailed for so long might break down.

Still, racial segregation in society disrupts the cross-race contact that people would ordinarily have. Segregation can distort people's ideas about each other and make them more susceptible to accepting the racial stereotypes present in areas such as popular culture, with consequences for both White people and people of color. With racial segregation, there is less sharing of culture, histories, ideas, and caring across racial lines. One recent study shows that, among other things, friendships are affected by segregation. A team of sociologists studied cross-race friendships in several high schools around the nation and found that cross-race friendships increase when school populations are more diverse (Quillian and Campbell 2003).

One of the hallmarks of a free society is that people are able to freely associate with others—as peers, friends, neighbors, lovers, marriage partners, or in any other relationship. But this has not always been the case in the United States. Historically, laws in thirty states prohibited White people from marrying someone of a different race. These *anti-miscegenation* laws prohibited so-called "race mixing." For example, the state of California passed a law in 1880 prohibiting any White person from marrying a "negro, mulatto, or Mongolian." This was designed to prevent marriages between White people and Chinese immigrants (Takaki 1989). Specific laws against intermarriage varied from state to state. Most southern states prohibited White people from marrying Negroes, while some western states, like California, were also anti-Asian. Laws did not prohibit non-White groups from marrying each other, however. Thus in Mississippi the Chinese married so-called Negroes, although neither group was allowed to marry White people.

In order to enforce anti-miscegenation laws, states had to devise ways to define race. States varied in this practice as well. "Alabama and Arkansas defined anyone with one-drop of 'Negro' blood as Black; Florida had a one-eighth rule," and other states varied in their racial definitions (Lopez 1996: 118). If someone wanted to marry a person of another race, they had to do so in a state that did not prohibit the union. However, they risked having their marriage denied if they moved to a state where such arrangements were illegal.

Not until 1967 were such laws declared unconstitutional by the decision in a U.S Supreme Court case, *Loving v. Virginia*—a case taken to the court by an interracial couple, Mildred Jeter (a Black woman) and Richard Loving (a White man) who had been married in the District of Columbia in 1958. When they returned to their home in Virginia after marrying, they were indicted and charged with violating Virginia's law banning interracial marriage. They were convicted and sentenced to one year in jail—a term they never served because the judge suspended the sentence on the condition that they leave Virginia. They returned five years later to appeal the decision. The Supreme Court decided the case in 1967 based on the argument that laws against intermarriage violated the 14th Amendment, which states: "No State shall make or enforce any laws which shall abridge the privileges or immunities of citizens of the United States; nor shall any State deprive any person of life, liberty, or property, without due process of law; nor deny to any person within its jurisdiction the equal protection of the laws" (U.S. Constitution, Amendment 14, Section 1).

Now interracial marriage is legal, although it is still relatively rare. As more people live in neighborhoods and attend schools with people of different races and learn about each other, intermarriage has increased, but slowly. The U.S. Census Bureau estimates that 3 percent of all marriages involve interracial couples, the most common being between a White person and someone of another race who is not Black. Marriage between Black men and White women is the second most common (U.S. Census Bureau 2004: 59). Despite its infrequency, public opinion polls show that people are now more accepting of interracial marriage than was true in the past, and interracial marriage is increasing.

The articles in this Part each explore different dimensions of racial identity and interracial relationships. A brief report from the U.S. Census Bureau ("Overview of Race and Hispanic Origin," by Elizabeth M. Grieco and Rachel C. Cassidy) shows the specific decisions the U.S. Census Bureau made for the 2000 census to enumerate the racial and ethnic character of the U.S. population.

As shown, the growth in population of those who have a multiracial identity has changed the categories that count as race and ethnicity. Beginning with the 2000 U.S. census, people were allowed to choose more than one racial identity for the first time (2 percent of the population did so). People could identify their "race" as "Black," "White" "American Indian or Alaska Native," one of various "Asian American" categories, or "other." (If "other" was chosen, the respondents were asked to write in what they considered their race to be.) The census defines race and Hispanic origin to be separate concepts. This means that, in the census data, Hispanics can be considered of any race, depending on their self-definition. As Clara Rodriquez noted in Part I, racial identities, at least as measured by official agencies, shift and fluctuate over time as societies change in how they think about race and ethnicity.

Harry Kitano's narrative ("A Hyphenated Identity") situates his identity in the historical experiences of Japanese Americans in the twentieth-century United States. His poignant essay, including his experience as a young man in the internment camps to which Japanese Americans were confined during World War II, shows how racial identity emerges in a context in which groups are unjustly treated. Within this context, Kitano defines himself as having a hyphenated identity—American and Japanese American—just as many other groups have formed unique identities based on their particular racial and ethnic experiences in the United States.

Many people have complex racial-ethnic identities, especially when they sit on the border between different racial-ethnic identities. Judy Scales-Trent ("Notes of a White Black Woman") is a Black American women often mistaken for a White person. She uses this biracial identity to explain race as a social construction, reflecting an earlier theme in our book. Her concept of *skinwalkers,* borrowed from the Navajo culture, provides a metaphor for how racial identity can change, emerge, and form in the context of social relations. Despite the multiplicity of biracial identities, Scales-Trent shows how racialized societies spend a lot of time trying to fit people into fixed racial categories.

Likewise, Heather Dalmage ("Tripping on the Color Line") shows how multiracial people have to negotiate the boundaries that racist societies produce. Multiracial (or biracial people) cross racial borders, negotiate relationships across such borders, and shape identities in the context of complex race relations. The burden of having to negotiate these barriers, Dalmage argues, makes its harder to form interracial relationships. Together, the articles in Part IV challenge fixed definitions of race, and show how race and ethnic identities are emerging in an increasingly diverse social context.

White people typically have not been considered to have a racial identity. Because they are the dominant group, their identity has been considered transparent, taken-for-granted, not marked as are the identities of racial and ethnic minorities. As Mark Chesler and his colleagues ("Blinded by Whiteness") point out, White people actually do have a racial identity, but it is often not salient until they encounter experiences wherein that identity is brought to light. Thus, White college students may confront their own racial experiences for the first time when they interact with students of color on campus. Chesler and his colleagues analyze the different phases that constitute the development of White identity and, in so doing, show how identity can develop to produce a commitment to a more racially just society.

REFERENCES

Berger, Peter L. 1963. *Invitation to Sociology: A Humanist Perspective.* New York: Doubleday-Anchor.

Lopez, Ian f. Haney. 1996. *White by Law: The Legal Construction of Race.* New York: New York University Press.

Quillian, Lincoln, and Mary E. Campbell. 2003. "Beyond Black and White. The Present and Future of Multiracial Friendship Segregation." *American Sociological Review* 68 (August): 540-566.

Takaki, Ronald. 1989. *Strangers from a Different Shore: A History of Asian Americans.* New York: Penguin.

Tatum, Beverly Daniel. 1997. *Why Are All the Black Kids Sitting Together in the Cafeteria? And Other Conversations about Race.* New York: Basic.

U.S. Census Bureau. 2004. *Statistical Abstract of the United States 2003.* Washington, DC: U.S. Department of Commerce.

16

Overview of Race and Hispanic Origin

Census 2000 Brief

BY ELIZABETH M. GRIECO AND RACHEL C. CASSIDY

Every census must adapt to the decade in which it is administered. New technologies emerge and change the way the U.S. Census Bureau collects and processes data. More importantly, changing lifestyles and emerging sensitivities among the people of the United States necessitate modifications to the questions that are asked. One of the most important change for Census 2000 was the revision of the questions on race and Hispanic origin to better reflect the country's growing diversity. . . .

UNDERSTANDING RACE AND HISPANIC ORIGIN DATA FROM CENSUS 2000

The 1990 Census Questions on Race and Hispanic Origin were Changed for Census 2000

The federal government considers race and Hispanic origin to be two separate and distinct concepts. For Census 2000, the questions on race and Hispanic origin were asked of every individual living in the United States. The question on Hispanic origin asked respondents if they were Spanish, Hispanic, or

"Overview of Race and Hispanic Origin, March 2001," U.S. Census Bureau. Website: www.census.gov

Latino.[1] The question on race asked respondents to report the race or races they considered themselves to be. Both questions are based on self-identification.

The question on Hispanic origin for Census 2000 was similar to the 1990 census question, except for its placement on the questionnaire. For Census 2000, the question on Hispanic origin was asked directly before the question on race. For the 1990 census, the order was reversed—the question on race preceded questions on age and marital status, which were followed by the question on Hispanic origin.

The question on race for Census 2000 was different from the one for the 1990 census in several ways. Most significantly, respondents were given the option of selecting one or more race categories to indicate their racial identities.[2]

Census 2000 Used Established Federal Guidelines
to Collect and Present Data on Race and Hispanic Origin

Census 2000 adheres to the federal standards for collecting and presenting data on race and Hispanic origin as established by the Office of Management and Budget (OMB) in October 1997.

The OMB defines Hispanic or Latino as "a person of Cuban, Mexican, Puerto Rican, South or Central American, or other Spanish culture or origin regardless of race." In data collection and presentation, federal agencies are required to use a minimum of two ethnicities: "Hispanic or Latino" and "Not Hispanic or Latino."

Starting with Census 2000, the OMB requires federal agencies to use a minimum of five race categories:

- White;
- Black or African American;
- American Indian or Alaska Native;
- Asian; and
- Native Hawaiian or Other Pacific Islander.

For respondents unable to identify with any of these five race categories, the OMB approved including a sixth category—"Some other race"—on the Census 2000 questionnaire. The category Some other race is used in Census 2000 and a few other federal data collection activities. As discussed later, most respondents who reported Some other race are Hispanic. For definitions of the race categories used in Census 2000, see the box.

---

[1] Hispanics may be of any race. The terms "Hispanic" and "Latino" are used interchangeably in this report.

[2] Other changes included terminology and formatting changes, such as spelling out "American" instead of "Amer." For the American Indian and Alaska Native category and adding "Native" to the Hawaiian response category. In the layout of the Census 2000 questionnaire, the Asian response categories were alphabetized and grouped together, as were the Pacific Islander categories after the Native Hawaiian category. The three separate American Indian and Alaska Native identifiers in the 1990 census (i.e., Indian (Amer.), Eskimo, and Aleut) were combined into a single identifier in Census 2000. Also, American Indians and Alaska Natives could report more than one tribe.

BOX 1   How Are the Race Categories Used in Census 2000 Defined?

"White" refers to people having origins in any of the original peoples of Europe, the Middle East, or North Africa. It includes people who indicated their race or races as "White" or wrote in entries such as Irish, German, Italian, Lebanese, Near Esterner, Arab, or Polish.

"Black or African American" refers to people having origins in any of the Black racial groups of Africa. It includes people who indicated their race or races as "Black, African Am., or Negro," or wrote in entries such as African American, Afro American, Nigerian, or Haitian.

"American Indian and Alaska Native" refers to people having origins in any of the original peoples of North and South America (including Central America), and who maintain tribal affiliation or community attachment. It includes people who indicated their race or races by marking this category or writing in their principal or enrolled tribe, such as Rosebud Sioux, Chippewa, or Navajo.

"Asian" refers to people having origins in any of the original peoples of the Far East, Southeast Asia, or the Indian subcontinent. It includes people who indicated their race or races as "Asian Indian," "Chinese," "Filipino," "Japanese," "Vietnamese," or Other Asian," or wrote in entries such as Burmese, Hmong, Pakistani, or Thai.

"Native Hawaiian and Other Pacific Islander" refers to people having origins in any of the original peoples of Hawaii, Guam, Samoa, or other Pacific Islands. It includes people who indicated their race or races as "Native Hawaiian," "Guamanian or Chamorro," "Samoan," or "Other Pacific Islander," or wrote in entries such as Tahitian, Mariana Islander, or Chuukese.

"Some other race" was included in Census 2000 for respondents who were unable to identify with the five Office of Management and Budget race categories. Respondents who provided write-in entries such as Moroccan, South African, Belizean, or a Hispanic origin (for example, Mexican, Puerto Rican, or Cuban) are included in the Some other race category.

The Census 2000 question on race included 15 separate response categories and three areas where respondents could write in a more specific race group (see Figure 1[on page 132]). The response categories and write-in answers can be combined to create the five minimum OMB race categories plus Some other race. In addition to White, Black or African American, American Indian and Alaska Native, and Some other race, seven of the 15 response categories are Asian and four are Native Hawaiian and Other Pacific Islander.

HISPANIC ORIGIN IN CENSUS 2000

According to Census 2000, 281.4 million people resided in the United States, and 35.3 million, or about 13 percent, were Latino (see Table 1[on page 133]). The remaining 246.1 million people, or 87 percent, were not Hispanic.

Reproduction of Questions on Race and Hispanic Origin From Census 2000

→ **NOTE: Please answer BOTH Questions 5 and 6.**

5. **Is this person Spanish/Hispanic/Latino?** *Mark* ☒ *the "No" box if **not** Spanish/Hispanic/Latino.*

☐ No, not Spanish/Hispanic/Latino.   ☐ Yes, Puerto Rican

☐ Yes, Mexican, Mexican Am., Chicano   ☐ Yes, Cuban

☐ Yes, other Spanish/Hispanic/Latino — *Print group* ↗

6. **What is this person's race?** *Mark* ☒ **one or more races** to *indicate what this person considers himself/herself to be.*

☐ White

☐ Black, African Am., or Negro

☐ American Indian or Alaska Native — *Print name of enrolled or principle tribe* ↗

☐ Asian Indian   ☐ Japanese   ☐ Native Hawalian

☐ Chinese   ☐ Korean   ☐ Guamanian or Chamorro

☐ Filipino   ☐ Vietnamese   ☐ Samoan

☐ Other Asian — *Print race* ↗   ☐ Other Pacific Islander — *Print race* ↗

☐ Some other race — *Print race* ↗

Source: U.S. Census Buerau Census 2000 questionnaire.

FIGURE 1 Reproduction of Questions on Race and Hispanic Origin From Census 2000.

RACE IN CENSUS 2000

The Race Data Collected by Census 2000
Can Be Collapsed Into Seven Categories

People who responded to the question on race by indicating only one race are referred to as the race *alone* population, or the group that reported[3] *only one* race category. Six categories make up this population:

- White *alone;*
- Black or African American *alone;*

---

[3]In this report, the term "reported" is used to refer to the response provided by respondents as well as responses assigned during the editing and imputation processes.

Table 1 Population by Race and Hispanic Origin for the United States: 2000

RACE AND HISPANIC OR LATINO	NUMBER	PERCENT OF TOTAL POPULATION
RACE		
Total population	281,421,906	100.0
One race	274,595,678	97.6
White	211,460,626	75.1
Black or African American	34,658,190	12.3
American Indian and Alaska Native	2,475,956	0.9
Asian	10,242,998	3.6
Native Hawaiian and Other Pacific Islander	398,835	0.1
Some other race	15,359,073	5.5
Two or more races	6,826,228	2.4
HISPANIC OR LATINO		
Total population	281,421,906	100.0
Hispanic or Latino	35,305,818	12.5
Not Hispanic or Latino	246,116,088	87.5

SOURCE: U.S. Census Bureau, Census 2000 Redistricting (Public Law 94-171) Summary File, Tables PL1 and PL2.

- American Indian and Alaska Native *alone;*
- Asian *alone;*
- Native Hawaiian and Other Pacific Islander *alone;* and
- Some other race *alone.*

Individuals who chose more than one of the six race categories are referred to as the *Two or more races* population, or as the group that reported *more than one* race. All respondents who indicated more than one race can be collapsed into the *Two or more races* category, which combined with the six *alone* categories, yields seven mutually exclusive and exhaustive categories. Thus, the six race *alone* categories and the *Two or more races* category sum to the total population.

The Overwhelming Majority of the U.S. Population Reported Only One Race

In Census 2000, nearly 98 percent of all respondents reported only one race (see Table 1). The largest group reported White alone, accounting for 75 percent of all people living in the United States. The Black or African American alone population represented 12 percent of the total. Just under 1 percent of all respondents indicated only American Indian and Alaska Native. Approximately 4 percent of all respondents indicated only Asian. The smallest race group was the Native Hawaiian and Other Pacific Islander alone population, representing

0.1 percent of the total population. The remainder of the "one race" respondents—5.5 percent of all respondents—indicated only Some other race.[4]

Only 2.4 Percent of All Respondents
Reported Two or More Races

The Two or more races category represents all respondents who reported more than one race. The six race categories of Census 2000 can be put together in 57 possible combinations of two, three, four, five, or six races; *see Table* Less than 3 percent of the total population reported more than one race. Of the 6.8 million respondents who reported two or more races, 93 percent reported exactly two. The most common combination was "White *and* Some other race," representing 32 percent of the Two or more races population.[5] This was followed by "White *and* American Indian and Alaska Native," representing 16 percent, "White *and* Asian," representing 13 percent, and "White *and* Black or African American," representing 11 percent. Of all respondents reporting exactly two races, 47 percent included Some other race as one of the two. Of all respondents who reported more than one race, 7 percent indicated three or more. Most of these (90 percent) reported three races.

The Office of Management and Budget Identified
Four Combinations of Two Races for Civil Rights
Monitoring and Enforcement

In March 2000, the OMB established guidelines for the aggregation and allocation of race responses from Census 2000 for use in civil rights monitoring and enforcement. These guidelines included the five OMB race categories and identified four specific combinations of two races.[6] These four OMB race combinations, which were the largest combinations reported in recent research, are:

- "White *and* American Indian and Alaska Native"
- "White *and* Asian"
- "White *and* Black or African American"
- "Black or African American *and* American Indian and Alaska Native."

---

[4]The Some other race alone category consists predominantly (97.0 percent) of people of Hispanic origin, and is not a standard OMB race category.

[5]The Two or more races categories are denoted by quotations around the combinations with the conjunction *and* in bold and italicized print to indicate the separate race groups that comprise the particular combination.

[6]*Guidance on Aggregation and Allocation of Data on Race for Use in Civil Rights Monitoring and Enforcement.* Office of Management and Budget Bulletin Number 00-02, March 9, 2000. Also included in the guidelines was the inclusion of any multiple race combinations [excluding Some other race] that comprise more than 1 percent of the population of interest. For more information, see www.whitehouse.gov/omb/bulletins/b00-02.html.

In fact, these four combinations are the largest categories, when combinations that include Some other race are excluded. Combined, these four combinations accounted for 43 percent of the population reporting Two or more races and 1 percent of the total population.

DISCUSSION QUESTIONS

1. Were you to develop categories to describe the racial and ethnic makeup of the U.S. population, would you create them in the same manner as did the U.S. Census Bureau? Why or why not?

2. Why do you think the Census Bureau decided to make race and ethnicity separate categories in the 2000 census?

17

A Hyphenated Identity

BY HARRY KITANO

rowing up as a Japanese American means a constant search for an identity. Even if one is monolingual and can speak only English, and knows nothing about the Japanese language and culture, one quickly learns that there are questions concerning an American identity. Comments such as, "Where did you learn how to speak English so well?" or "Why did your people bomb Pearl Harbor?" are reminders that acculturation and identifying as an American are insufficient criteria for belonging to the host society.

The reality of being different came about very early in my life. My parents were from Japan and they opened a hotel in San Francisco's Chinatown, so that in the beginning, what mattered were not racial differences, but those of nationality. China and Japan were enemies, ergo, the Chinese and the Japanese in America were also hostile to each other. When I attended elementary school, there was this Chinese girl, Dorothy Lee, who would mutter "Jap," and try to pinch me. I learned to suffer these taunts in silence, I knew that my teachers would be of no help, and that my parents would only laugh and scoff at the idea that I was afraid of a Chinese girl. I was unaware of the word prejudice at that time, but whatever it was, it was uncomfortable. I tried to avoid any contact with Dorothy, and counted my days as successful if I could escape without being pinched.

From *Names We Call Home: Autobiography on Racial Identity*, ed. by Becky Thompson and Sangeeta Tyagi (New York: Routledge, 1996), pp. 111–118. Copyright © 1996. Reprinted by permission of Routledge/Taylor & Francis Books, Inc.

There were only a few Japanese families living in or near Chinatown. They were concentrated in the small business sector; a dry cleaners, an art goods store, a pool hall, a restaurant, and small hotels and rooming houses. Most of their clientele were other minorities, especially Blacks, Filipinos, and other Asians. As far as I can remember, the racial tensions that are characteristic of present-day race relations among minority groups were not present then.

There was a closeness among the Japanese families, so that although we lived apart, we formed a community. One central gathering place was the Japanese school; most of the children attended this school after the regular school day. The Japanese classes met every day from 4 to 6, so for most of us, it was an extremely long school day. However, it was not all work and study; on the contrary, it was often disorganized chaos. One lone teacher had to deal with over 20 students, ranging from the very young to some who were in high school. Motivation to learn was not very high; I found it a chance to meet with my peers, to tease and nag the teacher, and to behave in a way that would not be tolerated in the regular school. I was quiet, obedient, conforming, studious, and turned in my homework in the regular school—I was the opposite in the Japanese school.

It was a situational orientation; I learned early in life that at least for me and my peers, the situation shaped behavior. We learned that we were to behave with much more reserve when dealing with the dominant group, and less so among fellow ethnics. However, I deeply regret that I did not take the opportunity to learn the Japanese language more thoroughly; my attendance for over ten years was not a total waste, but because I can speak only limited Japanese, it was obviously not a total success. The school involved the families, and ceremonies, dinners, meetings, and speeches fostered community cohesion. There were year-end pictures so that even today, looking back and wondering who went where remains a part of one's history. The school that I attended closed during World War II and never opened again.

The close network of the Japanese families constituted my first world. Parents would help each other; the reputations of all of the children were known, and the primary reference group was my family and the small Japanese American community in Chinatown. We shared experiences and were exposed to similar values, which were reinforced by the families and the ethnic community. Not all of us behaved according to the values, but we had a clear idea when norms were violated. I don't remember any dramatic clashes between the ethnic way and the American way, primarily because of the situational orientation. I did things the American way when dealing with the host society, and the ethnic way when dealing within the family and community. This was not too difficult since there was often a congruence between the two value systems. I had minimal voluntary contact with people who were not fellow ethnics. I went to school with children of Chinese, Italian and Mexican ancestry (it was the North Beach area of San Francisco), but never got to know them intimately. The public school was integrated, but my early life was within a structurally pluralistic ethnic group. For example, I was never invited to visit or play at any homes except by Japanese Americans, and never invited other than Japanese Americans to my home.

GROWING UP IN THE 1930s AND 40s

Life for a Japanese American growing up in the 1930s and 40s was, to use a current phrase, "not a piece of cake." I remember discrimination and domination quite well. We lived just below Nob Hill in San Francisco; there was a public playground next to the Mark Hopkins Hotel on the hill. I was told by my parents that I should not play there; no one would tell me why, so that when I was about 8 years old, I trudged up the hill with several of my ethnic companions. For a few times nothing happened, but one day a stern, white lady came up and said we could not play there. Her presence was dominating—we never questioned her authority and we hurriedly left. We thought that she was right: that that playground was reserved exclusively for the whites who lived on Nob Hill. It strengthened the wisdom of my parents that certain places were off limits, and that I shouldn't raise questions, especially when there were no appropriate answers.

Another incident, at about the same time, remains strong in my memory. A close bachelor friend of the family used to take me and my brother fishing and to other activities. Once he took us to a swimming spa called Sutro Baths which was purported to have both hot and cold swimming pools. We were extremely excited—we endured a long streetcar ride across the city to the baths. However, when we got there, the cashier evidently told my friend that "Japs weren't allowed." My friend looked embarassed, never told us what was said, and muttered something about not wanting to go swimming anyway. I went along with his rationalization, even venturing that I didn't want to go swimming either, although I felt that the rejection had something to do with our ancestry.

PEARL HARBOR AND THE
CONCENTRATION CAMPS

If I had doubts that racism was a problem, they were quickly erased after the Japanese attack on Pearl Harbor in 1941. I had just entered Galileo High School; I remember attending the all school assembly where President Roosevelt gave his speech about the "day of infamy," and the declaration of war against Japan. There were just a few Japanese American students; we glanced at each other furtively, feeling that all eyes were upon us. It was a trying moment; I felt confused and even guilty that Japan had attacked the United States. But, nothing untoward happened; school continued and I thought that, aside from the war, I could live a relatively normal life.

However, soon afterwards, there appeared signs on the telephone poles addressed to persons of Japanese ancestry—that we were to register, assemble and be prepared for removal to "relocation centers." Newspapers, especially the Hearst Press, and politicians led the charge—that all Japanese were traitors; they were dangerous to the war effort so they should be locked up and shipped to Japan. We were inundated by rumors, ranging from benign treatment by the authorities to permanent incarceration. It was an introduction to the next several years of my life, since rumors were an integral part of my wartime experience.

The FBI came and took my father away; it was a mixed blessing, since, while it gave him the status of being acknowledged as a community leader by the government, it meant he had to leave his family under crisis conditions. We were not to see him for several years; he never talked about his experiences in his camp, which I believe was in Bismarck, North Dakota. One irony was that since it was an "official" camp, under the Geneva Convention for prisoners of war, so that their food and treatment was better than what we were to experience.

Our worst fears were realized—that all of us, whether citizens or aliens were to be incarcerated. Disorganization was the order of the day. My mother, who had seldom been outside of the home, was left in charge and she panicked. She was sure that we were going to be taken out and shot, so she was prepared to die. (I should add that at the time of this writing she has reached her 102nd birthday.) We heard a variety of rumors. One was that that we were going to be shipped to Japan and traded for American prisoners of war; another more optimistic one was that once the United States realized that it had made a mistake, the evacuation orders would be rescinded. We lived by rumors, primarily because the government provided very little information. Perhaps they didn't know what to do with us, or, more likely, they kept secrets in the name of national security.

My family, and all others of Japanese ancestry, were ordered to close our homes, either sell or store our furniture and other valuables, pack what we could carry and assemble at designated sites. Although my past experiences had shaken my faith in my American identity, the forced evacuation led to even more doubts about who I was. I still believed that I was an American; I belonged to the ROTC band at Galileo and played "The Stars and Stripes Forever," marched behind the flag, and had little identification with Japan. However, it was clear that my ancestry was viewed with suspicion, and that my government saw me, my family, and members of my community as enemies, or in simple terms, we were all "Japs."

To add to my discomfort, the initial assembly place was on Van Ness Avenue, just a few blocks from my high school. I tried to hide my face from the students who were walking by; several came to express their sympathy, but others came at us making airplane-like noises in order "to kill the Japs." An important-looking white man with a clip sheet began calling out family names to board the buses that were to take us to the train station. His pronunciation of Japanese surnames could be termed "state of the art," and perhaps would have come as a source of comic relief if the occasion were not so serious.

LIFE IN A CONCENTRATION CAMP

It is hard to recapture my feelings of being herded into railroad cars, drawing the shades, and seeing armed soldiers at every entrance, especially since over half of the "prisoners" were women and children. We didn't know where we were going, how long the ride would be, and what was at the end of the journey. The most vivid remembrance was that of crying children, confused parents, and the hot, stuffy car.

We quickly adapted to camp norms. We formed youth groups, a euphemism for gangs, and I belonged to the San Francisco gang. There were already established gangs from various parts of Los Angeles, so that the primary identification for adolescents was through groups from their former area of residence. The change for me was that instead of identification with the family, my primary identification was with my group. Food was one of the highest priorities; my gang quickly learned that if we gulped down our food, we could run to another mess hall and eat another meal. Not that the food was of gourmet quality; we quickly labeled it "slop suey," but for a growing teenager, a full stomach meant a degree of contentment. My adaptation was to "monku"; that is to develop griping to an art. There was plenty to gripe about—the food, the smelly horse stalls, the bedding, standing in line, and most important, the lack of freedom, and facing an uncertain future. My group argued constantly, even picking on whether jeans with a red tag (Levis) were superior to those without a tag.

Santa Anita represented a growing away from the family. Almost every waking moment was spent with my peers; we got up, washed, ate, went to the bathroom, played, and argued together. The small, crowded barracks were for sleeping only, so that the influence of the family became minimal. My primary identification was with the San Francisco group; we met other peers through their group identification such as those from San Diego, Hawthorne, San Pedro, San Jose, and the numerous gangs from Los Angeles. I heard of a few fights between gangs, but my San Francisco group was known to be relatively peaceful, so that we stayed out of any serious conflict.

Looking back, I marvel at the ability of the Japanese to organize a coherent community under concentration camp conditions. There were about 20,000 of us, hemmed in by barbed wire; Santa Anita was to be only temporary until more permanent inland camps were developed; we came from different parts of California and yet, the community was able to offer a semblance of coherent life. I don't know who provided the organizational impetus, but there were softball and basketball leagues, dances, talent shows, and other activities that kept our mind off living behind armed wire. There were some riots and some beatings, but Santa Anita was remarkably peaceful, given the unusual circumstances.

Camp life exposed me to a wide variety of identities. I met those with strong Japanese identities—so strong that they were convinced that Japan would win the war. They advised me to prepare myself for the future by learning the Japanese language, its values and its culture. One can't miss the irony of that advice—knowledge of Japan is extremely valuable today, but not because they were victorious in World War II. Then there were others who were 150 percent American, who said that it was our patriotic duty to cooperate with the United States Government in order to win the war over fascist Japan, Germany, and Italy.

I vacillated in my identity, most often leaning towards an American identity. But I found it hard to answer the question, "If you say that you're an

American, why are you in the camp with the rest of us?" My only weak rejoinder was that there were imperfections in any system.

Although life in Santa Anita was short, from April to October, 1942, I had established some roots there. Even though it was temporary, there was an attempt to start a school, but since it was not mandatory and there was little organization, I did not attend school. Instead I roamed around and played most of the day, made friends from other parts of California, and began to have an interest in girls. The news that we were to be separated and sent to different permanent camps was met with tears and frustration. We were again to be pawns of governmental decisions; the feeling of being utterly powerless reinforced the cynicism that was a strong part of camp philosophy.

The move was to Topaz, a permanent camp set up in the middle of the desert in Utah. Here I got the feeling of being a real prisoner; at Santa Anita, one could see cars moving along the outside streets and some residents even had friends who would wave to them. But Topaz was in the middle of nothing, and the perimeters were sealed with barbed wire fences, guard towers, and armed soldiers. I lived in Topaz from 1942 to 1945 and graduated from high school there.

The relative shortness of the war meant that much of what I learned in camp was not internalized and permanently incorporated into my lifestyle. Nevertheless, many of the former residents still remember the feeling of being totally unwanted by the host society, and the hard struggle to re-establish themselves, once the camps were closed in 1946.

RELEASE FROM CAMP

There was a procedure that with governmental clearance we were free to relocate anywhere in the United States aside from the barred zones along the Pacific Coast. After graduating from high school in 1945, and talking over the decision with the remnants of my family (three of my sisters and my older brother had already left camp), I decided to apply for my leave. Armed with something like fifty dollars, a high school diploma, a one-way train ticket, and with fear and anxiety, I left my "home" with a vague idea that I would go to Milwaukee. There was no logical reason why I chose this city; I had no relatives, friends, or acquaintances there, although I could probably say the same of wherever I chose to go. I sometimes wonder if I would have the courage to make a similar move today, leaving a relatively safe "home" for an unknown, but probably very hostile society. But I was young, ignorant, and still believed in America.

I was extremely self-conscious when I boarded the train at Delta, Utah; I had not faced the outside world for over three years and the crowded train was filled with soldiers, sailors, and marines. I expected at any time to be spotted as a "Jap"; much to my surprise, I was totally ignored and even offered a seat. Everyone was more interested in their companions; some were even making love, which was a shock to one who had spent the last several years in a strict environment.

The War Relocation Authority (WRA) had set up an employment office in Milwaukee, and they referred me to a farmer whose specialty was erecting silos. Aside from disliking heights and hard physical labor, the job was mine for a short time. One of the things that I brought with me to camp was an old trombone, and I played in a number of dance bands that were organized in both Santa Anita and Topaz. I decided to try my hand at being a professional musician; I answered an ad in *Downbeat,* a magazine for musicians, and found myself traveling to Worthington, Minnesota to play in the Tiny Little orchestra. I was hired sight unseen, and given my two-weeks notice (fired), the minute I was seen. The manager was nice enough to refer me to a new band that was being formed in Austin, and so I began my career as one of the few musicians of "oriental" extraction playing in what is known as the Midwest territory. I had changed my name to Harry Lee and identified myself as of Chinese extraction, from San Francisco. Aside from some skeptics who could not understand why one would leave California to play through a Minnesota winter, no one raised any questions about my identity, although I always feared that I would be discovered. Occasionally some dancer would pull his eyes to achieve a slant, and mouth the word "Jap," but such displays were few and far between.

In one band, I was told that I was hired to replace a black musician; the musician was much more talented than what I could ever be (Oscar Pettiford, who later played for Duke Ellington and is one of the foremost jazz bassists), and I realized there was a pariah group even more the target of prejudice and discrimination than mine. There were also heated discussions about Jewish musicians; one trumpet player decided that Harry James was no longer his idol because he was Jewish. I suspect that if I had not been around, Asians would also have been the target of snide remarks. Nevertheless, I realized that there were also other groups that were not accepted. I strove very hard to belong, so that I dressed, talked, walked, acculturated, and identified as a jazz musician with a Chinese surname.

BERKELEY

I returned to California to attend the University of California in 1946 and received my B.A., M.S.W., and my Ph.D. in 1958. Meanwhile, I continued as a part-time musician. On applying for membership in the San Francisco Musician's Union, I was faced with a dilemma. There was a white union, I believe it was Local 6, and a Black union, Local 669. I had a choice; I joined Local 6 but played primarily in black bands.

One incident stands out in my mind. I played for an all black band that had a "gig" in Redding, in Central California. On the way, we stopped for something to eat. Imagine my surprise when they turned to me and said, "Lee, could you find out if they'll serve us?" Here I was with a group of talented musicians acting as their spokesperson, even though I had been recently released from a concentration camp. I think that the restaurant sold us sandwiches "to go."

ACADEMIC EXPERIENCES

I regret being a quiet, nonaggressive student during my years at the University of California at Berkeley. I stuck around mostly with fellow ethnics and did not participate in campus life. It was the post World War II era where all-white fraternities and sororities ruled campus social life, and Asian Americans were viewed as outsiders. As a consequence, I felt that my role was to remain in the background and to be anonymous, even to professors. I was satisfied to sit and listen, and to not raise any questions. It was only after I graduated with my Ph.D. that I felt that I had not taken full advantage of the opportunities that were available on the campus at Berkeley.

I was trained primarily as a psychologist, with a major interest in Social Psychology. In a course in Clinical Psychology, we were to write our own psychological histories; I don't remember the details, but I know that I wrote about myself without once mentioning my years in a concentration camp. I thought that it would not be of interest to the professor, although by that time I was also acquainted with such terms as repression and shame.

EMPLOYMENT

There were two jobs, prior to my appointment to the University of California, Los Angeles in 1958, that were important influences on my way of thinking. One was as a caseworker at the International Institute in San Francisco where the emphasis was on working with immigrant groups and helping them adapt to the American society. My major task was to work with a variety of Asian immigrants, wherein I developed a detailed knowledge of the kinds of problems that were associated with moving, with old country values and experiences, to a new country. The problems of the immigrants' children, the second generation, were especially interesting, since I had also gone through these conflicts in my own upbringing.

I also worked as a therapist in the Child Guidance Clinic of the San Francisco Schools. Here I was exposed to psychoanalytic thinking; we were supervised by consulting psychiatrists with a heavy psychoanalytic orientation. My clientele was minority children who were acting out. Although I would have preferred to work with Asian American children, almost none were referred to the Clinic. While it was exciting to deal with some of the intricacies of psychoanalytic thinking, concepts such as discrimination and racism were seldom given high priority in our psychiatric conferences. It was difficult for me to believe that insight gained through verbal interaction would lead to a significant change in the behavior of my child clients.

Sociological input and sociological thinking came primarily after my employment at UCLA. While conducting research on Japanese American Crime and Deliquency, I worked closely with sociologists and their perspectives opened up newer ways of thinking for me. I received a joint appointment in sociology and taught courses on ethnic groups and race relations. The most influential sociologist for me was Milton Gordon. His view of acculturation in his book, *Assimilation in American Life,* was extremely helpful in providing an organizational framework to my knowledge of my ethnic group. The idea of hyphenated Americans, the different stages in acculturation, the effects of discrimination and prejudice, and the role of structural variables went far beyond the social psychological and psychoanalytic models that were a part of my early training.

In 1969 I published *Japanese Americans*—a difficult book to write since it was the first book about my ethnic group after World War II and I had to cover material from many disciplines, history, economics, psychology, anthropology, sociology, psychiatry to name a few, and yet provide a coherent account of the experiences of an ethnic group. I used a sociological perspective to provide the basic framework, while my own life experiences were used to illustrate and add flesh to the frame.

The reaction to the book was very positive. I was especially pleased that members of my ethnic group reacted so well. Many called or wrote, with comments such as for the first time they began to understand something about their parents, themselves, and the Japanese American culture. It provided the impetus for lectures throughout the United States. The book was also translated into Japanese, so that on visits to Japan, I enjoyed a degree of celebrity status.

EXPERIENCES ABROAD

There were a number of teaching and research experiences abroad which were helpful in clarifying my ethnic identity. While serving as the Director of the University of California Tokyo Study Center at the International Christian University in the early 1970s, I quickly discovered that I was more American than Japanese. I was different from my Japanese colleagues, and walked, talked, dressed, and behaved more as an American than a Japanese. Conversely, while spending a summer at the University of Bristol in 1979 it was difficult to escape from my ancestral heritage. Questions about Japan were more common than questions regarding Los Angeles. The category of Japanese Americans was not readily understood and as a consequence, my identity as a Japanese, rather than as an American, was reinforced.

DOMINATION

My past experiences have lead me to develop my Domination Model in order to explain race relations. I tried to make sense of what had happened to the Japanese Americans, and whether such an incident could happen again in a democratic society.[1] The dominant group (D) has the power to erect barriers which limits the ability of the dominated group (d) to participate equally in the host society. The primary actions are prejudice, discrimination, and segregation, through such mechanisms as stereotypes, laws, and norms which result in the dominated group being avoided, placed at a competitive disadvantage, and isolated. These are the shaping factors, and the more severe actions, such as concentration camps, expulsion, and extermination can best be explained when these prior actions have been in place. The more severe acts can occur when the host society perceives a crisis—that its very existence will be in danger unless more drastic steps are instituted. The Japanese attack on Pearl Harbor and World War II were seen as such "triggers" on those of Japanese ancestry residing in America.

Once a group is incarcerated, there is always the danger of expulsion, exile, and extermination. Some Japanese Americans were deported to Japan; there was talk of using the internees as guinea pigs for "experiments," such as lowering caloric requirements and separating the sexes so that there would be no more "Jap bastards." Fortunately, more sane and humane forces prevailed.

There is also stratification in the lower part of the system (small d). I experienced these differences—that specific dominated groups are less acceptable than others. These are the conditions that lead to "middleman minorities,"[2] and model minorities.[3]

I suspect that if I were to write on my gravestone today, my epitaph would simply say, "a Japanese American who served as a Professor of Social Welfare and Sociology at UCLA." While changes in the dominant society have allowed a shift in what Ogawa (1971) calls *From "Japs" to Japanese,*[4] I also realize that it is my education that has enabled me to be comfortable with all of my identities— as a Japanese American, a college professor, a member of UCLA, a Los Angeleno, and a Californian. I have lived in Japan, Great Britain, and Hawaii and have retained all of my above identities.

NOTES

1. Harry H. L. Kitano. 1991. *Race Relations.* Englewood Cliffs, New Jersey: Prentice Hall.

2. Harry H. L. Kitano. 1974. "Japanese Americans: The Development of a Middleman Minority," *Pacific Historical Review,* November, XLIII, 4, 500–519.

3. Stanley, Sue and Harry H. L. Kitano. 1973. "The Model Minorities." *Journal of Social Issues,* 29, 2, 1–10.

4. Dennis Ogawa. 1971. *From "Japs" to Japanese.* Berkeley: McCutchan Publishing.

DISCUSSION QUESTIONS

1. How did the particular history of Japanese Americans in the United States influence Harry Kitano's biography? How does this reveal the connection between biography and history, and how does it suggest the importance of knowing the histories of diverse racial and ethnic groups?

2. Are there any experiences that you have had that make you especially empathetic with Kitano's account of his life?

3. What did you learn most about the experiences of Japanese Americans in the United States from his narrative?

18

Notes of a
White Black Woman

BY JUDY SCALES-TRENT

Because I am a black American who is often mistaken for white, my very existence demonstrates that there is slippage between the seemingly discrete categories "black" and "white." This slippage is important and can be helpful to us, for it makes the enterprise of categorizing by race a more visible—hence, a more conscious—task. It is at this point, then, that we can pause and look carefully at what we are doing. It is at this point of slippage that we can clearly see that "race" is not a biological fact but a social construct—and a clumsy one, at that. Stories about my life as a white black American also show that creating and maintaining a racial identity takes a lot of effort on my part, and on the part of other Americans. "Race" is not something that just exists. It is a continuing act of imagination. It is a very demanding verb. . . .

SKINWALKERS, RACE, AND GEOGRAPHY

In Navajo cosmology there exist certain powerful creatures who, although they appear to be mere humans, can change shape whenever they wish by taking on animal form. These are supernatural beings, not like you and me. They are called "skinwalkers."

From Judy Scales-Trent, *Notes of a White Black Woman* (University Park, PA: Pennsylvania State University Press, 1995), pp. 476–481. Reprinted by permission.

And I think about them, and this name, when I think about how we all "skinwalk"—change shapes, identities, from time to time, during the course of a day, during the course of our lives. I think about how we create these identities, how they are created for us, how they change, and how we reconcile these changes as we go along.

A young woman leaves her family on the farm and goes to medical school, where she learns a new language, a new culture. She tells me that she feels like an immigrant in a new land. She feels as if she is changing skin, shifting shape, and will be forever shifting as she travels back and forth between these two worlds. A young man goes to visit his parents with a wife and new child. He visits, however, in a complicated way, as he is now not only a son but also a husband and a father. And he shifts identities during the visit, mediating between these different roles within his newly structured family. A child whose parents have different religious beliefs—the father Methodist, the mother Episcopalian; the mother Roman Catholic, the father Greek Orthodox; the father Reform Jew, the mother Buddhist—this child learns to change shape as she communicates in two languages with her parents, as she visits with different sets of grandparents, aunts and uncles, cousins, during the holidays. And I, when I moved from a predominantly black civil rights community in Washington, D.C., to a predominantly white university in Buffalo, I too was a "skinwalker." In Washington, in the black community within a black city, I was a woman who just happened to be black. But in the Buffalo academic world, in this white community within a white city, I became a black person who just happened to be a woman.

All of these examples involve moving from one place to another, from one life to another, from one culture, one role, to another. But sometimes you can change identities while you are doing absolutely nothing at all. Things change around us. Society changes its rules and its boundaries, and suddenly you take on a different form: you become a heron, or forsythia, or your ancestor.

I have been thinking for a long time about two young girls, girls who were skinwalkers, sisters who never met.

The first, Marie, lived in Thionville, a small village in eastern France, in 1871. Like all good French girls, she went to Mass with her family, went to confession, and wore a beautiful white dress to her first Communion. Because her family had a prosperous farm, she also went to a small school, where she studied the French kings and queens, read the plays of Corneille and Racine, and learned the old songs of the region. Then one bright fall day, . . . while she was in the kitchen with her mother putting bread on the table for the midday meal, somewhere far away, in some office or lawmaking place, one of the people who get to draw the lines wrote something down on a piece of paper—and suddenly Marie had a different identity. She was no longer French. She was German.

How could she comprehend this? Was she really supposed to unlearn everything she had ever learned about who she was? About who her people were? That must be so, because Germany immediately installed German schools in its new territory, the former French province of Lorraine. And suddenly Marie, now German, was required to speak only German and study the glory of German history.

The second young girl, Hannah, lived on her family farm in the Tidewater area of Virginia in 1785. Both her parents were free Negroes. Her father worked as a carpenter, and the whole family worked their small farm, a farm that provided a good life for Hannah and her two sisters. With the other black children in the community, the girls learned reading and writing and Bible verses in classes held in the nearby black church—the center of religion, culture, and community for all Negroes, free and enslaved, who lived in the Tidewater area in those days. And after classes, the children played circle games together, and they sang the old work songs and spirituals their grandparents had taught them. Then one day, one muggy summer's day while Hannah was sitting at the kitchen table with her mother stringing beans for dinner, something happened miles away, something that would change her life forever. The Virginia legislature changed the line between black and white. Now before this time, Hannah and her sisters were black, because one of their great-grandparents was black. They thus met the statutory definition of "Negro" in 1784. But in 1785 the legislators redefined "Negro" to mean anyone who had one black grandparent. And this, Hannah didn't have. All her grandparents were white. So on that simple summer's day, while she was at her home on the farm, someone somewhere wrote three sentences on a piece of paper and, magically, supernaturally, Hannah, a "skinwalker," became "white."

Thinking about the lives of these two young girls—one whose life was thrown into disarray by lines drawn on a map; the other, who was turned inside out by lines marked down in a book of rules—thinking about these two girls makes me think about the relationship between race and geography.

In both instances, we are talking about an exercise in drawing lines, lines to separate Here from There. The line-makers are marking boundaries, borders, creating Insiders and Outsiders. They are creating an "us" and a "them." They are creating the Other. Also, in both instances, those who draw these lines are drawing pictures of the world. They are showing what the world looks like, how the world *should* look, what looks right to them. So if you study their picture, you will know who you are—black or white, French or German, "us" or "them."

All this means, of course, that the line-drawers have the authority to describe the world for everyone in it. They are exercising enormous power, power they have grabbed or earned or received or simply found. But they have it, this power to locate the line, to decide who stands where in relationship to the line, and to divide community resources based on that decision.

Thus, whether we are talking about race or geography, marking boundaries creates property rights, for it is the boundaries that define who gets what—who gets the most, who gets less, who gets nothing—who takes, and who gets taken. And whether we are talking about race or geography, both imply war, as property rights always do. For those who somehow have the power to draw these lines, the power to say that they will get the most, will then have to, in fact *must,* fight to maintain those lines. And that means war.

Race and geography have one more important trait in common. They are both equally arbitrary systems of (dis)organization. Whether a person is sitting at a desk drawing lines on a piece of paper that represents the surface of this planet, or putting marks on a piece of paper that form words telling how to separate humans one from the other—no matter which task one is engaged in, it is simply and only a task. It is not a given, not a fact, not an eternal truth. Also, it is a task that leads to other tasks. For after creating this idea and drawing this line, the line-drawers must then convince a lot of people that this is the right line to draw and an important line to draw. And then they must develop a system to maintain these lines.

This is a lot of work.

Just think of the time, energy, and resources that a country uses to create and maintain the lines between its tribe and the tribe on the other side of the line. First, there are probably wars to establish the lines. Then you have to have guards on patrol at all times to make sure that only certain people cross the line. And there are immigration rules and lawyers and border patrols and enforcement agencies and soldiers and sailors and pilots and planes and bombs. The country also spends enormous sums of money teaching its youth about the importance, the "rightness" of the tribal line, so they will be eager to guard the line when it is their turn to patrol.

Then realize that this is the same amount of time and energy we expend to maintain the lines of racial purity in this country. And it is done the same way. There have been fights on the battlefield. And there are still fights—in the courts, in legislatures, on the streets of America—a continuing struggle to maintain the line between black and white, to reinforce validity and power. The country has published rules, drafted forms, hired census-takers, created grandfather clauses and gerrymandering and segregated water fountains and back-of-the-bus and "cordons sanitaires."* It has deputized all Americans who are not black to engage in this battle as soon as the boat brings them here from Peru, from Ireland, from Japan. Similarly, white America expends enormous resources in school and in the media to teach its youth about the intrinsic rightness of this line, so that they will not question its value when they reach the age to stand guard.

One task is overt; the other, covert. But line-drawing is line-drawing. It is the same task. So the next time you say "black" or "white," the next time you hear someone use a racial designation, think about geography. And when you think about geography, see this picture:

> There is a small group of men in a tent, and it is night. These men, lieutenants and cartographers, are sitting, standing around a small table, trying to calculate where to put the line. An oil lamp on the table reflects its yellowish light on their tired faces, on papers strewn about. The men are concentrating on one of these papers—a heavy parchment scroll, its red wax seal broken. The scroll is from the general at the front, claming victory, and it tells the mapmakers what to do. The scroll says this:

> Draw the line here.
> We have taken more land.
> This much is ours.

---

*Eds. note: meaning, exclusionary zones

DISCUSSION QUESTIONS

1. How does Scales-Trent use the analogy of "skinwalkers" to describe her biracial identity?
2. How does Scales-Trent's account show how race is both imaginary and real?

19

Tripping on the
Color Line

BY HEATHER M. DALMAGE

To live near the color line, in the space Gloria Anzaldúa calls the border-lands, means to contend constantly with what I call borderism.[1] Borderism is a unique form of discrimination faced by those who cross the color line, do not stick with their own, or attempt to claim member-ship (or are placed by others) in more than one racial group. Like racism, bor-derism is central to American society. It is a product of a racist system yet comes from both sides. The manner in which people react to individuals who cross the color line highlights the investment, the sense of solidarity, and per-haps the comfort these observers have with existing categories. Perhaps most important, the reaction shows the wide acceptance of racial essentialism as the explanation for the color line. When the color line is crossed, the idea of immutable, biologically based racial categories is threatened. The individual who has crossed the line must be explained away or punished so that essential-ist categories can remain in place. Ironically, if race were natural and essential, individuals would not have to engage in borderism. The act of boderism is one of the many ways in which individuals construct or "do" race.

Multiracial family members contend with borderism in many aspects of life. It is both part of the workings of larger institutions and the outcome of individual actions. When families are unable to find accepting places of worship

From Heather M. Dalmage, *Tripping on the Color Line: Black-White Multiracial Families in a Racially Divided World* (Piscataway, NJ: Rutgers University Press, 2000), pp. 41–63. Copy-right © 2000. Reprinted by permission of Rutgers University Press.

and comfortable neighborhoods, they contend with examples of institutional borderism. A nefarious individual is not responsible for creating these situations. Rather, this borderism has developed in the context of a deeply racist and segregated society. Some borderism, however, does play out on an individual level and is meant to be hurtful. A family disowns a child for not sticking with his own. Peers tell a multiracial child that she is not black enough. An interracial couple is physically accosted in the street. Such borderism may stem from hostility, hatred, or feelings of betrayal and is grounded in ideas about how people ought to act. But it may also reveal concern. For example, before returning to graduate school I was working for a corporation that sent me to the Baltimore–Washington, D.C., area for a summer. It was a racially tense season. The Klan was active, hanging notices all over one town declaring it a "nigger-free zone." In another town white men killed a black man because he was walking with a white woman in a white neighborhood. When my husband, Philip, came to visit, we were cautious. One weekend went without incident until I dropped him off at the train station. We were talking when an elderly black woman walked up to Philip and scolded, "Get away from her; she's going to get you killed. You need to stick with your own." I believe she meant to be helpful to Philip. After all, another black man had just been killed. Nonetheless, these moments can be paralyzing. In such situations there is no way for me to say, "Please, my skin color belies my politics. I am likewise outraged and live with the fear of violence." For me to speak at these moments is to belittle the history of racial oppression and the fact that I do represent, in my physical being, the object and the symbol that has been used to justify so much oppression. Protecting white women's bodies has long been the justification for abusing and lynching thousands of black men and women.[2]

All members of multiracial families face borderism, although as individuals we face specific forms of discrimination based on our race, physical features, gender, and other socially significant markers. . . .

BORDER PATROLLING

The belief that people ought to stick with their own is the driving force behind efforts to force individuals to follow prescribed racial rules. Border patrollers often think (without much critical analysis) that they can easily differentiate between insiders and outsiders. Once the patroller has determined a person's appropriate category, he or she will attempt to coerce that person into following the category's racial scripts. In *Race, Nation, Class: Ambiguous Identities,* Etienne Balibar and Immanuel Wallerstein observe that "people shoot each other every day over the question of labels. And yet, the very people who do so tend to deny that the issue is complex or puzzling or indeed anything but self-evident."[3] Border patrollers tend to take race and racial categories for granted. Whether grounding themselves in essentialist thinking or hoping to strengthen socially constructed racial categories, they believe they have the right and the need to patrol. Some people, especially whites, do not recognize the centrality and problems of the color line, as evinced in color-blind claims that "there is only one race: the human race" or "race doesn't really matter any more." Such thinking dismisses the terror and power of race in society. These individuals may patrol without being aware of doing so. In contrast, blacks generally see patrolling the border as both problematic and necessary.

While border patrolling from either side may be scary, hurtful, or annoying, we must recognize that blacks and whites are situated differently. The color line was imposed by whites, who now have institutional means for maintaining their power; in contrast, blacks must consciously and actively struggle for liberation. . . .

White Border Patrolling

Despite the institutional mechanisms in place to safeguard whiteness, many whites feel both the right and the obligation to act out against interracial couples. If a white person wants to maintain a sense of racial superiority, then he or she must attempt to locate motives and explain the actions of the white partner in the interracial couple. A white person who crosses the color line threatens the assumption that racial superiority is essential to whites. The interracially involved white person is thus often recategorized as inherently flawed—as "polluted." In this way, racist and essentialist thinking remains unchallenged.

Frequently white families disown a relative who marries a person of color, but several people have told me that their families accepted them again once their children were born. The need to disown demonstrates the desire to maintain the facade of a pure white family.[4] By the time children are born, however, extended family members have had time to shift their racial thinking. Some grant acceptance by making an exception to the "rule," others by claiming to be color blind. Neither form of racial thinking, however, challenges the color line or white supremacy. In fact, both can be painful for the multiracial family members, who may face unending racist compliments such as "I'll always think of you as white." . . .

Privileges granted to people with white skin have been institutionalized and made largely invisible to the beneficiaries. With overwhelming power in society, why do individual whites insist on border patrolling? As economic insecurity heightens and demographics show that whites are losing numerical majority status, the desire to scapegoat people of color, especially the poor, also heightens. As whites lose their economic footing, they claim white skin as a liability. Far from recognizing whiteness as privilege, they become conscious of whiteness only when defining themselves as innocent victims of "unjust" laws, including affirmative action.[5] In their insecurity they cling to images that promote feelings of superiority. This, of course, requires a racial hierarchy and a firm essentialist color line. Border patrolling helps to maintain the myth of purity and thus a color line created to ensure that whites maintain privileges and power.

Black Border Patrolling

Some blacks in interracial relationships discover, for the first time, a lack of acceptance from black communities. Others experienced border patrolling before their marriage, perhaps because of hobbies and interests, class, politics, educational goals, skin tone, vernacular, or friendship networks. Patrolling takes on new proportions, however, when they go the "other way" and marry a white person. While all relationships with individuals not seen as black are looked down on, relationships with whites represent the gravest transgression. Interracially married black women and men often believe they are viewed as having lost their identity and culture—that they risk being sees as "no longer really black." Before their interracial marriage, most called black communities their home, the place from which they gained a sense of humanity, where they gained cultural and personal affirmation. During their interracial relationship many discovered black border patrolling. . . .

An overwhelming percentage of black-white couples involve a black male and a white female at a time when there are "more single women in the black community than single men."[6] Many black men are hindered by a racist educational system and job market that make them less desirable for marriage. Many others are scooped into the prison industrial complex. High-profile athletes and entertainers who marry white women confirm for many that black men who are educated and earn a good living sell out, attempting to buy white status through their interracial relationship.[7] Beyond issues of money and status, many black women see black male-white female interracial relationships "as a rejection of black women's beauty, [and] as a failure to acknowledge and reward the support that black women give black men."[8] . . .

Black women and men may both feel a sense of rejection when they see an interracial couple, but for each that sense of rejection comes from a different place. In a society in which women's worth is judged largely by beauty— more specifically, Eurocentric standards of beauty—black women are presumed to be the farthest removed from such a standard. Men's worth is judged largely by their educational and occupational status, two primary areas in which black men are undermined in a racist system. Black men with few educational and job opportunities lack status in the marriage market. Thus, when black men see a black woman with a white man, they may be reminded of the numerous ways in which the white-supremacist system has denied them opportunities. The privilege and power granted to whites, particularly to white males, is paraded in front of them; and they see the black women in these relationships as complicit with the oppressor. . . .

Today there are no longer legal sanctions against interracial marriage, but de facto sanctions remain. At times, family and friends exert pressure to end the interracial relationship; at other times, pressure may come from the border patrolling of strangers. Even if the relationship is clandestine, thoughts of how friends, family members, co-workers, employees, and the general public might respond can deter people from moving forward in a relationship. . . .

Border patrolling plays a central role in life decisions and the reproduction of the color line. As decisions are made to enter and remain in an interracial relationship, the color line is challenged and racial identities shift. Many blacks spoke of the growth they experienced because of their interracial relationship and border patrolling. Parsia explains, "I used to be real concerned about how I would be perceived and that as an interracially married female I would be taken less seriously in terms of my dedication to African American causes. I'm not nearly as concerned anymore. I would hold my record up to most of those in single-race relationships, and I would say, 'Okay, let's go toe to toe, and you tell me who's making the biggest difference,' and so I don't worry about it anymore." Identities, once grounded in the presumed acceptance of other black Americans, have become more reflective. Acceptance can no longer be assumed. Definitions of what it means to be black are reworked. Likewise, because of border patrolling, many whites in interracial relationships began to acknowledge that race matters. Whiteness becomes visible in their claims to racial identity.

NOTES

1. Gloria Anzaldúa, *Borderlands/La Frontera: The New Mestiza* (San Francisco: Spinsters/Aunt Lute, 1987).

2. Grace Elizabeth Hale, *Making Whiteness: The Culture of Segregation in the South, 1890–1940* (New York: Vintage, 1998).

3. Etienne Balibar and Immanuel Wallerstein, *Race. Nation. Class: Ambiguous Identities* (New York: Verso, 1991), 71.

4. Naomi Zack, *Race and Mixed Race* (Philadelphia: Temple University Press, 1994).

5. Charles Gallagher, "White Reconstruction in the University," *Socialist Review* 24, 1 and 2 (1995): 165–188.

6. Michael Eric Dyson, "Essentialism and the Complexities of Racial Identity," in *Multiculturalism: A Critical Reader,* ed. David Theo Goldberg (Cambridge, Mass.: Blackwell, 1994), p. 222.

7. Paul C. Rosenblatt, Terri A. Karis, and Richard D. Powell, *Multiracial Couples: Black and White Voices* (Thousand Oaks, Calif.: Sage, 1995), 155.

8. Dyson, "Essentialism," p. 222.

DISCUSSION QUESTIONS

1. Dalmage uses the concept of *borderlands.* What does she mean and how does she explain that borders are "patrolled?"

2. In what different ways do Black people and White people do their *border patrolling,* according to Dalmage, and how has this affected her experience and that of others who live along "borders?"

20

Blinded by Whiteness

The Development
of White College Students'
Racial Awareness

BY MARK A. CHESLER,
MELISSA PEET,
AND TODD SEVIG

R acial identity is the meaning attached to self as a member of a group or collectivity in racial situations, and individuals may express this identity differently in different circumstances (Cornell and Hartman 1998). Since identity is formed by class and gender as well as race, there are many ways of being white or any other race/ethnicity. Racial attitudes and changing attitudes are the statements of a person's preferred views or positions about others and about contemporary (or historic) policies and events. Attitudes are also shaped by one's social location and are expressed differently in different circumstances. Social and institutional structures and cultures provide the limits and opportunities for both the creation of racial identities and the formation and expression of racial attitudes.

Throughout, we present the voices of white students attending the University of Michigan, a university with a tradition of student, faculty, and administrative engagement with issues of racism and affirmative action. Recently, Michigan has become one of the nation's battlegrounds for competing narratives and institutional policies around racial matters. The data reported here were gathered from white students of varied backgrounds in individual and small-group interviews conducted between 1996 and 2000. Although they

From *White Out: The Continuing Significance of Racism*, ed. by Ashley W. Doane and
Eduardo Bonilla-Silva (New York: Routledge, 2003), pp. 215–227. Copyright © 2003.
Reprinted by permission of Routledge/Taylor & Francis Books, Inc.

are not geographically, temporally, or in terms of cohort representative of other white students' racial consciousness, they are useful windows into the ways in which racial processes become visible and are expressed.

BACKGROUND

The social and cultural context of the modern university is one of racial plurality but also of racial separatism and tension. Students come to these settings from racially separated and often segregated neighborhoods and communities (Bonilla-Silva and Forman 2000; Massey and Denton 1993). For many, the university is the first place in which they have sustained contact with a substantial number of students of another race. Although there are more numerous formal and informal opportunities for racial interaction and growth in the university than in most secondary educational environs, white students' lives in these environs are often not very different from their separated lives in previous home and school communities (Hurtado et al. 1994).

In these collegiate circumstances, white students are often confronted for the first time with the need to think about their own racial location. Having been socialized and educated at home, in their neighborhoods, through the media, and in previous schooling to expect people of color to be different, less competent, and potentially threatening, most young white people are ignorant, curious, and awkward in the presence of "others." Some may be aware of their racial group membership and identity, but others may be relatively unaware. Furthermore, during this developmental stage of late adolescence and early adulthood, students' identities as racial beings, as well as their racial attitudes, are subject to challenge and change. Hence it is important to understand the potential developmental trajectory of students' views as they move from their communities of origin to and through diverse collegiate experiences.

Recent explorations of whiteness suggest that changes in the economic, political, and cultural landscape have promoted greater self-consciousness about race. As a result, for many students the invisibility of whiteness, the notion that white is normal and natural, has become harder to sustain. Challenges to white ignorance and/or privilege have also increased some whites' sense of threat to their place in the social order and to their assumptions about their lives and society (Feagin and Vera 1995; Pincus 2000; Winant 1997). Discussions of historic privilege, structural inequality, and racial oppression have caused some white students (and college administrators and faculties as well) to question their enmeshment in pervasive (if unintended) patterns of institutional discrimination. In addition, institutions that now see the education of a diverse citizenry as integral to their missions of education and public service are struggling to make changes in the demographics of their faculty and student bodies, curricular designs, pedagogical tactics, student financial aid programs, and support services.

CONTEMPORARY THEORIES OF WHITE
RACIAL ATTITUDES AND IDENTITIES

In the context of these shifts and struggles, scholars have described and explained the genesis and nature of whiteness and white racial attitudes and experiences as well as the developmental aspects of white racial identity and consciousness. When understood in the context of larger patterns of institutional racism and changing cultural narratives about race, these identity and attitude frames are useful guides—heuristic devices—to understanding white racial consciousness and conceptions of whiteness itself. However, almost all interpretations and typologies of white attitudes and identities focus on their views of "the other" rather than on views of oneself or one's own racial group. That is, surveys of racial attitudes generally ask white people about their views of or prospective behavior toward people of color or race-related policies, seldom inquiring into whites' views of their own racial selves or of their earned/unearned status (i.e., privileges).

Similarly, most white identity development models focus on how whites view people of color rather than themselves; thus their racial identity is conceived as a reflection of their views of "the other." The stance that overlooks one's own race and focuses on others' can itself be seen as a manifestation of the "naturalness" and dominance of whiteness. Certainly one's views of the other and of the self are interactive, and people learn about their racial identity and attitudes in an interactive context, but one's views of others (or of the meaning of others' race) and one's view of themselves (or of the meaning of their own race) are not the same thing. . . .

The notion of white racial "identity stages" suggests a developmental process that generally proceeds as follows (Helms 1990; Rowe et al. 1994; Tatum 1992): 1) from racial unawareness or conformity to traditional racial stereotypes, sometimes called an "unachieved" racial identity; 2) through questioning of these prior familial and societal messages, with attendant confusion, dissonance, and perhaps even "overidentification" with the other and attendant rebellion; 3) to retrogressive reintegration, where white culture is idealized, others are rejected, and a racially "dominative" ideology holds forth; 4) into a generally liberal (sometimes called pseudo-) acceptance or tolerance of people of color, often accompanied by adherence to notions of "color blindness" or denial and conflict around remaining prejudices; and 5) it is hoped to an antiracist stance, wherein understanding of others' oppression and one's own privilege is (more or less) fully integrated into a personal worldview called an "autonomous" or "integrative" white racial identity. . . .

Given increased collegiate attention to racial injustice and the desire of some people and advocacy groups to challenge institutional racism, it is not surprising that some young white collegians are becoming more conscious of their racial membership and its privileges. Such consciousness is likely to be painful, as it requires acknowledging both systemic advantage and personal privilege and enmeshment (historically and contemporarily) in structural or institutional discrimination and oppression. A few scholars have pointed to the emergence of a "liberationist" or "antiracist" form of white racial attitudes, wherein white people acknowledge and grapple with their accumulated racial privilege and their role (intentional or not) in sustaining white advantage and the domination over people of color. The racial identity literature refers to this belief/action system as an integrated, autonomous, introspective, or antiracist racial consciousness. . . .

WHAT DO WHITE STUDENTS BRING WITH THEM TO THE UNIVERSITY?

In interviews, white students [we interviewed] discussed the neighborhood and schools in which they grew up and the effect these largely segregated experiences had on their conceptions of themselves, race, and racism. The major themes that characterize their precollege experience are lack of exposure, subtle and overt racism, racial tokenism, and lack of successful role models of people of color:

> I never really think about the fact that I am white. I just think that it is fortunate that we don't have to think about it, you know what I mean? It is one of the perks of being white.

> I consider myself white, but I don't think about it. The only time I think about it is when we have to do these dumb forms and think about what race we are.

According to Janet Helms (1990:3), racial identity is "a sense of group or collective identity based on one's perception that he or she shares a common racial heritage with a particular group." If the students above never thought about being white and didn't feel a sense of shared racial heritage, they could not possibly develop a self-conscious racial identity; they were at the unaware stage.

White students consistently indicated that their lack of prior contact with people of color, even in the midst of liberal rhetoric, failed to prepare them to engage meaningfully about race:

> I grew up in a very white community, and the church was really white. We talked about other cultures, but it was all about boys and girls are equal and worthy and so are people of different colors. It was all about "everything's OK."

> Where I grew up, everybody was white, and even though I knew (on some level) that not everyone was white, we never really had to deal with it, and so we didn't.

A few students reported coming from more diverse neighborhoods and schools, but they too indicated a relatively low level of sustained interaction or conscious educational attention to issues of diversity and intergroup relations.

In these "more diverse" settings, racial segregation was still the normative experience for white students (as well as for students of color):

> [The city] is very segregated in terms of housing, and there's all different kinds of people who live here. But there isn't a tremendous amount of communication and social interaction between the groups . . . unless you played sports or you were involved in something else, because it was tracked. Almost all of the kids on the college track were white and almost all of the kids on the other tracks were black . . . and then there were also Asian kids and they were generally in the white track.

This lack of meaningful contact with people from other races was often coupled with various forms of both subtle and overt racism. If fact, many students' comments indicate that intergroup separation supported the home and media-based racism they were exposed to, creating and sustaining conditions wherein remnants of "old-fashioned racism" and an identity stage of unawareness and acceptance of stereotypes could be maintained:

> So I grew up with my dad particularly being really racist, he didn't really say much about any other group except Black people. "Nigger" was a common word in my family. I knew that that was not a good thing in terms of race. I knew that there was the black side of town, there was the black neighborhood, and then the rest of it was white, and that's what I grew up in. . . . But we never had any personal interactions with anybody [from the black neighborhood].

> My whole town was white except for a few families who migrated from Mexico to work. I had the clear sense that they weren't supposed to be there. They were like some unspoken exception that was supposed to be invisible.

In addition to the lack of contact in school and neighborhood and the various forms of racism that students were exposed to, several students indicated that when they did learn about people from other races, they were usually token efforts of inclusion:

> The only thing I learned in school was that [George] Washington Carver was a black man and he discovered peanuts or something like that. I think we might have peripherally dealt with Martin Luther King. But four years, two years of history, two years of government, we really didn't touch on African-American or any other issues at all . . . that just didn't even exist as far as anybody was concerned. In elementary school we dealt with the Indians. You know, you put your hand on a piece of paper and you draw around it and you cut it out and you make a turkey, or you make little Indian hats and things like that with feathers.

Finally, even students who experienced token efforts of inclusion as unsatisfactory found little opportunity to formulate openly meaningful questions about race. Several students commented that when they did have racial questions and concerns during their high school years, they were simply told that there was "nothing to talk about":

> The message that I got from the white teachers at the school and other people was that the way not to be racist was to just pretend that you don't see any differences between people. And so everybody had feelings about race, but nobody talked, there was no place to talk about those things. And you only have to just treat everybody as an individual and everything will be fine.

> In my high school government class I asked a question about the Civil Rights movement and racism. The answer I got was basically that it was bad back then, but now everything was fine.

Growing up with everyday processes of segregation, lacking contact with racially (or socioeconomically) different peers, being exposed to various forms of racism and racial tokenism, and not being educated meaningfully about race and racism deeply affect white students' social identity—their sense of themselves as well as their relations with others. In their homes, schools, and communities these students acquired habitual attitudes, expectations, and ways of making meaning about their world. White students were socialized to not see themselves as having a race and did not understand their own (and their communities') exclusionary attitudes and behaviors. . . .

WHITE STUDENTS' EXPERIENCES ON CAMPUS: NEW CHALLENGES TO WHITENESS

Students' precollege socialization forms a grid of attitudes and expectations about race and whiteness that is often reenacted and reified through their collegiate experience. As several white students reported, once in college they still did not think about themselves as being white—even in the presence of diversity; no one and no program invited or required them to. Hence, as the racial majority on campus and the dominant group within the larger society, the experience of knowing themselves as white was primarily reactive. That is, white students' numerical and cultural dominance protected them from having to know or understand others' experiences. Consequently, in order to "see" their race, they had to have a critical encounter or be consciously challenged to think and reflect about the particular experiences (perhaps privileges) that they had as a result of their racial position. Unless this challenge occurs at a conscious level, their own racial identity remains unknown and invisible during their college years.

Even when white students do have a critical encounter that raises their awareness of their race, they may not have the skills and consciousness (or instructional and experiential assistance) to deal with or act on it productively. Compare, for instance, the level of insight conveyed in these two excerpts:

> I don't understand why all the black students sit together in the dining hall. They complain about people being racist, but isn't that racist?

> Something I see is that the different races tend to stick with people like themselves. Once, in a class, I asked why all the black students sit together in the dormitory cafeteria. A black student then asked: "Well, why do you think all the white kids sit together?" I was speechless. I thought that was a dumb question until I realized that I see white people sitting together as normal and black people sitting together as a problem. . . .

These comments reflect larger social assumptions about race relations on campus, wherein the prevailing myth has been that minority students are "self-segregating" and the exclusionary behaviors of the majority white group remain unseen (Elfin and Burke 1993; Tatum 1997). However, longitudinal research with over 200,000 students from 172 institutions found that it was white students who displayed the most exclusionary behaviors—particularly when it came to dating (Hurtado et al. 1994). Thus the view that minority students are self-segregating is clearly a skewed perspective that does not take into consideration the separatist and/or exclusionary behaviors of white students. It also fails to account for the ways in which institutional norms and cultures help students misinterpret patterns of interracial interaction.

Other white behaviors took the form of promoting or reacting to patterns of racial marginalization and separation in daily interactions in classrooms, social events, or casual encounters. The result, of course, continues to be minimal opportunity for sustained interaction:

> My black friend invited me to a party with her. And the first thing I could think of was how many white people are usually there. I remembered thinking, this is probably going to be uncomfortable, and I would rather just go out with my white friends. I'm feeling apprehensive about meeting their friends and therefore spending time with them.

> I used to feel very guilty thinking I don't have many diverse friends. I thought: "I have to go out and get a black friend."

Some white students reported finding these and other situations so discomforting that they began to express resentment against students of color. This type of resentment is supported by the discourse of whites as victims:

> I think white males have a hard time because we are constantly blamed for being power-holding oppressors, yet we are not given many concrete ways to change. Then we just feel guilty or rebel.

> I think that black people use their race to get jobs. I've seen it happen. My friend should have had this job as a resident advisor, but a black guy got it instead. There's no way the black guy was qualified.

The particular reference to "my friend" in the excerpt above is referred to by Eduardo Bonilla-Silva as one of the main "story lines of color blindness

(2001:159)." Views such as these, expressing the emergence of a self-interested form of racial awareness, are consistent with Lawrence Bobo's (1999) discussion of the group-position frame of racial attitudes.

Hence we encounter the view of the white person as the "new victim" of racism or as the target of "reverse discrimination" (Gallagher 1995; Pincus 2000). Victimhood, like all racial identities and views, is historically situated, and current public discourse about affirmative action and other race-based remedies stimulates and supports its development and expression. A lack of understanding of one's own prejudices, the realities of racial discrimination, and the advantages whites have leads to the view that minority advance is unmerited and a reflection of special privilege. The result often is aversive or self-interested racism that facilitates the interpretation of interracial encounters or circumstances as overprivileging minorities and victimizing whites. This also is referred to as the reintegrative or dominative stage of white racial identity.

The inability to understand racial membership is compounded by denial of any racial prejudice or racism. As a result of professed innocence about the meaning and implications of their own racial status and privileges, white students are often "blind" to the reality and status of students of color and regard themselves as "color-blind." If white students do not understand the personal or structural implications of being white and are unable to see how their racial behaviors affect others, they blindly negotiate racial encounters with the sense that all that matters is their good intentions. Their structural position of racial dominance, together with precollege socialization and color-blind ideology, makes it very difficult to distinguish between good intentions (or innocence) and a reflective consciousness that can enact just racial encounters:

> I am a pretty open person and someone who wouldn't even think about race, who would try to be color-blind.

> When I was asked in a class to describe my beliefs about race, it was easy. I said that I think that the whole idea of race has gone too far, that we need to stop thinking about race and start remembering that everyone is an individual.

Robert Terry (1981) identifies this pervasive color-blind ideology as an attempt to ignore or deny the relevance of race by emphasizing everyone's "humanness." Others have pointed out that the changing discourse of affirmative action—from a need to remediate past injustice to a concern about reverse discrimination—has affected how white people construct racial meaning (Gamson and Modigliani 1987). The new discourse of white victimhood not only acts to obscure the experiences of students of color but also further reinforces barriers to white students' ability to acknowledge their own racial identity as members of the dominant or privileged group.

Despite these reports of unawareness, negativity, blindness, and victimhood, there are also signs that some white students develop more sophisticated and progressive views of race. As they encounter themselves and others, some white students report moving out of the stage of "conformity" or "dissonance," going beyond "color blindness," and acknowledging their racism, prejudices, and stereotyped assumptions or expectations. This occurs partly as a function of structured educational experiences and informal contacts:

> It took me a long time to be able to get to a point where I can say that I have prejudices.

> Something I learned is that people have stereotypes. I learned that having stereotypes about other groups is part of the environment that we grow up in.

For a number of white students, these realizations led to a sense of shame or guilt: several scholars have also referred to these responses as the symbolic or emotional "costs" of white racism (Feagin and Vera 1995; Rose 1991);

> But I was so guilt-ridden, just horribly liberal guilt-ridden, paralyzed and unable to act. I was totally blowing every little minor interaction that I had with people of color way out of proportion and thinking that this determines whether or not I'm a good white person or a bad white person, and whether I'm racist or not. I saw how hard it was for me to stop doing that and start being more productive. And how hard it was for me to not be scared.

Such strong feelings, when combined in sensitive ways (as contrasted with self-pitying or defensive ways) with new educational input, helped some white students understand some of the privileges that were normally accorded them as a function of their white skin color (and associated socioeconomic and educational status):

> I learned that being white, they're so many privileges that I didn't even know of . . . like loans from the bank, not being stopped by the police, and other things me and white kids can get away with.

> I had not noticed the extent to which white privilege has affected and continues to affect many aspects of my everyday life. I thought "I" had accomplished so much, but how much of where I am is due to my accumulated privilege—my family, economic status, school advantages? . . .

Innovative educational programs must be designed and implemented to address these issues in students' racial identities and attitudes. However, even such innovations will not be effective or sustained without parallel changes in the operations of departments and the larger collegiate or university environment. Without changes in this broader organizational landscape, it is unlikely that individual white students' attitudes will change or that their racial identities will continue to "progress"—or that such change programs, if initiated, can be maintained. Moreover, students' consciousness and the academy itself are enmeshed in our society's continuing struggle with racial discrimination and racial privilege. There are real limits for any change toward more liberationist or antiracist white identities or racial attitudes within a highly racialized and racist society and higher educational system.

REFERENCES

Bonilla-Silva, Eduardo, and Tyrone A. Forman, 2000. "'I Am Not a Racist But . . .': Mapping White College Students' Racial Ideology in the USA." *Discourse & Society* 11: 50–85.

Cornell, Stephen and D. Hartman, 1998. *Ethnicity and Race: Making Identities in a Changing World.* Thousand Oaks, CA: Pine Forge.

Helms, Janet E., ed. 1990. *Black and White Racial Identity: Theory, Research and Practice.* New York: Greenwood.

Hurtado, S., Dey, E., & L. Trevino, 1994. "Exclusion or Self-Segregation: Interaction Across Racial/Ethnic Groups On Campus." Presented to meetings of the American Educational Research Association. New Orleans, LA.

Massey, Douglas S., and Nancy A. Denton, 1993. *American Apartheid: Segregation and the Making of the Underclass.* Cambridge, MA: Harvard University Press.

Terry, Robert W. 1981. "The Negative Impact on White Values." Pp. 119–151 in *Impacts of Racism on White Americans,* ed. Benjamin P. Bowser and Raymond G. Hunt. Beverly Hills, CA: Sage.

DISCUSSION QUESTIONS

1. What are the different phases that Chesler et al. identify as affecting the development of White people's racial identity?

2. Chesler and his colleagues identify definite patterns in the attitudes of White students as they encounter new racial experiences in the college setting. Have you seen evidence of these same patterns in your observations of White students on your campus? How are they similar and/or different?

InfoTrac College Edition:
Bonus Reading

http://infotrac.thomsonlearning.com

You can use the InfoTrac College Edition feature to find an additional reading pertinent to this section. You can also search on the author's name or keywords to locate the following reading:

O'Neil, John. 1997. "Why Are the Black Kids Sitting Together? (Interview with Author Beverly Daniel Tatum on Discrimination in Education)" *Educational Leadership* 55 (December): 12-17.

Beverly Tatum argues that understanding the formation of racial identity is important to an understanding of why students of color often group themselves with other students of color. Because White students usually don't understand how racism works, they can misinterpret this pattern as so-called "self-segregation" (even though *they* may associate mostly with other White students). What does Tatum suggest is necessary to create more interracial understanding?

STUDENT EXERCISES

1. Design a brief study of interracial relationships for which you interview students on your campus about their attitudes toward interracial marriage. If possible, interview students from different racial-ethnic backgrounds. What degree of support do you find for *dating* someone of a different race? Is there a different level of support for *marrying* someone of a different race? Do these attitudes vary between men and women and/or between those from different racial-ethnic backgrounds? What factors do you think most influence people's attitudes toward interracial relationships?

2. Think back to the first time you remember recognizing your own racial identity. What were the circumstances? What did you learn? Now ask the same question to someone whose race is different from your own. How do the experiences of the two of you compare and contrast? How do the answers illustrate how racial identity is formed in different contexts and with different meanings depending on the group's experience?

PART V

The Political Economy
of Race

Until now, we have been looking at race largely in terms of beliefs, relationships, and identity. But race is also part of larger social structures that form the organization of society. **Political economy** refers to the links that exists between the economic and political systems of society—systems whose evolution is based on the exploitation of racial-ethnic labor and the denial of citizenship rights to people of color. The political economy is a system in which those with the most economic resources tend to have political power—power to shape the organized systems of social behavior within a society. These organized systems of social behavior are called **social institutions.** They include the *family*; the ways work and commerce are organized (*economy*); the schools where people learn their places in the society (*education*); and the *media* that communicate ideas to the public. Social institutions are not fixed, but change over time. Because the political economy of race is a huge topic, we cover it in three sections: race, citizenship and labor; immigration, ethnicity, and migration; and opportunity structure, class formation, and social mobility.

Power enables a group to shape institutions for its own benefit, while groups without power have to work within social organizations they did not design. In terms of the economy, those with power have historically appropriated the labor of others for their own benefit. As the nature of the economy shifts, a few more people might gain access to political power—such as the extension of the vote to White men without property in the mid-nineteenth century. However, the practice of denying racial-ethnic groups the rights of citizenship was not challenged until the twentieth century. It was hard for these groups to earn a living and support their families, as well as change unfair economic and political arrangements.

When the United States was an agricultural nation, landowners, banks, and budding industries depended on the labor of enslaved and indentured people, as well as people who had no land and had to work for others. In the U.S. Constitution and many state constitutions, these workers, along with women, were denied the right to vote. Newspapers and documents of the time presented these arrangements as appropriate and justified by the assumed ability of propertied White men to rule over others, just as fathers ruled in families. In other words, not all people were considered full citizens of the nation.

You can think of citizenship in two ways. First, there is an actual system of rights (the right to vote, the right to sit on a jury, the right to own property, the right to be counted for purposes of representation, and so forth) by which people are able to play a role in shaping institutions. Sometimes you can succeed within the system; other times you do not, but you are still a participant. Citizenship is also symbolic—meaning the right of belonging to a nation or a community (Glenn 2002). For example, at the ceremony that opened the Lincoln Memorial in Washington, D.C. in 1922, Black Americans had to witness the event from behind a rope. The pattern of segregation, like separate schools for people of color and the denial of access to libraries and museums, communicated that they were not full citizens in the nation.

To understand the political economy and its impact on people, we have to look closely at discrepancies in particular groups' access to power. Inequity can be seen in many arenas—labor, education, housing, and cultural resources, to name a few. Many people immigrate to the United States with hopes of earning a decent wage, raising and educating their children, and securing social resources to practice their religions and cultural rituals, goals also shared by native-born people of color. Such goals are not easy to achieve, and vary over time. Thus, when and how groups arrive, and the social, economic, and political conditions they face shape the likelihood of economic opportunity and political freedom. Race plays a key role in shaping those opportunities.

After our nation had been established, the 1790 Naturalization Act offered citizenship to White immigrants to the new nation, while American Indians, African Americans, and other people of color had to change laws to become citizens. Rather than enjoying life, liberty and the pursuit of happiness, these groups had their labor and land appropriated and used for the benefit of others. Europeans who entered the United States after the War of 1812 found the

Native White population a little distant, but most of the immigrants were able to settle in rural and opening sections of the country like Ohio and Illinois, where they could replicate the lives they left in Europe. Although these new-comers were foreign to native-born White Americans, they were mostly farm-ers and craftspeople whose way of life did not threaten much of the established population. Over time and after the Civil War these residents were considered Americans. The Irish, the majority poor men and women who settled in cities in the 1820s to 1850s when there were few jobs, faced harsh discrimination. The men worked in the harbors and traveled to work on canals and other indus-trial sites, while the women did domestic work. With little money and few work opportunities, these immigrants clustered in urban centers like New York and Boston, and eventually gained citizenship and played a role in urban and state politics. By the time new immigrant groups came from eastern and southern Europe during the development of our industrial economy (1880s to 1920s), the Irish were a group with some degree of political power in cities.

In contrast, the Japanese immigrants in California, mostly a literate popula-tion, could not become citizens; they thus lacked the voting power to shape their lives. They found their employment options limited, so they served their own community and farmed outside of cities, where they provided fruits and vegetables for the growing urban population. As their settlements grew, so did opposition from the native and immigrant White population. Californians first passed laws in 1913 limiting Japanese to leasing land for three years. Later the Alien Law Act of 1920 prohibited them from leasing or owning land, so the Japanese either changed occupations or had White people purchase land for them (Takaki 1989). Their labor was critical for the growth of the nation, but the denial of rights made them vulnerable to exploitation and limited their occupational choices.

Immigration has always been essential to the economic well-being of the United States, but the immigrants themselves have not always been welcome. Today, we see many of the descendants of eastern and southern Europeans as White people, but when they first arrived, their ancestors had a difficult adjust-ment and faced many barriers to acceptance. In Part II, Karen Brodkin ("How Did Jews Become White Folks?") discusses this transition for her own ethnic group. Nancy Foner ("From Ellis Island to JFK: Education in New York's Two Great Waves of Immigration") provides more details of the experiences of those early immigrants and how their children advanced. Immigrant labor was essential to the industrial owners whose factories made the United States a manufacturing center. Yet, these newcomers were perceived as racially differ-ent from native-born White Americans. Their participation in industrial work changed the nation from a land of independent farmers and craftsmen to a nation state where economic power was concentrated in the hands of a few robber barons. At the time, many native-born White Americans wanted to limit immigration as a means of holding onto their established status. Industrial employers, on the other hand, wanted to keep the borders open so that they would always be able to hire people at low wages, then replace them with newer immigrants when their current workers wanted higher wages. National leaders, including presidents, sided with the big industries because such businesses were important to growing the country.

The European newcomers, seen as racially inferior, were perceived as threats to many native-born White Americans who then used their political power to shape who could become future citizens. Alarmed at the prospect of downgrading the "stock" of White Americans, they passed laws that reflected their beliefs about racial hierarchies. First, they passed the Literacy Act in 1917, requiring that immigrants be literate. Second, in 1924 the Johnson-Reed Act limited immigration to 150,000 people per year and mandated a quota system that reflected the initial origins of the nation. This Act also barred the admission of people who were ineligible for citizenship, thus ending immigration for most Asians. In Part II, we learned about the difficulties Takao Ozawa faced regarding citizenship; he had arrived in California in 1894. After the 1920s, few Japanese immigrants would even gain access to this nation for many years.

The 1924 law, with many modifications, remained the framework of U.S. immigration policy until 1965. In 1928, when Congress established quotas, the groups with some political power could insure that the people from their nations of origin could continue to enter the United States. Thus, citizenship rights helped the Irish get a higher quota for those from Ireland. Meanwhile, the quotas for Italians, Polish, and those from other eastern European nations were lower than the numbers from prior to 1920. While initially defined as racially distinct, European immigrant groups still had more advantages than most people of color already in the United States, who were highly restricted in their social and economic options at the time.

During this industrial era, many European immigrants worked in light industries (such as garment factories) and heavy industries (e.g., steel, oil, glass, and automobiles), often located in rapidly growing cities. People of color were thought unable to do industrial work and were largely confined to agricultural and service work. The majority of African Americans lived in the South, where many labored as sharecroppers, working land they did not own for a portion of the crop. They were frequently in debt, with money coming into households only from the labor of mothers and daughters doing domestic work in the homes of White families (Marks 1989). Asian immigrants and the small Asian American population also faced limited employment options—many worked only in agriculture and service jobs (Takaki 1989). Mexican Americans, citizens since the United States had won the Mexican-American War and annexed the majority of the southwest in 1848, found their experiences shaped by wealth. Those with land and resources had a measure of power, but as more Anglos dominated the region, many lost their rights and resources. Long-term residents and new immigrants found employment in agriculture and the hard labor of building new industries like mining and railroads (Acuña 1981).

Often the only industrial job African Americans and other people of color could gain was work as strikebreakers when employers wanted to prevent White workers from unionizing. Over time, groups like African Americans, Mexican Americans, and some Chinese Americans would find their way into industrial jobs. Ideologies meant to exclude were modified to give these minority groups access to industrial work, especially because their labor was needed, but there were ceilings on their advancement. Men of color doing industrial work earned lower wages than White men in the same industries and few had access to union membership. The practice of a dual wage system was common in many industries, and it had many implications for families. White workers could earn enough to support their wives and enable their children to remain in school, while many families of color required the labor of many family members to survive. Ideas about the latter's assumed inferiority were kept alive by residential segregation and segregated schools, segregated hospital wards, and segregated public accommodations, including seating on buses and trains. Yet, people of color resisted and worked in various ways to change their situations, a subject that we address more in Part VII.

After World War II, there was much rethinking about racial hierarchies and citizenship, but it would still take decades for oppressed groups to challenge ideologies that presented them as racially distinct and inferior. In the 1960s, Congress passed civil rights legislation and revised immigration policy. The Hart-Celler Act of 1965 opened the door to immigrants without racial or ethnic bias. Many Italian and Eastern European families were reunited, and other groups now had access to this democratic nation. At the end of the twentieth century, the majority of immigrants were coming from Latin America, Asia, and the Caribbean. They entered a political economy that was again changing—this time from a manufacturing-based economy to one focused on service work—both highly skilled (such as doctors, financial planners, and college professors) and less skilled (such as janitors, domestic workers, and fast-food workers). Because workers are needed, issues of citizenship and race have emerged as key to economic and civic opportunities. As adults cross borders to secure employment, they either bring their children and other family members with them, or face years of separation while their earnings in the United States support children and other family members in their native lands. Many of these new immigrants are *transnationals*—that is, some family members may work and reside in one nation while other members of the immediate family remain in the country of origin. The technologies in the global economy enable them to maintain ties with their nations of origins.

The articles in Section A explore issues of citizenship and patterns of exclusion and their implications for various communities in terms of access to political participation, the labor market, housing, and education. Today, many people born in the United States take citizenship for granted, but many groups struggled for decades to acquire essential rights that gave them new employment opportunities—opportunities critical for economic stability and essential for improving life chances for the next generation.

Efforts by immigrants and people of color to gain political and economic rights began early in our nation's history, continued into the twentieth century, and are ongoing. Evelyn Glenn ("Citizenship and Inequality") explores how people of color battled the limitations of the Constitution and opened up options for others. Even when laws have been changed, she finds that race structures rights and responsibilities in policies and practice, such as in contemporary disputes about entitlements. Glenn's article shows how citizenship is integrally linked to the labor people provide. This can be seen in a different context, including an international one, in the article by Joanne Mariner ("Racism, Citizenship, and National Identity"). Mariner argues that citizenship is a basic human right, but that people have been denied it because of race and ethnic identification in various locales and at different points in history. Who counts as a citizen and who does not is also linked to a sense of national identity and the history of exclusionary national practices. Suzanne Oboler ("It Must Be a Fake!") also sees the impact of our historical legacy in contemporary relationships, since racial ideologies serve as barriers to building a nation that is a community of equals. She discusses the shift from an ideology of racism based in biology to a more social form, in which new ethnic labels, in particular Hispanic, monitor group progress but also limit full participation and rights.

The indigenous groups present when European settlers arrived on the continent have faced difficulties gaining economic resources in the new political economy. Russell Thornton ("American Indians in the United States") examines the many laws that were supposed to help but in the end harmed

American Indian groups and individuals. Only in the twentieth century have tribes gained a measure of control and real sovereignty to improve their status, but their rights are still contested in many states.

We like to celebrate equality in light of the passage of civil rights laws in the 1960s, but the goal of "one person, one vote" has not been achieved. The Voting Rights Act of 1965 made disenfranchising people illegal, and it set up ways to monitor voting practices. Poll taxes, literacy tests, and other barriers that had been used to keep people of color and poor White people from voting have been removed, but, as we learned during the 2000 election, voting can still be compromised and in new ways. Revathi Hines ("The Silent Voices: 2000 Presidential Election and the Minority Vote in Florida") analyzes the difficulties minorities faced when they exercised their right to vote in Florida. Voting is essential for full citizenship, thus voting irregularities impede building a community of equals.

After decades of limited immigration, new policies and a new economy made the United States a major destination for immigrants from around the globe. Section B looks at immigration from different angles, but with attention to the importance of the political economy. Leo R. Chavez ("Covering Immigration: Popular Images and the Politics of the Nation") shows the dramatic changes that resulted from the 1965 immigration law. The law not only increased the number of immigrants, but also opened the doors to people from nations that had previously been excluded. Now, the growth of the Latino and Asian American populations is slowly shifting the racial and ethnic composition of the nation. Nancy Foner ("From Ellis Island to JFK: Education in New York's Two Great Waves of Immigration") compares the pursuit of education of earlier immigrants (1880–1920) and those immigrants coming to New York City since the 1965 law. The majority of new immigrants, from Latin America, Asia, and the Caribbean, are often compared to the earlier waves of Europeans entering during the industrialization period. Foner clarifies how the economic progress of earlier, uneducated immigrants translated into educational advantages for their children. Those immigrants entering the country in a postindustrial economy are arriving when higher education is essential for economic security. Many make sacrifices so that their children gain those credentials, a trend that promotes their children's incorporation into the mainstream.

Looking at immigrants, we see that some fare better than others, even though all struggle to adjust to the social structural conditions they encounter. Some immigrants earn high wages, while others survive in the low-wage segment of the job market. The resources immigrants bring and the complexity of employment options, housing markets, schools, and the degree of hostility directed at groups from others all matter in the immigrants' well-being, even though the media frequently attribute any group's success or failure to cultural traits. Alejandro Portes ("Immigration's Aftermath"), himself an immigrant from Cuba, has studied the experiences of others who come to these shores. In his article, Portes focuses on the children of the recent wave of immigration, the new "second generation," whose mobility is strongly linked to the economic prospects of their parents.

Min Zhou ("Are Asian Americans Becoming 'White'?") raises many questions about the economic diversity of Asian Americans, the majority of whom are foreign-born. Zhou unpacks the "model minority" stereotype that attributes the economic success of Asian Americans to cultural traits. Like most racial immigrant groups, Asian Americans face many obstacles and have had to work hard to overcome the prejudice directed against them. Many of the newer immigrants are successful because they arrived in the United States with educational credentials, thus providing for their children's mobility in the school system.

In Section C, we examine social mobility and the formation of social class and its connection to race. In the new, postindustrial era, race relations have increased in complexity but continue to be framed by opportunity structures and access to full civil rights. *Social mobility* is the movement of individuals or groups between social classes, up or down. In the eighteenth century, this would have meant going from being an indentured servant to becoming an independent farmer. During the industrial era, a time when many people lost their farms and worked for others, many experienced downward mobility. At that same time, going from working for others to owning your own business would be considered upward mobility. After World War II, the movement of people from service jobs to industrial jobs with benefits gave upward mobility to some groups. With an education, some segments of the population could advance from working-class origins to middle-class status, especially as access to education and white-collar jobs increased for White Americans. But race continues to affect career paths, as well as the educational opportunities required to become upwardly mobile. Members of groups that had advantages in the past are able to pass them on to their own children—creating options to perhaps remain middle class or even to become upper middle class.

The complexities of social class today mean that the paths to mobility are more complicated and require more resources. In the new economy, it is not just what a person earns, but the assets and wealth that shape the lifestyle and options of individuals and their families. By securing political and civil rights early in the history of the nation, some groups accumulated advantages, while others struggled simply to maintain a secure foothold. Melvin Oliver and Thomas Shapiro ("Wealth and Racial Stratification") introduce the importance of wealth to the study of racial inequality, *wealth* meaning the assets that people have—homes, cars, and stocks and bonds—once debt is subtracted. Formerly, most scholars studying racial inequality looked only at earnings. Civil rights laws and freer access to work environments have reduced the earnings gap between White people and people of color, but wealth produces even more persistent inequality.

Knowing about differences in wealth is part of understanding the development of Black urban middle-class residential communities. Mary Pattillo-McCoy ("The Making of Groveland") researched the circumstances in Groveland, a Black middle-class urban community in Chicago. As urban communities have experienced a decline of jobs, there is more and more poverty surrounding even these middle-class enclaves. Civil rights legislation opened up educational options to those with resources. This enabled people to enter professional and managerial positions, particularly in formerly segregated areas of the labor market. As a consequence, the African American community now is more economically diverse than in the past because of access to new jobs. Yet, achieving a middle-class social position does not signal the end of struggle for equality in the United States. Middle-class residents still have to contend with residential segregation and the decline of industrial jobs in urban communities.

Opportunities for advancement are shaped not only by race and class but also by generation. Different age groups face varied opportunities, depending on the changed conditions of their times. The baby boomer generation (those born roughly between 1946 and 1967—both native born and immigrants) gained opportunities during an expanding economy, and as civil rights legislation opened new doors. Those born later now find deindustrialization dividing the work world into either high- or low-paying service work. Celeste Watkins ("A Tale of Two Classes: The Socio-Economic Divide among Black Americans Under 35") addresses current political and economic trends that affect the opportunities for Black Americans under 35.

Vilna Bashi Bobb and Averil Clarke ("Experiencing Success: Structuring the Perception of Opportunities for West Indians") also look at the differences between generations. They challenge the myth that culture is the key to explaining the success of West Indians' immigration. The authors compare the experiences of West Indian immigrants in New York City to those of the second generation, those born here or who came as young children. The experiences with racism of the children of immigrants have created different perceptions from those of their parents. The children are less likely to sacrifice and to believe in education as a strategy for advancement. Bobb and Clarke's article highlights how the persistence of racism challenges the idea that working hard is all newcomers need to do to achieve the American Dream.

Altogether, the articles in Part V focus our attention on citizenship rights and how those given such rights early in the nation's history have enjoyed more opportunities to secure economic and political resources. Groups that faced obstacles in the past to securing such rights are still more vulnerable to exploitation, and much work is required to change the laws that limit them. We find this old pattern repeating itself in the twenty-first century, as the lack of recognition of particular groups of people as full citizens is again a barrier to opportunities. Such practices keep race and ethnicity alive as key ingredients in the paths to mobility.

REFERENCES

Acuña, Rudolfo. 1981. *Occupied America: A History of Chicanos,* 2nd Edition. New York: Harper and Row.

Glenn, Evelyn Nakano. 2002. *Unequal Freedom: How Race and Gender Shaped American Citizenship and Labor.* Cambridge: Harvard University Press.

Marks, Carole. 1989. *Farewell—We're Good and Gone: The Great Black Migration.* Bloomington, IN: Indiana University Press.

Takaki, Ronald. 1989. *Strangers from a Different Shore.* Boston: Little, Brown and Company.

21

Citizenship
and Inequality

BY EVELYN NAKANO GLENN

HISTORICAL DEBATES ABOUT
CITIZENSHIP IN THE UNITED STATES

... The concept of citizenship is, of course, historically and culturally specific. The modern, western notion of citizenship emerged out of the political and intellectual revolutions of the seventeenth and eighteenth centuries, which overthrew the old feudal orders. The earlier concept of society organized as a hierarchy of status, expressed by differential legal and customary rights, was replaced by the idea of a political order established through social contract. Social contract implied free and equal status among those who were party to it. Equality of citizenship did not, of course, rule out economic and other forms of inequality. Moreover, and importantly, equality among citizens rested on the inequality of others living within the boundaries of the community who were defined as non-citizens. The relationship between equality of citizens and inequality of non-citizens had both rhetorical and material dimensions. Rhetorically, the "citizen" was defined and, therefore, gained meaning through its contrast with the oppositional concept of the "non-citizen" as one who lacked the essential qualities needed to exercise citizenship.

From *Social Problems* 47(1): 1–20, 2000. Copyright © 2000 by The Society for the Study of Social Problems, Inc. All rights reserved. Reprinted by permission.

Materially, the autonomy and freedom of the citizen were made possible by labor (often involuntary) of non-autonomous wives, slaves, children, servants, and employees. . . .

A specifically sociological conception of citizenship as membership is offered by Turner (1993:2), who defines it as "a set of practices (judicial, political, economic, and cultural) which define a person as a "competent" member of society. . . ." Focusing on social practice takes us beyond a juridical or state conception of citizenship. It points to citizenship as a fluid and decentered complex that is continually transformed through political struggle.

Membership entails drawing distinctions and boundaries for who is included and who is not. Inclusion as a member, in turn, implies certain rights in, and reciprocal obligations toward, the community. Formal rights are not enough, however; they are only paper claims unless they can be enacted through actual practice. Three leading elements in the construction of citizenship, then, are membership, rights and duties, and conditions necessary for practice.

These three elements formed the major themes that have run through debates, contestation, and struggles over American citizenship since the beginning. First, membership: who is included or recognized as a full member of the imagined community (Anderson 1983), and on what basis? Second, what does membership mean in terms of content: that is, what reciprocal rights and duties do citizens have? Third, what are the conditions necessary for citizens to practice citizenship, to actually realize their rights and carry out their responsibilities as citizens?

MEMBERSHIP

Regarding membership, there are two major strains of American thought regarding the boundaries of the community. One tradition is that of civic citizenship, a definition based on shared political institutions and values in which membership is open to all those who reside in a territory. The second is an ethno-cultural definition based on common heritage and culture, in which membership is limited to those who share in the heritage through blood descent (Smith 1989; Kettner 1978).

Because of its professed belief in equality and natural rights (epitomized by the Declaration of Independence), the United States would seem to fit the civic model. However, since its beginnings, the U.S. has followed both civic and ethno-cultural models. The popular self-image of the nation, expressed as early as the 1780s, was of the United States as a refuge of freedom for those fleeing tyranny. This concept, later elaborated in historical narratives and sociological accounts of America as a nation of immigrants, blatantly omitted Native Americans, who were already here, Mexicans, who were incorporated through territorial expansion, and Blacks, who were forcibly transported. This exclusionary self-image was reflected at the formal level in the founding document of the American polity, the U.S. Constitution. The authors of the

Constitution, in proclaiming a government by and for "we the people," clearly did not intend to include everyone residing within the boundaries of the U.S. as part of "the people." The Constitution explicitly named three groups: Indians, who were identified as members of separate nations, and "others," namely slaves; and finally, "the people": only the latter were full members of the U.S. community (Ringer 1983).

Interestingly, the Constitution was silent as to who was a citizen and what rights and privileges they enjoyed. It left to each state the authority to determine qualifications for citizenship and citizens' rights, e.g., suffrage requirements, qualifications for sitting on juries, etc. Individuals were, first, citizens of the states in which they resided and only secondarily, through their citizenship in the state, citizens of the United States. The concept of national citizenship was, therefore, quite weak.

However, the Constitution did direct Congress to establish a uniform law with respect to naturalization. Accordingly, Congress passed a Naturalization Act in 1790, which shaped citizenship policy for the next 170 years. It limited the right to become naturalized citizens to "free white persons." The act was amended in 1870 to add Blacks, but the term "free white persons" was retained. As Ian Haney Lopez (1996) has documented, immigrants deemed to be non-white (Hawaiians, Syrians, Asian Indians), but not Black or African, were barred from naturalization. The largest such category was immigrants from China, Japan, and other parts of Asia, who were deemed by the courts to be "aliens ineligible for citizenship." This exclusion remained in force until 1953. . . .

It was Black Americans, both before and after the Civil War, who were the most consistent advocates of universal citizenship. Hence, it is fitting that the Civil Rights Act of 1865 and the Fourteenth Amendment, ratified in 1867 to ensure the rights of freed people, greatly expanded citizenship for everyone. Section 1 of the Fourteenth Amendment stated that "All persons born or naturalized in the United States and subject to the jurisdiction thereof, are citizens of the United States and of the State wherein they reside. No State shall make or enforce any law which shall abridge the privileges or immunities of citizens of the United States; nor shall any State deprive any person of life, liberty, or property, without due process of law; nor deny to any person within its jurisdiction, the equal protection of the law."

In these brief sentences, the Fourteenth Amendment established three important principles for the first time: the principle of national citizenship, the concept of the federal state as the protector and guarantor of national citizenship rights, and the principle of *birthright* citizenship. These principles expanded citizenship for others besides Blacks. To cite one personal example, my grandfather, who came to this country in 1894, was ineligible to become a naturalized citizen because he was not white, but his daughter, my mother, automatically became a citizen as soon as she was born. Birthright citizenship was tremendously important for second and third generation Japanese Americans and other Asian Americans who otherwise would have remained perpetual aliens, as is now the case with immigrant minorities in some

European countries. Foner (1998) was on the mark when he said that the Black American struggle to expand the boundaries of freedom to include themselves, succeeded in changing the boundaries of freedom for everyone.

The battle was not won, once and for all, in 1867. Instead, the nation continued to vacillate between the principle of the federal government having a duty to protect citizens' rights and states' rights. By the end of the Reconstruction period in the 1870s, the slide back toward states' rights accelerated as all branches of the federal government withdrew from protecting Black rights and allowed southern states to impose white supremacist regimes (Foner 1988). In the landmark 1896, *Plessy v. Ferguson* decision, the Supreme Court legitimized segregation based on the principle of "separate but equal." This and other court decisions gutted the concept of national citizenship and carved out vast areas of life, employment, housing, transportation, and public accommodations as essentially private activity that was not protected by the Constitution (Woodward 1974). It was not until the second civil rights revolution of the 1950s and 1960s, that the federal courts and Congress returned to the principles of national citizenship and a strong federal obligation to protect civil rights. . . .

MEANINGS OF RIGHTS
AND RESPONSIBILITIES

Just as with the question of membership, there has not been a single understanding of rights and duties. American ideas on rights and duties have been shaped by two different political languages (Smith 1989). One, termed *liberalism* by scholars, grew out of Locke and other enlightenment thinkers. In this strain of thought, embodied in the Declaration of Independence, citizens are individual rights bearers. Governments were established to secure individual rights so as to allow each person to pursue private as well as public happiness. The public good was not an ideal to be pursued by government, but was to be the outcome of individuals pursuing their own individual interest. The other language was that of *republicanism,* which saw the citizen as one who actively participated in public life. This line of thought reached back to ancient Greece and Rome where republicanism held that man reached his highest fulfillment by setting aside self-interest to pursue the common good. In contrast to liberalism, republicanism emphasizes practice and focuses on achieving institutions and practices that make collective self-government possible.

There has been continuing tension and alternation between these two strains of thought, particularly around the question of whether political participation is essential to, or peripheral to, citizenship. In the nineteenth century, when many groups were excluded from participation, the vote was a mark of standing; its lack was a stigma, a badge of inferiority, hence, the passion with which those denied the right fought for inclusion. Three great movements— for universal white manhood suffrage, for Black emancipation, and for women's suffrage—resulted in successive extensions of the vote to non-propertied white men, Black men, and women, between 1800 and 1920 (Roediger 1991; Foner 1990; DuBois 1978). According to Judith Shklar (1991), once the vote became broadly available, it ceased to be a mark of status. Judging by participation rates, voting is no longer an emblem of citizenship. One measure of the decline in the significance of the vote has been the precipitous drop in persons voting. In 1890, fully 80% of those who were eligible, voted. By 1924, four years after the extension of suffrage to women, participation had fallen to less than half of those registered, and it has been declining ever since....

In the U.S., participation is discouraged by not making registration easy. Requiring re-registration every time one moves is only one example of the obstacles placed upon a highly mobile population. While Americans today would object if the right to vote was taken away, the majority don't seem to feel it necessary to exercise the vote. The U.S. obviously has the capacity to make registration automatic or convenient, but has not made major efforts to do so. Needless to say, it is in the interest of big capital and its minions that most people don't participate politically.

CONDITIONS FOR PRACTICE

This leads to the third main theme, the actual practice of citizenship and the question of what material and social conditions are necessary for people to actually exercise their rights and participate in the polity.

The short answer to this question for most of American history has been that a citizen must be *independent*—that is, able to act autonomously. . . . Independence has remained essential, but its meaning has undergone drastic transformation since the founding of the nation. . . . In the nineteenth century, as industrialization proceeded apace and wage work became common even for skilled artisans and white collar men, the meaning of independence changed to make it more consistent with the actual situation of most white men. It came to mean, not ownership of productive property, but nothing more than ownership of one's own labor and the capacity to sell it for remuneration (Fraser and Gordon 1994). This formulation rendered almost all white men "independent," while rendering slaves and women, who did not have complete freedom to sell their labor, "dependent." It was on the basis of the independence of white working men that the movement for white Universal Manhood Suffrage was mobilized (Roediger 1991). By the 1830s, all states had repealed property requirements for suffrage for white men, while simultaneously barring women and Blacks (Litwak 1961).

By the late 19th century, capitalist industrialization had widened the economic gap between the top owners of productive resources and the rest, making more apparent the contradiction between economic inequality and political democracy. Rising levels of poverty, despite the expansion in overall wealth, raised the question of whether low-income, unemployment, and/or lack of access to health care and other services, diminished citizenship rights for a large portion of the populace. Growing economic inequality also raised the issue of whether some non-market mechanism was needed to mitigate the harshness of inequities created by the market.

One response during the years between World War I and World War II was rising sentiment for the idea of what T. H. Marshall called social citizenship. In Marshall's words (Marshall 1964:78), social citizenship involves "the right to a modicum of economic security and to share in the full social heritage and to live the life of a civilized being, according to the standards prevailing in the society." Economic and social unrest after World War I spurred European states to institute programs to ensure some level of economic security and take collective responsibility for "dependents"—the aged, children, the disabled, and others unable to work. . . . The significance of the "redistributive" mechanisms of the Welfare State, according to Marshall, was that they enabled working class people to exercise their civil and political rights. . . . Compared to Western Europe, the concept of social citizenship has been relatively weak in the U.S. Welfare state researchers have pointed out that, although 1930s New Deal programs such as social security and unemployment insurance greatly increased economic security, they continued a pattern of a two-tiered system of social citizenship from the 1890s (Nelson 1990; Skocpol 1992; Fraser and Gordon 1993; Fraser and Gordon 1994). The upper tier consisted of "entitlements" based on employment or military service, e.g., unemployment benefits, old-age insurance, and disability payments, which were relatively generous and did not require means-testing. The lower tier consisted of various forms of "welfare," such as Aid to Dependent Children (changed to Aid to Families of Dependent Children—AFDC—in 1962), which were relatively stingy and entailed means testing and surveillance by the state.

White men, as a class, have drawn disproportionately on first tier rights by virtue of their records of regular and well-paid employment. White women, more often, had to rely on welfare, which is considered charity, a response to dependence, rather than a just return for contributions. Latino and African American men were generally excluded from employment-based benefits because of their concentration in agriculture, day jobs, and other excluded occupations; Latina and African American women, in turn, have often been denied even second tier benefits (Oliver and Shapiro 1995; Mink 1994). Moreover, in contrast to the situation in most European countries, there has been little sense of collective responsibility for the care of dependents. Thus, raising children is not recognized as a contribution to the society and, therefore, as a citizenship responsibility that warrants entitlements such as parental allowances and retirement credit, which are common in Europe (Glenn 2000; Sainsbury 1996).

As with the other issues of citizenship, the 1980s and 1990s [saw] a neoconservative turn with a concerted effort to roll back even attenuated social citizenship rights. Government funding of social services has been vilified for draining money from hardworking citizens to support loafers and government bureaucrats who exert onerous control over people's lives. And, after years of attacks on Black and other single mothers "dependent" on welfare, Congress passed the Personal Responsibility and Work Opportunity Act in 1996, which dismantled AFDC and replaced it with grants to the states that limited eligibility for federal benefits for a lifetime maximum of five years. The aim was, as Representative Richard Armey put it, that the poor return to the natural safety nets—family, friends, churches, and charities. States instituted stringent work requirements and limited total lifetime benefits. By limiting total years that they can stay on welfare, proponents argued the new regulations would wean single mothers from unhealthy dependence on the state. In contrast to an overly indulgent government, the market is expected to exert a moral force, disciplining them and forcing them to be "independent" (Roberts 1997; Boris 1998). . . .

Thus, we must ask: if people have a responsibility to earn, then, don't they also have a corresponding right to earn?—i.e., to have a job and to earn enough to support themselves at a level that allows them, in Marshall's (1964:78) words, "to participate fully in the cultural life of the society and to live the life of a civilized being according to the standards prevailing in the society."

REFERENCES

Anderson, Benedict. 1983. *Imagined Communities: Reflections on the Origins and Spread of Nationalism.* London: Verso.

Boris, Eileen. 1998. "When work is slavery." *Social Justice* 25, 1:18–44.

DuBois, Ellen Carol. 1978. *Feminism and Suffrage.* Ithaca: Cornell University Press.

Foner, Eric. 1988. *Reconstruction: America's Unfinished Revolution.* New York: Harper and Row.

———— 1990. "From slavery to citizenship: Blacks and the right to vote." In *Voting and the Spirit of American Democracy*, ed. Donald W. Rogers. Urbana and Chicago: University of Illinois Press.

———— 1998. *The Story of American Freedom.* New York: Norton.

Fraser, Nancy, and Linda Gordon. 1993. "Contract versus charity: Why is there no social citizenship in the United States?" *Socialist Review* 212, 3:45–68.

———— 1994. "A genealogy of dependence: Tracing a keyword of the U.S. welfare state." *Signs* 19, 2:309–336.

Glenn, Evelyn. 2000. "Creating a caring society." *Contemporary Sociology* 29, 1: 84–94.

Haney Lopez, Ian F. 1996. *White By Law: The Legal Construction of Race.* New York: New York University Press.

Kettner, James. 1978. *The Development of American Citizenship, 1608-1870.* Chapel Hill: University of North Carolina Press.

Litwak, Leon F. 1961. *North of Slavery: The Negro in the Free States, 1790-1860.* Chicago: University of Chicago Press.

Marshall, T. H. 1964. *Class, Citizenship, and Social Development.* Garden City, NY: Doubleday and Company.

Mink, Gwendolyn. 1994. *The Wages of Motherhood: Inequality in the Welfare State, 1917–1942.* Ithaca: Cornell University Press.

Nelson, Barbara. 1990. "The origins of the two-channel welfare state: Workman's compensation and mother's aid." In *Women, the State and Welfare*, ed. Linda Gordon, 123–151. Madison: University of Wisconsin Press.

Oliver, Melvin L., and Thomas M. Shapiro. 1995. *Black Wealth/White Wealth: A New Perspective on Racial Inequality*. New York: Routledge.

Ringer, Benjamin B. 1983. *We the People and Others: Duality and America's Treatment of Its Racial Minorities*. New York: Tavistock.

Roberts, Dorothy. 1997. *Killing the Black Body: Race, Reproduction, and the Meaning of Liberty*. New York: Pantheon.

Roediger, David. 1991. *The Wages of Whiteness: Race and the Making of the American Working Class*. London: Verso.

Sainsbury, Diane. 1996. *Gender, Equality and Welfare States*. Cambridge, UK: Cambridge University Press.

Shklar, Judith. 1991. *American Citizenship: The Quest for Inclusion*. Cambridge, MA: Harvard University Press.

Skocpol, Theda. 1992. *Protecting Soldiers and Mothers: The Origins of Social Policy in the United States*. Cambridge, MA: Harvard University Press.

Smith, Rogers M. 1989. "'One United People': Second-class female citizenship and the American quest for community." *Yale Journal of Law and the Humanities* 1:229–293.

Turner, Brian S. 1986. *Citizenship and Capitalism: The Debate over Reformism*. London: Allen and Unwin.

Woodward, C. Vann, ed. 1974. *The Strange Career of Jim Crow*, 3rd rev. New York: Oxford University Press.

DISCUSSION QUESTIONS

1. What elements does Glenn identify as constructing citizenship?
2. During what era did social citizenship rights expand and why?
3. Thinking about your own citizenship status, what do you see as the rights and obligations linked to your position in the nation?

22

Racism, Citizenship, and National Identity

BY JOANNE MARINER

THE RACE CONVENTION

At least as a matter of principle, there are few proscriptions as widely accepted and deeply felt as the prohibition against racial and ethnic discrimination. The experiences of slavery, colonialism, the holocaust, South African apartheid and America's white supremacist system of Jim Crow, among others, have left an enduring imprint on the global fabric of values. Racism and related forms of discrimination are condemned not only as unfair, but also as morally repugnant. Reflecting this perspective, the Race Convention, the primary international treaty relating to discrimination on racial and ethnic grounds, is categorical in its call to eradicate the practice. As it emphasizes: 'there is no justification for racial discrimination, in theory or in practice, anywhere.'[1]

The Race Convention, which extends to discrimination on the basis of race, colour, descent and national or ethnic origin, covers a wide range of government policies, from those relating to political rights, such as the right to vote, to those relating to economic, social and cultural rights, such as the rights to housing and education. But, strikingly, the convention shifts gears with regard to rules regulating citizenship. Despite its broad and unqualified language about the necessity of eliminating racial and ethnic discrimination in all

From *Development* 46(3): 64–70, 2003. Reprinted by permission of Palgrave Macmillan.

of its manifestations, the treaty contains an explicit exception for countries' citizenship and naturalization policies. This provision, set out in Article 1, specifically states that the convention's protections against discrimination do not generally extend to legal rules on citizenship and naturalization, although they do bar discrimination against particular nationalities.

To comply with the Race Convention, countries cannot, for example, allow European immigrants to naturalize while barring, say, Haitians. But in myriad other ways, countries are free to impose highly discriminatory citizenship rules, ones that favour one racial or ethnic group over everyone else. Practices that would, in short, merit the sternest reproach in nearly every other area of government policy are considered permissible in the area of citizenship.

Nor is there anything abstract about the possibility of racial and ethnic discrimination in states' regulation of citizenship. Historically, citizenship has often been restricted along racial or ethnic lines, with certain groups being excluded from the citizenry and others facing mass denationalization. Even now, a number of countries discriminate overtly on the basis of racial or ethnic criteria in granting citizenship. In Israel, for example, the Law of Return guarantees citizenship to anyone with at least one Jewish grandparent. In Germany, similarly, the law protects a 'Right of Return' that allows ethnic Germans, mostly from eastern Europe and the Soviet successor states, to obtain citizenship upon their arrival in the country. Both states, needless to say, apply more demanding rules to persons of other ethnicities who seek citizenship.

The permissiveness of international norms regarding racial and ethnic discrimination in citizenship policy is all the more striking when compared with analogous rules on gender discrimination. The Women's Convention, although similar to the Race Convention in its general approach to discrimination, is much more demanding than the latter text with regard to government regulation of citizenship. Rather than carve out an explicit exemption for citizenship rules, Article 9(1) of the treaty states emphatically that 'States Parties shall grant women equal rights with men to acquire, change or retain their nationality'.

Why, then, the relative tolerance for racial and ethnic discrimination in citizenship policy? The question is particularly timely now, as increased immigration flows magnify the impact of discriminatory citizenship rules, and raise the spectre of long-term migrant populations being excluded from full membership in their countries of residence.

THE REGULATION OF "HUMANITY'S BASIC RIGHT"

It is hard to overstate the crucial importance of citizenship. Citizenship has been called 'humanity's basic right': 'nothing less than the right to have rights.' This statement, though it exaggerates the vulnerability of non-citizens, usefully conveys the fundamental nature of the individual's right to citizenship.

But citizenship questions are also of profound significance to states. What is more fundamental to a state, after all, than sovereignty over people and territory?

Given the exceedingly sensitive nature of citizenship questions, it should not be surprising that the traditional rule is that citizenship policy falls within the 'reserved domain' of state authority. Citizenship policy is, in principle, a matter of domestic concern. Because countries' citizenship rules can come into conflict—particularly with regard to the right to diplomatic protection, understood as a corollary of citizenship—there have always been some limits to a state's unilateral power to determine who its citizens are, but the scope of state discretion with regard to such issues has historically been broad.

A person may obtain citizenship at birth in two possible ways: by descent, based on the citizenship of one or both parents (*jus sanguinis*), or by birth on a country's territory (*jus soli*). In general, states are free to recognize either method of acquiring citizenship, or to rely upon some combination of the two. As should be obvious, the two approaches tend to vary meaningfully in their impact on the racial and ethnic composition of a country's citizenry. The rule of *jus sanguinis* obviously limits a country's racial and ethnic diversity, as a population simply replicates itself in racial and ethnic terms. The rule of *jus soli* is more flexible in its impact; yet it still only undermines racial and ethnic uniformity to the extent that there is diversity in a country's immigration practices. Unless a country has a fairly open immigration policy, its racial and ethnic make-up will remain relatively static.

A person may obtain citizenship after birth via naturalization. In general, states enjoy enormous discretion in crafting substantive and procedural requirements for naturalization. In many countries, therefore, naturalization rules are extremely restrictive.

PAST EXAMPLES OF DISCRIMINATORY CITIZENSHIP POLICIES

There is no dearth of historical examples of discriminatory citizenship rules. In the U.S., for instance, the Naturalization Act of 1790 imposed explicitly racist criteria for eligibility for citizenship, limiting naturalization to 'free white person[s]'. From 1882 until 1943, under the Chinese Exclusion Act and subsequent laws, the U.S. specifically barred all Chinese from naturalization. In 1922, extending this exclusionary principle, the U.S. Supreme Court ruled that Japanese immigrants were ineligible for naturalization. The following year, in a case involving 'a high-caste Hindu, of full Indian blood', the Court ruled that all Asians were barred from naturalizing. It was only in 1946 that the U.S. Congress passed laws that abolished the last of the country's racially discriminatory citizenship rules.

In European countries during the 18th and 19th centuries, Jews, Roma and other minority populations were often excluded from citizenship. In the first part of the 20th century, as well, discriminatory citizenship policies were not uncommon. A number of treaties concluded at the end of the First World War covering the former Austro-Hungarian Empire contained what one commentator termed 'ethnic options,' granting persons of minority ethnic groups the right to acquire citizenship in other countries in which the majority population was of the same ethnicity (Donner, 1994: 259, citing Kunz, 1930–31). But it was the period of the Second World War, with the mass denationalization decrees of the Nazi regime that deprived German Jews of citizenship, that brought the most notorious and harmful examples of discriminatory citizenship rules.

Even after the Second World War, countries have continued to strip people of citizenship based on ethnic criteria. In 1945–46, Czechoslovakia and Poland passed laws that effected the denationalization of members of their country's German or Hungarian minorities (Donner, 1994: 157). In the early 1960s, according to reports, as many as 140,000 persons of Kurdish origin were deprived of their citizenship in Syria. These are not the only examples.

POST-WAR INTERNATIONAL LEGAL RULES

Since the Second World War, in reaction both to wartime abuses and to the post-war explosion in refugee populations, there has been a concerted international effort to impose stricter limitations on governments' power over citizenship. A primary concern of this effort has been the avoidance of statelessness. Unsurprisingly, therefore, international law has focused on limiting the deprivation of citizenship more than on rules regulating the granting or denial of citizenship.

International legal barriers have been constructed to prevent states from depriving people of citizenship on racial or ethnic grounds. Article 9 of the Convention on the Reduction of Statelessness, for example, specifically addresses this issue. Although the Convention is far from universally accepted, having been ratified by only 21 states, its prohibition on the discriminatory deprivation of citizenship arguably codifies a principle of customary international law. Supporting the prohibition's claim to customary law status is the Universal Declaration of Human Rights, which, in Article 15, provides that no person shall be arbitrarily deprived of citizenship. Given international norms against racial and ethnic discrimination, stripping a person of citizenship on racial or ethnic grounds would undoubtedly be considered arbitrary.

Yet states continue to enjoy surprising latitude to take racial and ethnic considerations into account in decisions regarding to whom citizenship should be granted. Besides the German and Israeli examples, mentioned above, numerous other countries, including Armenia, Bulgaria, Croatia, Estonia, Honduras, Hungary and Norway, have laws granting ethnic preferences in the conferral of citizenship. Most strikingly, the Liberian Constitution restricts

Liberian citizenship to 'persons who are Negroes or of Negro descent'. Moreover, countries such as Japan and Kuwait that strictly limit naturalization and do not recognize *jus soli* as a grounds for granting citizenship at birth are, in effect, limiting access to citizenship on ethnic grounds. In Japan, in fact, generations of foreigners—including well over half a million Koreans, many of whom are culturally and linguistically indistinguishable from Japanese citizens—are unable to obtain citizenship.

RIGHT OF SELF-DETERMINATION

Defenders of racial and ethnic discrimination in citizenship policy claim to honour a competing ideal: that of national self-determination. Enshrined in Article 1 of the UN Charter, the concept of 'self-determination of peoples' arguably militates in favour of discrimination. Unsurprisingly, countries that understand themselves to be the national home of a particular ethnic group are especially apt to enforce overtly discriminatory citizenship rules. Existentialist fears regarding the future of the national group, or the group's status as a majority population in the country's territory, further reinforce the link between discriminatory citizenship policy and national identity.

As a particularly illustrative example, consider the case of Israel. Zionism, which the United Nations General Assembly once notoriously branded as a form of racism, claims much of its legitimacy from the Jewish people's right of self-determination. As the official website of Israel's Ministry of Foreign Affairs puts it: 'the State of Israel is a Jewish state, established as the independent state of the Jewish people exercising the right of self-determination in the Land of Israel, its ancestral homeland'.

The Law of Return, which, together with the country's Citizenship Law, grants automatic citizenship to Jewish immigrants to Israel, is considered the 'statutory expression' of the Zionist ideal of Israel as the state of the Jewish people (Shachar, 2000: 394). Not only does the Law of Return give Jewish immigrants a preference over all other potential immigrants to Israel, it gives them priority over Palestinians who fled or were driven from the country during the 1948 and 1967 wars (Takkenberg, 1998: 245–6). While it is impossible to know what the ethnic breakdown of immigration to Israel would have been in the absence of this discriminatory rule, it is beyond dispute that the Law of Return has played a pivotal role in maintaining Israel's character as a Jewish state. In all, some 2.7 million Jews immigrated to Israel between 1948 and 1998, more than outweighing the growth in the Palestinian population during that period (Shachar, 2000: 396).

The Israelis are hardly unique in considering their state to be the embodiment of the national aspirations of a people. The Japanese, for example, whose restrictive citizenship policies were mentioned earlier, have a similarly robust sense of national identity. Indeed, even though these two countries appear to lie at the far end of the spectrum, there are probably few, if any, states in which the population's self-understanding is entirely free of racial and ethnic bias.

NOTES

1. International Convention on the Elimination of All Forms of Racial Discrimination, 660 UNTS 195, entered into force on 4 January 1969. Further attesting to the widespread acceptance of its underlying norms, the Race Convention has been ratified by 155 states, making it one of the world's most widely ratified international treaties.

REFERENCES

Barrington, L. (2000) 'Understanding Citizenship Policy in the Baltic States,' in T.A. Aleinikoff and D. Klusmeyer (eds.) *From Migrants to Citizens: Membership in a changing world.* Washington, DC: Carnegie Endowment for International Peace.

Cassese, A. (1995) *Self-Determination of Peoples: A legal reappraisal.* Cambridge: Cambridge University Press.

Department of State, Bureau of Democracy, Human Rights, and Labor (2001). *Country Reports on Human Rights Practices 2000.* Washington, DC: U.S. Department of State.

Donner, R. (1994) *The Regulation of Nationality in International Law.* Irvington-on-Hudson: Transnational.

Kunz, J. (1930–31) 'L' option de nacionalité', *Recueil des Cours* 31: 150.

Mariner, J. (2001) 'Citizen Dad: An international perspective on the Supreme Court's *Nguyen* Case', *FindLaw's Writ,* 11 January.

Oz, A. (2001) 'Israel Must Face Refugee Problem Before It Can Win Peace', *Sydney Morning Herald,* 11 January.

Shachar, A (2000) 'Citizenship and Membership in the Israeli Polity', in T.A. Aleinikoff and D. Klusmeyer (eds) *From Migrants to Citizens: Membership in a changing world.* Washington, DC: Carnegie Endowment for International Peace.

Takkenberg, L. (1998) *The Status of Palestinian Refugees in International Law.* Oxford: Clarendon Press.

DISCUSSION QUESTIONS

1. In the past, the United States had highly discriminatory laws on eligibility for citizenship. Why are such practices seen as questionable today?

2. Why does Mariner think that the discriminatory policies on granting citizenship found in many nations are problematic? Give one example.

3. Of what country are you a citizen? Would you be able to become a citizen if you migrated to another country? Do you have the "right of return" to become a citizen in another country? What does your own circumstance suggest about Mariner's concerns?

23

"It Must Be a Fake!"

Racial Ideologies, Identities, and the Question of Rights

BY SUZANNE OBOLER

About two and a half years ago Luis Gutiérrez, a Puerto Rican congress-man from Chicago, was standing in line with his sixteen-year-old daughter and his niece, waiting to get into the Capitol to show them his office there. They had just been to "a tribute to all the veterans of the all–Puerto Rican 65th Army Infantry Regiment of the Korean War, including the 743 soldiers who were killed and the 2,797 who were wounded in that conflict." As a result, his daughter and niece were carrying small Puerto Rican flags. Gutiérrez told them to roll the flags up, thinking (mistakenly, as it turned out) that they were not allowed to bring flags into the Capitol. The girls did roll up the flags, but "they got caught in the rollers of the conveyer belt and unfurled." A Capitol police security aide, Stacia Hollingsworth, saw the unfurled flags and, according to the congressman, "yelled in [my] ear: 'Those flags cannot be displayed!'"

The *Chicago Tribune* journalists reporting the incident, David Jackson and Paul de la Garza, tell us that "Gutiérrez was embarrassed, but told his daughter to get rid of the flags, saying, "You know what the rules are."[1]

Overhearing him, Hollingsworth asked: "Who are you that you know what the rules are?" When he told her he was Luis Gutiérrez, a member of Congress, she replied, "I don't think so."

From *Hispanics/Latinos in the United States: Ethnicity, Race, and Rights*, ed. by Jorge Gracia and Pablo De Greiff (New York: Routledge, 2000), pp. 125–139. Copyright © 2000. Reprinted by permission of Routledge/Taylor & Francis Books, Inc.

So Gutiérrez showed her his congressional ID card. Her immediate response was to say, "It must be a fake." And then she added, "Why don't you and your people just go back to the country you came from?"

According to Jackson and de la Garza, "Gutiérrez was stunned. 'It wasn't like on a side street in Chicago,' he said. 'This was in the middle of the gallery. In the Capitol. Where I work. Can you imagine how humiliating this was in front of my 16-year-old daughter?' At that point, a Capitol Police dignitary protection officer rushed over, recognized Gutiérrez and pulled the aide aside. He told Gutiérrez he saw what happened and suggested that Gutiérrez file a complaint."[2]

The exchange between Congressman Gutiérrez and the Capitol security aide raises at least four related issues that characterize the situation of Puerto Ricans and, more generally, of Latinos in the United States today. First—and although it is important to note that his racial characteristics are never explicitly stated in the article—Gutiérrez clearly did not look like a member of the U.S. Congress, which is largely made up of white males. Second, since he did not look like a congressman, he could not be trusted. Hence the aide assumed he had "faked" his congressional ID card. Third, Gutiérrez's visual features marked him as foreign to the image of people who belong in the United States. As such, he was told that he was neither recognized as a U.S. citizen nor welcome in this country. And finally, this relatively insignificant tale exemplifies the lack of awareness of the long historical presence and citizenship status both of Puerto Ricans (officially U.S. citizens since 1917) and, more generally, of the majority of the population today officially known as "Hispanics" in the United States.

Minimally, Congressman Gutiérrez's experience suggests the extent to which racism in the United States ensures that, unlike white Americans, Latinos constantly have to prove their citizenship and to insist on their rights—including their "right to have rights" as citizens of this society. In fact, the very symbolism of this exchange having taken place at the entrance to the building housing the U.S. Congress is illuminating, for it points to the ongoing emphasis on racial features and phenotype in defining membership in the nation's legislature, where the very meaning of national belonging is negotiated and the experience of representative democracy is recorded into the laws that reinforce the belief in a community of equals in the United States.

One conclusion we can draw from the Gutiérrez family's experience is that in spite of both the end of legal segregation brought about by the 1954 *Brown v. Board of Education* decision and of the civil rights movements of the 1960s and 1970s, racism continues to interfere with the possibility of creating a community of equals—and its modern synonym, citizenship—in the United States. Indeed, this encounter reminds us yet again that while citizenship may be commonly understood as a legal status, it is above all a political reality. As such, it cannot be fully understood without taking into account the specificity of the context within which it is understood and differentially experienced in people's daily lives. . . .

The aim of this essay is to describe and clarify the national and global context within which we can discuss the ethnic identity, culture, and group rights of Hispanics/Latinos. I argue that it is a context in which, increasingly, racism and xenophobia shape both the meaning and social value attributed to individuals' ethnic identities and to their lived experience of national belonging in contemporary U.S. society. Insofar as citizenship is the political expression of national belonging, my aim is to clarify the contemporary role of racism in the decline of citizenship and in ensuring the impossibility of belonging to a national community of equals, both in U.S. society and in the broader international context. . . .

CITIZENSHIP AS A POLITICAL
AND HISTORICAL CONSTRUCTION

In considering the experience of Congressman Gutiérrez or, more generally, of Latinos in the United States, it is important to keep in mind that the concept of citizenship was never defined by the founding fathers in the Constitution. Instead, they spoke of "the people of the United States" and rarely mentioned the word *citizen.* Therefore, the political reality and social value of citizenship became contingent on a series of laws and/or court cases that at various times in the nation's history either reinforced or challenged each other. At the same time, both the laws and the courts ultimately aimed at specifying the role and implications of "race" in determining who could be a citizen, as well as in clarifying the responsibility of the state to the citizenry.

The Dred Scott case of 1857, for example, argued that the founding fathers did not mean to include blacks when they spoke of "the people of the United States." The subsequent Civil Rights Act of 1866 was specifically designed to reverse the Dred Scott decision, and was followed by the Fourteenth Amendment, ratified in 1868, which, for the first time in U.S. history, created a *national* citizenry and established the principle of equality under the law for all people born in the United States. . . .

Less than two decades later, however, the *Plessy v. Ferguson* decision in 1896 effectively challenged that amendment. Ruling that "legislation is incapable of eradicating racial instincts," the Supreme Court established racial segregation as the law of the land for the next sixty years. The *Brown v. Board of Education* decision ended legal segregation in 1954, thus countering the *Plessy* ruling by pointing to the ways that the psychological damage created by segregation prevented black children's access to equal opportunity. But it took the subsequent civil rights movements of the 1960s to create the various civil rights acts specifically aimed at enforcing the *Brown* decision.

From this perspective, the policies enacted in the last thirty years of the twentieth century represent a new attempt to create a national community of equals—an attempt grounded in the explicit acknowledgement of the historical role of race in shaping the political reality and social value of citizenship. . . .[3]

Like citizenship, race is also not a static concept. As a result of the discrediting of scientific racism underlying Nazism during World War II, its contemporary meanings and social value have gradually changed both in the United States and abroad. Certainly at the end of the twentieth century, the world as a whole has witnessed the disappearance of legal discrimination and, consequently, the seeming attenuation of racism in the political sphere of every society. Yet, paradoxically, the end of legal discrimination has signaled the unchallenged entrenchment of racism in social relations, particularly in the private sphere.

In the United States, the dualistic black/white biological racism that justified legal segregation until 1954 has been undermined over the past four decades by the emergence of a new ideology of "social racism," embedded in a new kind of social relations that are reminiscent of those found in Latin America. The emergence of this ideology of social racism in the United States is particularly apparent in the growing adherence to the idea of racial mixture (or biracialism—in Latin America, it is called *mestizaje*), leading some to stress that social class rather than race is the key to understanding and solving the ongoing problem of poverty and deprivation of large sectors of the population, including racial minorities, in the United States. Reinforced by those who point to what one prominent mainstream magazine defined in 1990 as "the browning of America, this perceived belief in the "declining significance of race" has since been reinforced by the growing emphasis on the need to explicitly (re)define and stress American nationality as the basis for national unity.

The emphasis on national unity and the simultaneous insistence on the insignificance of biologically determined racial characteristics have long defined the meaning and social value of "race" and, hence, race relations in Latin America. . . .

In the course of the twentieth century, Latin American racial ideologues increasingly modified and eventually rejected the notion of scientific racism that their U.S. colleagues consistently articulated at various inter-American conferences on eugenics.[4] Contrary to the biological determinism that historically has pervaded U.S. race relations, Latin American intellectuals and scientists alike understood "race" in social terms—specifically in terms of the belief in the existence of higher and lower cultures, which could clearly be assimilated into a national sociracial hierarchy organized and (in)visibly marked by skin color and phenotype. While the choice of terms, like their connotations and uses, continues to be debated throughout the continent, recent Latin American scholarship leaves no doubt that this hierarchy of cultural differences, which ensures both that "everyone knows their place" and the impossibility of forging national communities of equal citizens, has historically been grounded in what the Peruvian anthropologist Marisol de la Cadena has defined as "silent racism."

As I suggest in the following pages, this Latin American ideology of social racism, with its emphasis on the unifying force of nationality to the detriment of racial considerations, appears to be increasingly accepted in U.S. society, superimposed on—although not replacing—the biologically based black/white dualism that has been dominant for much of the nation's history. . . .

THE LABEL "HISPANIC" AND THE QUESTION OF RIGHTS

. . . The effects of differentiating and, in effect, racializing the entire U.S. population through . . . ethnic categories have been contradictory. Undoubtedly it has allowed us to track the progress toward political inclusion of racial minorities, as well as of women, since the end of legal segregation. But it has simultaneously reinforced the belief in the superiority of whiteness and "white privilege," making explicit the continuing existence of a socioracial hierarchy in a society that historically, and to this day, proclaims its adherence to the belief in equality for all. In fact, unlike past perceptions and beliefs that U.S. society was a "melting pot," there is today an implicit acknowledgment of an organized socioracial hierarchy, with whites at the top and blacks and Latinos alternating at the bottom. . . .

In short, the official creation of these ethnic categories has ensured that, as in Latin America, everyone "knows his (or her) place" in U.S. society. And as in Latin America, the outcome is the impossibility of establishing an expanded community of equals in the United States. This assertion is reinforced by the following five interrelated points.

1. The vagueness of the census definition has led to many debates in this country concerning who is a Hispanic and on what grounds. This debate includes questions such as whether citizens from Latin America's sovereign nations currently living in the United States are "as Latino" as those born in the United States. Should this distinction be made? Given the vagueness of the wording and its consequences for public policy, social and race relations, and individuals' daily lives, it is essential that we acknowledge that in the United States, the term *Hispanic*—as originally conceived by the state in the 1970s and currently understood—is first and foremost a bureaucratic invention, used for census data collection. Like its grassroots alternative designation *Latino*, the term *Hispanic* does not refer to, and is in no way tied to, an actual historical, territorial, or cultural background or identity of any of the national-origin groups or ethnic populations it encompasses in the United States. Instead, it comprises the populations of all the Spanish-American nations and of Spain. . . .

2. The term *Hispanic*, like other ethnic labels, is here to stay. And from this point of view, the Hispanic (or Latino) experience and identity in the United States cannot be understood outside of the context of the relations that colonized citizens (such as the Puerto Ricans) and conquered peoples

(such as sectors of the Chicano population) have historically had with the U.S. government. This context conflated race and nationality and, in 1977, allowed for the official designation of the ethnic label Hispanic which homogenized all people of Latin American descent. Nor can it be understood outside of the context of the historical and very specific differences that mark U.S. relations with each of the various Latin American nations. These relations invariably differentiate the sociopolitical experiences of each national population in this country. But it is also important to note that this is an unprecedented historical moment in the history of the hemisphere and of its populations, for it is the first time that there has been a significant meeting of the various national populations of Latin America in one country, which, perhaps ironically, happens to be the United States.

3. The emphasis on ethnicity, and more particularly on ethnic labeling, is directly related to the distribution and withdrawal of resources and opportunities. Yet the establishment of these official categories has not significantly improved either the social and economic conditions or society's attitudes and perceptions toward people of Latin American descent. Indeed, according to a recent news release by the National Council of La Raza, "Hispanics now have the highest poverty rate of any major ethnic or racial group in the U.S."—albeit still closely followed by African Americans.

    This points to the contradictory role that labels are playing today. On one hand, these labels do allow us to track and compare poverty and illiteracy rates among racial groups, to measure the nation's progress toward what Johnson called "equality as a fact and as a result." On the other, the labels are not improving the social or economic conditions in which people live. Instead, the label "Hispanic" marks all Latinos as culturally and socially inferior, as having "bad values" that are perceived to be related to their "foreign"—un-American—origins. . . . Hence, as the case of Congressman Gutiérrez suggests, on the basis of their "un-American" cultural and linguistic difference, as well as of their racial markings, the label is in fact serving to locate all "Hispanics" as a group in a hierarchy in which . . . inequalities are naturalized on the basis of racial, gendered, and cultural characteristics.

4. The label "Hispanic," like the categories "Asian American" and "African American," exemplifies the impact of globalization in "minoritizing" all populations from Asia, Africa, and Latin America in the United States. The minoritization of the Third World and the simultaneous emphasis on ethnic-group belonging rather than on citizenship (that is, Hispanic first, American second) has resulted in a variety of complex responses by both Latinos and non-Latinos to the growth of the Latino population in the United States. These are visible in the heated and often acrimonious debates on, and subsequent passage of, anti-immigrant, anti-affirmative-action, and anti-bilingual-education propositions in California and elsewhere, as well as in the proposals for similar bills in Congress. One of the consequences of this is that it now seems natural that the burden and responsibility of

protecting both the human and the citizenship rights of individuals lie solely with the particular "ethnic group" to which they ostensibly belong, rather than with the national society as a whole, or with the state, for that matter. . . .

5. Finally, it is important to note that the growth of the populations of Latin American descent in the United States and its racialization as a homogeneous "Hispanic ethnic group" are taking place in a larger global context, which I believe frames the entire debate on the ethnic identity, culture, and rights of Latinos. Clearly, the international context of this post–Berlin Wall decade has immersed all democracies in a process of expanding the scope of citizenship. Yet there has been relatively little, if any, sign of a significant and structured general debate within or among the older democracies about how to define the very notion of a collectivity—of a national citizenry—in the new global context. Instead, the historically inherited structures of citizenship rights—like the very political reality of citizenship itself—are being brought into question, with little effort made toward creating new international agreements and institutions (or at least reinforcing those that exist) in order to fully guarantee the human and political rights of the world's population. . . .

CONCLUSION

Ethnic labels such as "Hispanic" allow us to identify a racial hierarchy that, now rationalized in essentializing "cultural" terms, accounts for the ongoing (in)visibility of people of Latin American descent. In so doing, it is reinforcing inequalities not only within the United States, but also—as a result of the consequent minoritization of the entire Third World—between the Western developed nations and the developing world. From this perspective, while the state-imposed categories increasingly undermine the possibility of constructing a community of equals, they simultaneously highlight the process by which the United States is moving in the direction of a rigid, class-based society—a society in which, as in Latin America, the lack of social mobility, like the concomitant widening gap between the rich and the poor, can be explicitly rationalized along ethnic and racial lines.

Ultimately, the persistence of racism and of the ongoing racialization practices in the United States and abroad has put us in a quandary. On the one hand, we are confronted with the question of the very viability of focusing the analysis of the concept of rights—whether we are referring to group or individual rights—exclusively within the old parameters of national boundaries. On the other, given the absence of legitimized international institutions that protect the human rights of all individuals, regardless of citizenship, we need to find new ways of reinforcing the institutions of citizenship, even while we simultaneously create new ways of safeguarding human rights in an increasingly transnational world.

NOTES

1. David Jackson and Paul de la Garza, "Rep. Gutiérrez Uncommon Target of a Too Common Slur," *Chicago Tribune*, April 18, 1996, 1.

2. I first read a summary of this story in Kevin R. Johnson's thought-provoking essay "Citizens as Foreigners," in Richard Delgado and Jean Stefancic, eds., *The Latino/a Condition: A Critical Reader* (New York: New York University Press, 1998), 198–201.

3. Meta Mendel-Reyes, *Reclaiming Democracy: The Sixties in Politics and Memory* (New York: Routledge, 1995); William H. Chafe, "The End of One Struggle, the Beginning of Another," in Charles W. Eagles, ed., *The Civil Rights Movement in America* (Jackson: University Press of Mississippi, 1986), pp.127–48.

4. Nancy Leys Stepan, "*The Hour of Eugenics*": *Race, Gender and Nation in Latin America* (New York: Cornell University Press, 1996), 171–96.

DISCUSSION QUESTIONS

1. What is the difference between biological racism and social racism?

2. What are the advantages and disadvantages of the national government employing ethnic labels?

24

American Indians
in the United States

BY RUSSELL THORNTON

T he idea of American Indian tribal sovereignty within the United
States and the related issue of political participation within the larger
. . .   American society have long been important issues for American
Indians. They have, however, achieved new prominence in recent decades.

SOVEREIGNTY: MYTH OR REALITY?

Chief Justice John Marshall described American Indian tribes as "domestic
dependent nations" with "aspects of sovereignty" (Strickland, 1998). As
Strickland pointed out (Strickland, 1998):

> [F]rom the beginning of the Republic, the courts have acknowledged that
> Native American government is rooted in an established legal and histori-
> cal relationship between the United States and Native American tribes or
> nations. This is at the heart of Native American constitutionalism and
> grows from precontact tribal sovereignty. [Moreover] the rights and obli-
> gations of Native Americans, unique to Indian law, derive from a legal

From *America Becoming: Racial Trends and Their Consequences*, vol. 1, ed. by Neil Smelser,
William J. Wilson and Faith Mitchell (Washington, DC: National Academies Press, 2000),
pp. 143–147. Copyright © 2001 by the National Academy of Sciences, courtesy of the
National Academies Press.

status as members or descendants of a sovereign Indian tribe, not from race. [Nevertheless] for the Native American, law and the courts have been seen alternatively as shields of protection and swords of extermination, examples of balanced justice and instruments of a conquering empire (p. 248).

The federal government has a long history of defining, and thereby determining, the tribal status of both American Indian groups and American Indian individuals (Thornton, 1987). In 1871, Congress enacted legislation that basically destroyed tribal sovereignty, by ending the rights of American Indian groups to negotiate treaties with the United States. It said, "Hereafter no Indian Nation or Tribe within the Territory of the United States shall be acknowledged or recognized as an independent nation, tribe, or power with whom the United States may contract by treaty" (Blackwell and Mehaffey, 1983:53). Between then and 1934, American Indian tribes "became increasingly disorganized, in part because of other legislation passed in the late 1800s calling for the allotment of tribal lands" (Thornton, 1987:195). In 1934, the Indian Reorganization Act was passed, allowing that an American Indian group had "rights to organize for its common welfare," and delineated steps whereby this might occur (Cohen, 1982). Subsequently, though, "the U.S. government adopted policies more or less aimed at ending the special legal status of American Indian tribes, and in fact, 61 tribes were officially terminated" (Thornton, 1987:195)—i.e., no longer recognized by the federal government . . .

President Richard Nixon rejected the idea of terminating American Indian tribes, and in 1976 the Federal Acknowledgment was created, specifying seven mandatory criteria for an American Indian group to achieve federal recognition. It also placed the "burden of proof" on the American Indian group itself (Thornton, 1987). The seven criteria are:

1. A statement of facts establishing that the petitioner has been identified from historical times until the present on a substantially continuous basis, as "American Indian," or "aboriginal."

2. Evidence that a substantial portion of the petitioning group inhabits a specific area or lives in a community viewed as American Indian and distinct from other populations in the area, and that its members are descendants of an Indian tribe which historically inhabited a specific area.

3. A statement of facts which establishes that the petitioner has maintained tribal political influence or other authority over its members as an autonomous entity throughout history until the present.

4. A copy of the group's present governing document, or in the absence of a written document, a statement describing in full the membership criteria and the procedures through which the group currently governs its affairs and its members.

5. A list of all known current members of the group and a copy of each available former list of members based on the tribe's own defined criteria.

6. The membership of the petitioning group is composed principally of persons who are not members of any other North American tribe.

7. The petitioner is not, nor are its members, the subject of congressional legislation which has expressly terminated or forbidden the federal relationship (U.S. Bureau of Indian Affairs, 1978).

Given that a tribe is federally recognized, however, "the courts have consistently recognized that one of an Indian tribe's most basic powers is the authority to determine questions of its own membership. A tribe has power to grant, revoke, and qualify membership" (Cohen, 1982). . . .

Today, American Indian tribes as entities are healthy, if not thriving. Both tribes and individuals, however, are dominated by a maze of laws and their interpretation. Strickland (1998) notes:

> Much contemporary confusion results from the duality of traditional tribal law and federally enforced regulations. . . . The courts have powers of life-and-death proportion over tribal existence. The nature of U.S. constitutional law and public policy is such that legal issues loom large in even the smallest details of Native American cultural, economic, and political life. More than four thousand statutes and treaties controlling relations with Native Americans have been enacted and approved by Congress. Federal regulations and guidelines implementing these are even more numerous. The tribe's own laws, and some state statutes dealing with Indians, further complicate this legal maze (p. 252).

Importantly, American Indian tribes and individuals are unique in American society—they are the only segment of the U.S. population with a separate legal status, both as groups and as individuals.

> As Native American peoples prepare to move into the twenty-first century, the issues facing tribes are not substantially different from those faced over the last five centuries. . . . The miracle of the past 500 years is that Native American people and their values have survived in the face of the most unbelievable onslaughts. There is little question that the law and the courts have been, and will continue to be, a major battlefield in the struggle for sovereign survival (Strickland, 1998:255). . . .

Until the late nineteenth century, American Indians were the dominant "minority group" the U.S. government had to deal with on the national, political scene. From the Civil War until the 1980s, however, American Indians were a "moral" but not "powerful" minority political group.

With the reaffirmation and reestablishment of American Indian tribes as legal entities since the 1970s, and the accompanying economic well-being of some of these tribes, however, American Indian tribes are becoming increasingly important and increasingly sophisticated political actors, something we have not seen since the subjugation of the great Sioux Nations around 1890.

ECONOMIC DEVELOPMENT
AND ECONOMIC WELL-BEING

One of the most intriguing developments since the 1970s is the increased economic development of American Indian tribes and the increased control of American Indian tribes over this development. As Snipp (1988) noted,

> Historically, American Indians have been one of the most economically deprived segments of American society. Joblessness and the accouterments of poverty, such as high infant-mortality rates and alcoholism, have been a traditional plague among Indian people (p. 1). . . . [A]s internal colonies, Indian lands are being developed primarily for the benefit of the outside, non–Indian economy (p. 3). [Thus] the tribes have been relatively unsuccessful in capturing the material benefits of development, and some observers claim that Indians are now exposed to subtle forms of economic exploitation, in addition to the political dominance they have experienced as captive nations.

Since Snipp made his arguments, the situation has changed partially; certainly not totally.

What Is Tribal Economic Development?

Tribal economic development is generally conceived of as an increase in economic activities, particularly successful ones, on the part of the tribe itself as an entity, rather than increased economic well-being of tribal members per se. Individual economic well-being, nevertheless, is an important objective of tribal economic development; and American Indian tribes and individuals engage in virtually the entire spectrum of economic activities available in modern society, ranging from small service industries to manufacturing to extraction of natural resources—fishing, logging, hunting, etc. In some instances, the ability to exploit such resources has involved extensive legal issues engendered by American Indians' unique legal status in American society (Olson, 1988).

Tribes are also engaged in activities more specifically related to American Indian culture and themselves as American Indian peoples or peoples in rural areas. As is the case with many indigenous peoples worldwide, American Indian tribes are often involved in tourism, as objects of tourism or providers of facilities for tourists in tribal areas or both. Activities related to tourism on tribal lands include running museums, gift shops, gas stations, hotels, and restaurants; providing transportation and other direct services; and performing cultural plays, pow wows, dances, and, sometimes, ceremonies.

American Indian tribes have also engaged in economic activities available to them because of (rather than in spite of) their unique legal status in American society. First and foremost is legal gambling. In some instances, tribes have built and/or operate large, successful casinos that have brought some degree of prosperity to them and their members. Some of the more successful ones are operated by the Mississippi Choctaw in Philadelphia and Mississippi and by the Pequot in Connecticut.

Generally, tribal and individual economic well-being go hand in hand. It is not, however, always a simple, straightforward matter. . . .

Important in the decision to engage in economic activities is the issue of the type of activity to engage in and how chosen activities may or, typically, may not fit into the traditional cultural values of the tribe. Nowhere does more conflict occur than in considering the issue of gambling. Some tribes have explicitly decided not to engage in such activities—as profitable as they might be—because they conflict with important values. Wilma Mankiller, the former principal chief of the Cherokee Nation of Oklahoma, said that one of the most difficult decisions she made as principal chief was the decision that the Nation would not engage in gaming. "I literally cried when I made the decision," she said (personal conversation with the author). Gaming could have been very profitable for the Cherokee Nation and could have improved the economic well-being of tribal members, but it is also against Cherokee values.

There are American Indian communities who see economic development either as a return to old subsistence practices or as simply a reaffirmation of such practices. The attempt by the Makah Nation of Neah Bay, Washington, to return to traditional whaling practices is a case in point. Similarly, there are Inuit communities in Alaska who still cherish their traditional, subsistence lifestyles and are determined to preserve them.

REFERENCES

Blackwell, C., and J. Mehaffey. 1983. American Indians, trust and recognition. In *Nonrecognized American Indian Tribes: An Historical and Legal Perspective*, F. Porter, III, ed. Occasional Papers Series, no. 7. Chicago: The Newberry Library.

Cohen, F. 1982 [1942]. *Handbook of Federal Indian Law* (reprint). New York: AMS Press.

Olson, M. 1988. The legal road to economic development: Fishing rights in western Washington. Pp. 77–112 in *Public Policy Impacts on American Indian Economic Development,* C. Snipp, ed. Albuquerque: Institute for Native American Development, Development Series No. 4, University of New Mexico.

Snipp, C. 1988. "Public policy impacts on American Indian economic development." In *Public Policy Impacts on American Indian Economic Development,* C. Snipp, ed. Albuquerque: Institute for Native American Development, Development Series No. 4, University of New Mexico.

Strickland, R. 1998. The eagle's empire. In *Studying Native America: Prospects and Problems,* R. Thornton, ed. Madison: University of Wisconsin Press.

Thornton, R. 1987. *American Indian Holocaust and Survival: A Population History since 1492.* Norman, OK: University of Oklahoma Press.

U.S. Bureau of Indian Affairs. 1978. Guidelines for Preparing a Petition for Federal Acknowledgment as an Indian Tribe. Washington, DC, pp. 3, 8–11, 17.

DISCUSSION QUESTIONS

1. When did American Indians gain access to the political system?
2. How does a group of American Indians secure federal recognition as a tribe?
3. What does tribal sovereignty mean in the contemporary United States, and how is it linked to economic development?

25

The Silent Voices

2000 Presidential Election and the Minority Vote in Florida

BY REVATHI I. HINES

". . . No person acting under color of law shall fail or refuse to permit any person to vote who is entitled to vote under any provision of this Act or is otherwise qualified to vote, or willfully fail or refuse to tabulate, count, and report such person's vote"
—Voting Rights Act, 1965, Section 11

Historically, the right to vote was not a guaranteed right to African Americans in the United States. Efforts to enfranchise all African American males (women achieved the right to vote in 1920 with the 19th amendment) resulted in the ratification of the fifteenth amendment to the Constitution of the United States of America. In theory, the fifteenth amendment stated that no citizen could be refused the right to vote simply because of race, color, or previous condition of servitude. However, enforcement of this amendment was not automatic. Many states adopted barriers that prevented African Americans from exercising this right. Devices such as poll taxes, literacy tests, white primaries, intimidation, and grandfather clauses were effectively used to render the fifteenth amendment moot (Barker and Jones, 1994).

It was not until the passage of the Voting Rights Act of 1965 (VRA) did Congress intervene to ensure compliance by the states of the fifteenth

From *Western Journal of Black Studies*, 26 (2): 71–74, 2002. Reprinted by permission.

amendment. VRA was designed to bring coherence in voter eligibility requirements and eliminate discrimination of eligible voters. It rendered devices such as poll taxes and literacy tests discriminatory. It also set forth procedures against states found in direct violation of the VRA provisions (Rueter, 1995).

RESEARCH OBJECTIVES

This research examines the disenfranchisement of African Americans and other minorities during the 2000 presidential election in Florida. Specifically, it studies problems with voter list purges, accessibility to poll booths, outdated machinery, and lack of translators and then examines its impact on ex-felons, the disabled, Blacks, and Hispanics. Hopefully, this study will provide some insight on the voting day irregularities that occurred during this election in Florida. Some general conclusions can be drawn on the efforts that should be undertaken in subsequent elections to address these types of issues.

BACKGROUND

The 2000 Presidential Election

The 2000 presidential election was unique in many ways. Both candidates were well known in the political circle and had political experience. The election was the first presidential race of the new millinneum. It also proved to be a presidential race that was extremely close in votes. When the final counts of the states were reported, the fate of the two presidential candidates, Albert Gore and George Bush, hung on one state: Florida.

As the United States and the world watched the news media make projections and retractions of their projections of the Florida vote count, another problem was unfolding in the state. Allegations of voting day irregularities such as outdated machines, improper counts and tabulations, inadequate access to individuals with disabilities, lack of translators for immigrants, and inability of eligible individuals to exercise their right to vote further clouded the already uncertain election (Pierre and Morello, 2000).

THE SILENT VOTES OF THE MINORITIES

The 2000 presidential election saw a dramatic increase of African American voter registrants in the state of Florida. There was an increase of both African American Republicans and Democrats. However, the increase of Black Democrats was ten-to-one to that of the Republicans. The increase in the number of Black Democrats could be attributed to the discontent of Black

Floridians on the many conservative actions taken by Governor Jeb Bush. One of them included his decision to remove the state's Affirmative Action laws (Walton, Jr., 2001). However, the increase in the number of African American voters did not mean an automatic increase in the number of votes cast by them. No, the problem was not apathy on the part of Black voters and other minorities, but the silence of their votes. Civil rights groups, African American leaders, and Democrats have accused the state with a litany of charges ranging from confusing ballots, error-prone machines, lack of trained poll workers, accessibility to poll sites for the disabled, and improper purges.

Outdated Machinery
and the Black Vote

The relationship between race and rejected votes in the 2000 Florida elections is evident in many studies conducted since the November 7th election. A study undertaken by the United States Commission on Civil Rights (2001) found that there was a positive relationship between race and voter disenfranchisement. Counties with large minority populations were more likely to have voting systems with higher ballot rejection rates than the more affluent counties with large white populations. Nine of the ten counties with the highest percentage of African American voters had rejection rates above the Florida average. However, only two of the ten counties with the highest percentage of white voters had rejection rates above the state average. Moreover, seventy percent of African American voters lived in counties where the error-prone punch cards were used to tally the votes. Such a relationship between race and ballot rejections was also obvious when three specific counties, Duval, Miami-Dade, and Palm Beach, were studied. Precinct data, where available, indicated that 83 of the 100 precincts with the higher number of spoiled ballots were also Black-majority precincts.

Similar studies by the *Miami-Herald, New York Times,* and *Washington Post* indicated a strong relationship between race and rejected ballots in Florida. The *New York Times* study indicated that the majority of Florida's African American voters cast their ballots on error-prone punch card machines and thereby had a higher chance of having their vote rejected. Black precincts in Miami-Dade county saw a rejection rate nearly twice that of Hispanic precincts and four times the rate of white precincts.

An analysis by the *Washington Post* on voided votes in Florida indicated that ballots cast by African Americans were voided at a greater rate than those of whites. The analysis indicated that precincts with a majority of minority population were more likely to have outdated and faulty voting machines (Mintz and Keating, 2000).

Another analysis by the *Miami Herald* showed similar findings. Presidential ballots in nearly all of Florida's Black precincts were invalidated at a higher rate than the ballots in mostly White neighborhoods. While the study did conclude that both Black and White poor and less-educated voters were more likely to have their ballots voided, the balloting problem in Florida did have a greater impact on Blacks. While majority-black districts had rejection rates of one in every ten, majority-white precincts had a rejection rate of one in every thirty-eight. Also, 82 percent of the state's 463 majority-black precincts had a rejection rate that was higher than the statewide average of three percent. However, only 41 percent of the state's majority-white districts had a rejection rate higher than the statewide average. Also, a high percentage of Black voters lived in the twenty-four counties that used the punch-card machines (Lane, 2000).

Governor Bush's Select Task Force on Election Procedures, Standards, and Technology (2001), which studied the 2000 presidential election in Florida, had somewhat similar conclusions. The report stated that the error rates for different types of voting systems were directly related to the type of machine being used. It also stated that the difference in error rates among the different types of machines being used could not be accounted for solely by "uneducated, uninformed, or disinterested voters."

Purged Lists and Ex-felons

Eligible voters faced yet another hurdle during the 2000 presidential elections in Florida. Thousands of voters faced unnecessary purges on November 7th. These purges struck ex-felons the hardest. The state of Florida had been legally required to recognize the rights of ex-felons according to a 1998 court decision which ruled in favor of the full faith and credit clause as it pertained to weapon permits from other states. This ruling applied to voting rights as well. However, officials did not fully comply with the court decision during the November 7th election (March, 2001).

During the 2000 election, Floridians convicted of felonies in other states, were purged off the voter rolls even if they had their rights restored in those states. In order to be able to vote in Florida, they had to prove in writing that their rights had been restored or they could appeal and receive clemency from Florida officials. During the 2000 elections, the ex-felons were given as little as five months before November 7th to obtain the proof necessary to have their rights to vote restored in Florida (March, 2001). This policy applied to 2,834 Floridians in the Gore-Bush presidential election.

Civil rights groups called this a direct violation of the full faith and credit clause of the Constitution which requires a legal ruling in one state to be honored by another. Civil rights groups also noted that the policy was unconstitutional since Florida's purging laws had a disproportionate impact on African Americans who constitute a large percentage of those convicted and sentenced. Similar concerns were also echoed in the findings by the United States Commission on Civil Rights (2001). Their report stated that African Americans were placed on purge lists more erroneously than Hispanics or White voters. There were no clear guidelines to protect eligible voters from being wrongly removed.

Accessibility and Special Needs

Another issue that took center stage during the November 7th election was the lack of accommodations on behalf of those physically disabled and those needing language assistance. The Voter Accessibility for the Elderly and Handicapped Act of 1984 requires that polling sites be made accessible to disabled voters. The Act states that each state should be responsible for making sure that all polling places for Federal elections are accessible to handicapped and elderly voters. Section 2A of the United States Code states that when a polling site is not accessible by a disabled or elderly voter, it must be relocated or made temporarily available to the voters in need. However, many disabled and elderly Floridian voters found themselves not being able to vote or in some cases being forced to cope with the situation because of inaccessibilty to the voting places (Hill and Seeley Jr., 2000).

According to testimonies obtained by the United States Commission on Civil Rights (2001) and other civil rights organizations, some in wheelchairs simply turned back or faced humiliation by requesting help to be carried to the polling place. Some visually impaired voters also claimed that they were not provided with the proper tools to assist them in reading the ballots. Some of the visually impaired voters requested help from the poll workers to cast their ballot, thereby, losing the right to a secret ballot.

The inability to exercise the right to vote also extended to some voters who needed assistance with the English language. The 1975 amendments to the VRA addressed the multilingual requirement of voting laws. According to the amendments, if more than five percent of the voting age population of a state are individuals of a single language minority and have limited proficiency in the English language, the state has to provide language assistance. Some Hispanic and Haitian voters were not provided ballots in their native languages (Pierre and Slevin, 2001). A report by the Commission on Civil Rights indicated that in some parts of Florida, voters needing language assistance neither received bilingual assistance or bilingual ballots as required under the VRA.

NECESSITY OF COMPREHENSIVE
ELECTION REFORM IN FLORIDA

A 2001 study conducted by the *Washington Post* and other media organizations simply affirmed the many issues that arose during the 2000 election. They studied 175,010 Florida ballots that were not counted in the November 7th election. Evidence indicated that the ballots of voters in Black neighborhoods were most likely uncounted for. In large Black neighborhoods, thirteen of every thousand ballots went uncounted for, compared to six in every thousand ballots in predominantly white neighborhoods (Keating and Mintz, 2001).

The 2000 presidential elections in Florida had a great impact on Blacks and other minorities. Blacks in Florida were at least ten times more likely than other voters to have had their ballots rejected. Reports from the myriad of sources indicate that Black voters and Hispanics had higher rates of problems than other voters in Florida. Blacks were assigned to polling sites that lacked resources to confirm voter eligibility; used defective and complicated ballots that caused overvotes and undervotes; used defective election equipment in poor precincts; failed to provide bilingual assistance to Hispanic and Haitian voters; failed to ensure access for the disabled: and erroneously purged ex-felons off the voters lists (United States Commission on Civil Rights, 2001).

As the battle over chads and dimples evolved, many minority votes were silenced. These votes fell victim to error-prone machines, language and physical barriers, and faulty purge systems. Blacks, Hispanics and other immigrants, the disabled, and ex-felons have the most to gain from subsequent election reforms in Florida.

In 2001, the Florida legislature passed election reform legislation. It established funds for a statewide and uniform voting system, voter education programs, a database of registered voters and a requirement that the Division of Elections develop uniform recount procedures (*Tampa Tribune*, 2001). However, the reforms passed by the legislature have been criticized by many civil rights and civil liberties organizations. The two major objections that have been cited are the continued purging of convicted felons from voter rolls and voter responsibility signs at polling places that places undue burden on the voters who might not be able to read. For the disabled and minority voters in Florida, comprehensive election reform in Florida is not yet a reality.

REFERENCES

Barker, L. and Jones, M. (1994). *African Americans and the American political system.* Englewood Cliffs, NJ: Prentice Hall.

March, W. (2001). "Policy mix-up kept ex-felons from polls." *The Tampa Tribune*, 1.

Hill, E., and Seeley, J. (2000). "Now see Florida voting through eyes of disabled." *The Los Angeles Times*, 9B.

Keating, D., and Mintz, J. (2001). "Florida Black ballots affected most in 2000." *The Washington Post*, 3A.

Lane, C. (2000). "Miami Herald finds voting irregularities." *The Washington Post*, 8A.

Mintz, J., and Keating, D. (2000). "A racial gap in voided votes; precinct analysis finds stark inequity in polling problems." *The Washington Post*, 1A.

Pierre, R. and Morello, C. (2000, December 12). "Irregularities cited in Florida Voting." *The Washington Post*, 38A.

Pierre, R., and Slevin, P. (2001). "Florida vote rife with disparities, study says: rights panel finds Blacks penalized." *The Washington Post*, 1A.

Rueter, T. (1995). *The politics of race: African Americans and the political system.* New York: M. E. Sharpe.

"The limits of election reform." (2001). *The Tampa Tribune*, 8.

United States Commission on Civil Rights. (2001). *Voting irregularities in Florida during the 2000 presidential election.* Washington, DC: Government Printing Office.

Walton Jr., H. (2001). "The disenfranchisement of the African American voter in the 2000 presidential election: The silence of the winner and loser." *The Black Scholar* 31(2), 21–24.

DISCUSSION QUESTIONS

1. What were the goals of the Voting Rights Act of 1965? Why does this law have to be modified periodically?

2. How do practices that occurred during the Florida 2000 Presidential election reinforce the racial inequality of citizens?

26

Covering Immigration

Popular Images and the Politics of the Nation

BY LEO R. CHAVEZ

On 5 July 1976, *Time* magazine published an issue celebrating the nation's bicentennial birthday. The cover image was a mosaic of words printed in red, white, and blue, with the bold text "The Promised Land" forming a protective semicircle above the text "America's New Immigrants." Inside the magazine was another mosaic of images made from photographs of immigrants from different periods in U.S. history and from different countries. *Time's* 1976 birthday issue was an affirmative rendition of "the nation of immigrants" theme that is a central part of the story America tells about its history and national identity.

On 17 October 1994, the cover of the *Nation* told a different story about immigration. Its cover text proclaimed "The Immigration Wars." The cover is a collage of overlapping images. The central image appears to be the western border of the United States on the circular globe of the earth as seen from space. To the left of the continental border, where the Pacific Ocean would normally fill in the rest of the globe, is a multitude of people, a mass of heads and partial bodies, many wearing hats and scarves, evoking the mass movement of refugees or migrants. Walking north across the globe, with one foot on the border of the North American continent, is a man with a knapsack on

From Leo R. Chavez, *Covering Immigration: Popular Images and the Politics of the Nation* (Berkeley, CA: University of California Press, 2001), pp. 1–8. Copyright © 2001 The Regents of the University of California. Reprinted by permission.

his back and a Mexican sombrero on his head. A barking dog pulls tightly on its leash, right above the Statue of Liberty, which has an upside down American flag sticking out of her head. In the background is another line (border?) with grass beyond it and a rectangular frame that appears to be engulfed in flames.

The *Nation*'s cover used images that evoked a sense of the prevailing climate toward immigrants at the time, a climate filled with a sense of alarm about the perceived negative impact of immigration on the nation. The sentiments clearly elicited by the cover's image did not necessarily represent the editorial stance of the magazine itself, and in this case the *Nation* offered up such an image in critique of what it perceived as pervasive anti-immigrant views in U.S. society. But the point is that the *Nation*'s cover stands in marked contrast to *Time*'s affirmative rendering of immigration and the nation on its cover of almost twenty years earlier. These two magazine covers reflect the demonstrable shift to an increasingly anti-immigrant public debate and public-policy initiatives that occurred during the last quarter of the twentieth century. Moreover, they represent two opposing and yet interlocked views of immigration, a double helix of negative and positive attitudes that have existed throughout America's history. Immigrants are reminders of how Americans, as a people, came to be, and immigration is central to how we view ourselves as a nation. As Oscar Handlin once wrote, immigration *is* the history of the nation. But immigrants are also newcomers whose difference and "otherness" do not go unquestioned or unremarked upon. Their very presence raises concerns about population growth, economic competition, and various linguistic and "cultural" threats. These polarized views constitute the immigration dilemma in American society. . . .

1965: A WATERSHED YEAR

I begin this investigation with 1965 because that was the year the U.S. Congress passed monumental immigration reform that radically changed the criteria used to admit immigrants to the United States [In the Immigration and Nationality Act Amendments of 1965] (Reimers 1985). Anti-immigration sentiments crystallized in the 1924 immigration law, which instituted the national origins quotas and virtually shut off immigration from eastern and southern Europe. The quota for each nation was defined as 2 percent of the number of foreign-born persons for each nationality present in the United States in 1890 (changed to 1920 in 1929). Because of the composition of the U.S. population in 1890 and 1920, this quota system was heavily weighted in favor of immigration from northern Europe. For example, after 1929, 82 percent of the visas for the Eastern Hemisphere were allotted to countries in northwestern Europe, 16 percent to southeastern Europe and 2 percent to all others (Pedraza 1996, 8). Asians had little to no chance of obtaining a visa.[1] By the 1950s, Europeans still dominated the flow of immigrants to the United States. Europeans accounted for 52.7 percent of all immigrants between 1951 and 1960. Relatively few Asian immigrants (6.1 percent) came during that time. Latin Americans accounted for 22.2 percent of the immigrants during that decade (Pedraza 1996, 4).

The 1965 immigration law abolished national origins quotas. David Reimers (1985) has called this a "cautious reform" with unintended consequences. Instead of national origins quotas, the new law allotted immigrant visas on a first-come, first-served basis within a system of preferences. The law established seven preferences, five for close relatives of U.S. citizens and legal residents and two for immigrants with professions, skills, occupations, or special talents needed in the United States (Reimers 1985, 72). The preference system was built on the principle of family unification, a principle that few policy makers at the time believed would lead to a change in the composition of the immigrants since it was assumed that close relatives would come from the same parent countries as the citizens (Reimers 1985, 75). The "unintended" consequence of the preference system was an increase in the proportion of Asian and Latin American immigrants. Between 1981 and 1990, Asians (37.3 percent) and Latin Americans (47.1 percent) accounted for 84.4 percent of all legal immigrants (Pedraza 1996, 4). In addition to these changes, the number of legal immigrants coming to the United States has increased from 2.5 million between 1951 and 1960 to 7.3 million between 1981 and 1990. During the 1990s, about 800,000 legal immigrants came to our shores each year. When undocumented immigrants (popularly called "illegal aliens") are added, the total number of newcomers was over one million a year. That about equals the number of immigrants that came during the peak years of immigration during the early 1900s. . . .

Although the total number of immigrants has grown, it must be seen as relative to general population growth. One way to consider this is the proportion of foreign-born residents in the total U.S. population. In 1960, the foreign born accounted for 5.5 percent of the U.S. population. In 1970, they actually went down to 4.7 percent of the population. In 1980, the foreign born accounted for 6.2 percent of the total population, and in 1990 they were 7.9 percent. These numbers are proportionally smaller than in earlier decades of this century, when immigrants made up much larger proportions of the total population. In 1910, for example, the foreign accounted for 14.7 percent of the U.S. population (Rumbaut 1996, 25; U.S. Immigration and Naturalization Service, 1990–93).

Although the year 1965 marks the beginning of a major demographic shift in immigration patterns, not all of these changes can be blamed on alterations in the nation's immigration laws. Since 1965, the nation has received refugees and unauthorized immigrants from various regions of the world. Southeast Asian refugees began migrating to the United States after its military withdrawal from the region in the mid-1970s. Cubans began fleeing the Castro regime in the 1960s, with various moments of increased refugee movements, such as the infamous Mariel boat exodus in 1980, when more than 125,000 Cubans made their way to Florida's coast. Central Americans fleeing conflict in the region migrated to the United States throughout the 1980s and early 1990s. Eastern Europeans came, escaping the economic collapse of their countries in the post–Cold War era. Undocumented immigration has also contributed to the flow of immigrants both before the 1965 law and after. All of these factors contributed to demographic shifts not just in the national origins of today's immigrants, but also to changes in the ethnic and racial makeup of the nation generally.

At the same time that the total number of immigrants has grown, the native white population in the United States has aged demographically. Relatively young immigrants and an aging (less fertile) white population has meant that immigration's impact on the nation's population growth rate has increased proportionately. In the 1951–60 decade, net immigration accounted for 10.6 percent of the nation's population growth. Between 1961 and 1970, immigrants still accounted for only 16.1 percent of population growth. There was little change between 1971 and 1980 in immigration's relative impact on population growth (it was still only 17.9 percent). But between 1981 and 1990, immigration's net impact on population growth rates doubled from previous decades, accounting for 39.1 percent of the nation's population growth. Moreover, the composition of the immigrants had also changed. Europeans, who accounted for almost 53 percent of legal immigrants between 1951 and 1960, accounted for only 12.5 percent of the immigrants between 1981 and 1990. Asians (37.3 percent) and Latin Americans (47.1 percent) were the major immigrant populations in the 1981–90 decade (Rumbaut 1996, 25; Pedraza 1996, 4). As these statistics suggest, demographic changes related to immigration became more evident during the 1980s and 1990s, at least compared to the previous two decades.

Given immigration trends and fertility rates, Latino and Asian American populations will experience significant growth over the next fifty years (Martin and Midgley 1994). Latino growth should increase from about 11 percent of the nation's population to about one quarter. Asian populations will more than quadruple in size, or grow from about 3.5 percent to about 16 percent. Whites will decrease from about 75 percent to about half of the U.S. population, and African American growth will remain fairly constant in relative numbers. As we shall observe, the implications of these demographic trends inform much of the debate over immigration and the nation.

America was once viewed as a great "melting pot" that blended many immigrant strains into a single nationality. While we may now assert that ethnic identities and traditions are not so easily lost by immigrants, and that becoming American is not always a simple linear process, the melting pot continues to retain its narrative power as a metaphor for American society, if only, for some, to parody. The power of America to absorb immigrants is both marveled at and questioned, but continues to be an important story we tell about ourselves as a people and as a nation. That we can call ourselves "a nation of immigrants" depends on the power of this common narrative about our history.

During the later decades of the twentieth century, the American public was noticeably uneasy with both undocumented and legal immigration and with the melting pot narrative (see Mills 1994). Tensions revolved around the way we think of ourselves as a nation and as a people. As historian David Hollinger

(1995) might put it, who is included in "the circle of we" is increasingly debated and narrowed as immigrants, both legal and unauthorized, are targeted as belonging outside the "we." The "rhetoric of exclusion" embedded in contemporary discourse on immigration runs the risk of arousing nativism (Perea 1997, Stolcke 1995). In his classic book, *Strangers in the Land,* John Higham (1985 [1955]) defined nativism as "intense opposition to an internal minority on the ground of its foreign (i.e., 'un-American') connections." Higham argued that nativism gets much of its energy from modern nationalism and that "nativism translates broader cultural antipathies and ethnocentric judgments into a zeal to destroy the enemies of a distinctively American way of life." Indeed, the proponents of restricting immigration often view today's immigrants as a threat to the "nation," which is conceived of as a singular, predominantly Euro-American, English-speaking culture. The "new" immigrants—the *trans*nationalists—threaten this singular vision of the "nation" because they allegedly bring "multiculturalism" and not assimilation (Martinez and McDonnell 1994). From this perspective, the pot no longer has the capacity to melt.

NOTES

1. The Chinese Exclusion Act of 1882 and the Gentleman's Agreement with Japan in 1907 had already severely restricted immigration from those two countries.

REFERENCES

Higham, John. 1985. *Strangers in the land.* New York: Atheneum.

Hollinger, David A. 1995. *Postethnic America: Beyond multiculturalism.* New York: Basic Books.

Martin, Philip, and Elizabeth Midgley. 1994. Immigration to the United States: Journey to an uncertain destination. *Population Bulletin* 49, 2 (September). Washington, D.C.: Population Reference Bureau.

Martinez, Gebe, and Patrick J. McDonnell. 1994. Prop. 187 forces rely on message—not strategy. *Los Angeles Times* 30 October, sec. A, p. 1.

Pedraza, Silvia. 1996. Origins and destinies: Immigration, race, and ethnicity in American history. In *Origins and destinies: Immigration, race, and ethnicity in America,* edited by Silvia Pedraza and Rubén Rumbaut, I-20. Belmont, Calif.: Wadsworth.

Perea, Juan F., ed. 1997. *Immigrants out! The new nativism and the anti-immigrant impulse in the United States.* New York: New York University Press.

Reimers, David M. 1985. *Still the golden door: The third world comes to America.* New York: Columbia University Press.

Rumbautn, Rubén G. 1996. Origins and destinies: Immigration, race, and ethnicity in cotemporary America. In *Origins and destinies: Immigration, race, and ethnicity in America,* edited by Silvia Pedraza and Rubén Rumbaut, 21–42. Belmont, Calif.: Wadsworth.

Stolcke, Verena. 1995. Talking culture: New boundaries, new rhetorics of exclusion in Europe. *Current Anthropology* 36:1–24.

U.S. Immigration and Naturalization Service. 1990–93. *Statistical yearbooks.* Washington, D.C.: U.S. Government Printing Office.

DISCUSSION QUESTIONS

1. Why does Chavez argue that the 1965 immigration law was so important?
2. Why are the Latino and Asian American populations growing in percentages of the U.S. population?
3. How would you characterize the reactions to the changing racial composition of the nation?

27

From Ellis Island to JFK

Education in New York's Two Great Waves of Immigration

BY NANCY FONER

N ew York City is in the midst of a profound transformation as a result of the massive immigration of the last four decades. More than two and a half million immigrants have arrived since 1965, mainly from Latin America, Asia, and the Caribbean, and they are now streaming in at a rate of over 100,000 a year. Immigrants already constitute over a third of the city's population. In the midst of these dramatic changes, commentators and analysts, popular and academic, in the press and in the journals, are comparing the new immigration with the old.

This is not surprising. Few events loom larger in the history of New York City than the wave of immigration that peaked in the first decade of the 20th century. Between 1880 and 1920, close to a million-and-a-half immigrants arrived and settled in the city, so that by 1910, fully 41 percent of all New Yorkers were foreign-born. The immigrants, mostly Eastern European Jews and southern Italians, left an indelible imprint on the city—indeed, a large and influential part of New York's current citizens are their descendants.

An elaborate mythology has grown around immigration of a century ago, and perceptions of that earlier migration deeply color how the newest wave is seen. For many present-day New Yorkers, their Jewish and Italian immigrant

From *Brandeis Review* 21(2): 32–37, 2001. Reprinted by permission of the author.

forebears have become folk heroes of a sort—and represent a baseline against which current arrivals are compared and, unfortunately, often fail to measure up.

Nowhere is the nostalgia for the past more apparent than when it comes to education. Sentimental notions about Eastern European Jewish immigrants' love affair with education and their zeal for the life of the mind have become part of our picture of the "world of our fathers." Jews are remembered as the "people of the book" who embraced learning on their climb up the social ladder. These memories set up expectations about what immigrants can and should achieve in the schools. If my grandparents and great-grandparents could succeed in New York City's schools a hundred years ago, without special programs to help them adjust, why—many people say—can't today's immigrants and their children do well when they get so much more assistance?

A comparison of New York's two great waves of immigration shows that inspirational tales about Eastern European Jews' rise through education and their success in New York's schools in the so-called golden immigrant age do not stand up against the hard realities of the time. Today, despite a dramatically different context and significant problems in the schools, many immigrant children are doing remarkably well.

In the years before World War I, most Eastern European Jews did not make the leap from poverty into the middle class through education. Those who made substantial progress up the occupational ladder in this period generally did so through businesses in the garment, fur, shoe, and retail trades and in real estate. It was only in later decades that large numbers of Eastern European Jewish children used secondary and higher education as a means of advancement.

A hundred years ago, most Jewish immigrant children left school with, at best, an eighth grade education; few went to high school, and even fewer graduated. In the first decade of the 20th century, well below five percent of the Russian Jewish children in New York City graduated from high school; less than one percent of Russian Jewish young people of college age ever reached the first year of college. By 1908, the City College of New York (CCNY) had already become a largely Jewish school, but Jewish undergraduates there and at other New York colleges were a select few. Only a tiny number graduated. In 1913, City College's entire graduating class had only 209 students, less than 25 of them Eastern European Jews. The 25 percent of Hunter College graduates who were Eastern European Jews in 1916 amounted to only 58 women. This was at a time when the Jewish population of New York City was almost a million! It was not until the 1930s that there were big graduating classes at City College that contained large numbers of Jews of Russian and Polish origin.

One reason so few Russian Jewish students went to high school or college is that there weren't many high schools or colleges at the time. In 1911, after a decade of expansion, the city had 19 high schools, but the high school student body was still only about a quarter of the size of the four preceding elementary school grades. CCNY and Hunter, the only two public colleges, had about 1,400 students in 1908. A high school degree wasn't necessary for the jobs employing most New Yorkers, and an eighth grade graduate could even get a white-collar job. A business career didn't require four years of college nor did teaching or the law. In any case, extended schooling was a luxury beyond the means of most new immigrant families, who needed their children's contributions to the family income. Even those who managed it rarely saw all their children go to high school.

The elementary school also did a poor job of educating immigrant children so that many were not prepared or motivated to continue on. The schools were severely overcrowded in the wake of the huge immigrant influx and the inability of school construction to keep up with demand. By 1914, enrollments had grown to almost 800,000, more than triple the figure for 1881. Educators of the time joked that teachers should have prior experience in a sardine factory before being hired to work in the New York schools.

There's a nostalgia for the "sink or swim" approach to learning English, but unfortunately, many in the past "sank" rather than "swam." Most non-English speaking children were placed in the lowest grade regardless of their age. (The special "steamer" classes introduced in 1904—in which students were totally immersed in English for a few months—catered to a mere fraction of students needing them, only 1,700 students throughout the city in 1908.) Teachers made promotion decisions, and many children who could not do the work were left back. In 1908, over a third of the Russian Jewish elementary school pupils in New York City were two or more years over age for their grade.

If Russian Jewish children in this early period were not the education exemplars often remembered, they did do much better than Italians, the other "greenhorns" at the time. This favorable comparison helps explain why Jewish academic achievements have stood out and received so much attention. Russian Jewish students' progress, however, was fairly similar to that of native white New Yorkers at the elementary school level—and they did less well than native whites in making it to high school and college.

It's difficult to compare immigrants' educational achievements today with those in the past because the context in which education is a path to mobility is so radically different: formal education, and a more extended education, is now more important in getting a job owing to educational upgrading and transformations in the world of work. In the last great immigration wave, high schools were just becoming mass institutions. Today, getting a high school diploma is the norm, achieved by more than 70 percent of New York City's adult population and essential for many low-skilled positions. College is no longer an institution for a tiny elite—in 1990, a quarter of New Yorkers 25 and older had a college degree. Today, college graduates compete for jobs that immigrants with a high school diploma could have obtained a century ago, and a college degree—or more—is required for the growing number of professional, technical, and managerial positions.

How are immigrant children doing today? Admittedly, there are many dropouts and failures—something more serious now when more education is needed to get a decent job. Many, perhaps most, immigrants attend New York City schools where student skill levels are low, dropout rates high, and attendance rates poor. Once again, the surge of immigrants has led to soaring public school enrollments—now over the million mark—and serious overcrowding. New language programs are inadequate to meet the enormous need. Whereas a hundred years ago, New York schools mainly had to cope with Yiddish- and Italian-speaking children, today they confront a bewildering array of languages; a Board of Education count indicates that more than 100 languages are spoken by students from over 200 countries. When high schools were institutions for a minority of the better and more motivated students, violence, crime, and student indifference—and hostility—were not issues the way they are today.

Despite these problems, a substantial number of students are making it, and some are doing exceptionally well. There are new educational opportunities: expanded, and more widely available, higher education; a host of new programs for teaching students English; and even special immigrant schools designed specifically for newcomer children. Perhaps most important, large numbers of immigrant children have highly educated parents, and some immigrant children themselves have previous experience in fine schools in their home country. This has translated into academic success for many newcomers.

Although data on immigrant students in New York City are woefully inadequate—the only Board of Education data on immigrants refer to "recent immigrant students" who have entered as U.S. school system for the first time in the past three years—they show immigrants comparing favorably with other students in several ways. Students who were recent immigrants to the public school system in middle school, graduate from high school on time by a slightly greater percentage than their native-born peers. They also have lower dropout rates. Although recent immigrants' median test scores in math, reading, and English are somewhat lower than those of other students, they improved their scores between 1989-1990 and 1990-91 more than the rest of the student body. At the City University of New York, immigrants make up a high, and growing, proportion of the student body: in 1998, 48 percent of CUNY's freshman class were foreign-born.

National studies based on large representative samples show immigrants often outperforming their native-born peers. In a study of eighth graders, children of immigrants—those born abroad and in the United States—earned higher grades and math scores than children of native-born parents even after the effects of race, ethnicity, and parental socioeconomic status were held constant. Another national study, which interviewed more than 21,000 10th and 12th graders in 1980 and followed them over a six-year period, compared immigrants and the native born in the aggregate as well as immigrants and the native born in four different ethnoracial groups (Asian, white, black, and Hispanic). Whichever way they were compared, immigrants were more likely to follow an academic track in high school than their native-born counterparts; once graduated, immigrants were also more likely to enroll in postsecondary education, to attend college, and to stay continuously through four years of college.

Like Jews of an earlier era, today's educational exemplars are Asians; white European immigrants are also doing comparatively well. Most striking is Asian (native as well as foreign born) overrepresentation in New York City's elite public high schools that select students on the basis of notoriously difficult entrance exams. In 1995, an astounding half of the students at the most selective high school of all, Stuyvesant, were Asian; at the Bronx High School of Science, 40 percent were Asian, and at Brooklyn Technical High School, 33 percent. This is at a time when Asians were 10 percent of the city's high school population. So many Korean students now attend Horace Mann School—one of New York's most competitive private secondary schools—that there is a Korean parents' group there.

Just how important is Asians' culture in accounting for their educational achievements? As among Jews in the past, it plays a role. But I would argue that social-class factors, then as well as now, are more important. And today, race must also be considered.

A major reason Eastern European Jews did so much better academically than Italians in the old days was their occupational head start. Jews were more urban and arrived with higher levels of vocational skills, which gave them a leg up in entering New York's economy. Because the Jewish immigrant population was, from the start, better off economically than the Italian, Jewish parents could afford to keep their children in school more regularly and for longer. The poorer, less skilled Italians were more in need of their children's labor to help in the family. That Jewish children were more likely to have literate parents was also a help; the children themselves often arrived with a reading and writing knowledge of one language, making it easier to learn to read and write English than it was for southern Italian immigrant children, who generally arrived with no such skills.

Today, educational background plays a much larger role in explaining why the children of Asian immigrants are doing so well. Relatively high proportions of Asian students have highly educated parents. Although they often experience downward occupational mobility in New York, highly educated parents have higher educational expectations for their children and provide family environments more conducive to educational attainment. If their children started school in the home country, they typically attended excellent—and rigorous—institutions. Well-educated parents, moreover, are usually more sophisticated about the way the American educational system works and have an easier time, and more confidence in, navigating its complexities—and steering their children into good schools—than those with less education. In large part because so many come from professional and middle-class backgrounds, Asian New Yorkers are also doing fairly well economically. A century ago, economic resources were important because they allowed immigrants to keep their children in school; now they make it possible (or easier) to send children to private schools or to move to areas in the city (and suburbs) with better schools.

As for culture, the high value Jews placed on education—and the fact that southern Italians' heritage made them less oriented to and more skeptical about the value of book learning—helps account for the different educational achievements of the two groups. Today, most immigrant parents, in all groups, arrive with positive attitudes to education and high educational expectations for their children. Asian immigrants have particularly high aspirations for their children, through it's hard to say how much these aspirations are due to the cultural values and resources they bring to America as opposed to social-class advantages. That several national studies show Asian students outperforming all other racial/ethnic groups even after taking into account family income, household composition, and parental education, strongly suggests that culture is a factor.

Among Chinese immigrant parents, for example, hard work and discipline, not innate intelligence, are the keys to educational success. If their children study longs hours, parents believe, they can get As, and they put intense pressure on their children to excel. Confucian teaching, it is said, emphasizes discipline, family unity, and obedience to authority, all of which contribute to academic success. Children who do poorly in school bring shame to Chinese families; those who do well bring honor. According to sociologist Min Zhou, Chinese immigrant parents denounce consumption of name-brand clothes and other "too American" luxuries, but do not hesitate to pay for books, after-school programs, Chinese lessons, private tutors, music lessons, and other educationally oriented activities. Chinese and Korean immigrants also have imported after-school institutions that prepare their children for high-school admissions and college-entry exams. In Chinatown, "school after school" has, according to Zhou, become an accepted norm. According to one survey, a fifth of Korean junior and senior high school students in New York City were taking lessons after school, either in a private institution or with a private tutor.

Finally, there is the role of race. At the turn of the 20th century, race was irrelevant in explaining why Jews did better academically than Italians. Both groups were at the bottom of the city's ethnic pecking order, considered to be inferior white races. Today, the way Asian, as opposed to black and Hispanic, immigrants fit into the racial hierarchy makes a difference in the opportunities they can provide their children. Because they are not black, Asian (and white) immigrants have greater freedom in where they can live and, in turn, send their children to schools. Asians have been able to move into heavily white neighborhoods with good schools fairly easily. Moreover, their children are less likely than black or Hispanic immigrants to feel an allegiance with native minorities and be drawn into an oppositional peer culture that emphasizes racial solidarity and opposition to school rules and authorities and sees doing well academically as "acting white."

What, then, in a broad sense, can be learned from this comparison? The remembered past is clearly not the same thing as what actually transpired, and it is wrong to judge today's immigrants by a set of myths rather than actual realities. We place an added burden on the newest arrivals if we expect them to live up to a set of folk heroes and heroines from a mythical golden age of immigration. As New York, and indeed the nation as a whole, continues to be transformed by the current wave of immigration, this is something important to keep in mind.

DISCUSSION QUESTIONS

1. List two of the major differences that Foner cites comparing immigrants entering New York City at the end of the nineteenth century and those entering now.

2. What factors account for the difference in educational success for Jewish and Italian immigrants to New York early in the twentieth century?

28

Immigration's Aftermath

BY ALEJANDRO PORTES

I t is well known by now that immigration is changing the face of America. The U.S. Census Bureau reports that the number of foreign-born persons in the United States surged to 28 million in 2000 and now represents 12 percent of the total population, the highest figures in a century. In New York City, 54 percent of the population is of foreign stock—that is, immigrants and children of immigrants. The figure increases to 62 percent in the Los Angeles metropolitan area and to an amazing 72 percent in Miami. All around us, in these cities and elsewhere, the sounds of foreign languages and the sights of a kaleidoscope of cultures are readily apparent. But the long-term consequences are much less well known.

A driving force behind today's immigrant wave is the labor needs of the American economy. While those needs encompass a substantial demand for immigrant engineers and computer programmers in high-tech industries, the vast majority of today's immigrants are employed in menial, low-paying jobs. The reasons why employers in agribusiness, construction, landscaping, restaurants, hotels, and many other sectors want this foreign labor are quite understandable. Immigrants provide an abundant, diligent, docile, vulnerable, and low-cost labor pool where native workers willing to toil at the same harsh jobs for minimum pay have all but disappeared.

From *The American Prospect*, April 8, 2002, pp. 35–37. Reprinted by permission.

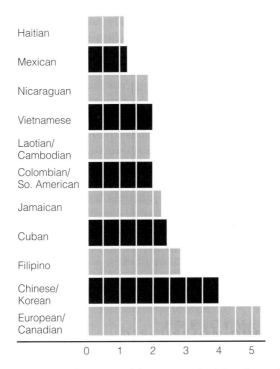

CHART 1 Relative Monthly Incomes of U.S. Immigrant Parents, 1996
(in thousands of dollars)

SOURCE: *Children of Immigrants Longitudinal Study (CILS) data*

The same agribusiness, industrial, and service firms that profit from this labor have extracted from Congress ingenious loopholes to ensure the continued immigrant flow, both legal and undocumented. Most notable is the requirement, created by the Immigration Reform and Control Act of 1986, that employers must certify that their employees have proper documents without having to establish their validity. Predictably, an entire industry of fraudulent papers has emerged. Would-be workers at construction sites and similar places often are told to go get "their papers" and return the following day. Through such subterfuges, firms demanding low-wage labor have continued to receive a steady supply, thus guaranteeing their profitability.

Defenders of this free flow portray it as a win-win process: Immigrants seeking a better life and the businesses that need their labor both gain. Opponents denounce it as a kind of invasion, as if employers did not welcome these workers. But this debate sidesteps a more consequential one: What becomes of the children of these immigrants? Business may think of them as nothing but cheap labor—indeed, that's why many business groups support pure bracero programs of temporary "guestworkers." But the vast majority of these immigrants want what everyone else wants: families.

So the short-term benefits of migration must be balanced against what happens next. The human consequences of immigration come in the form of children born to today's immigrants. Immigrant children and children of immigrants already number 14.1 million—one in five of all Americans aged 18 and under—and that figure is growing fast. A large proportion of this new second generation is growing up under conditions of severe disadvantage. The low wages that make foreign workers so attractive to employers translate into poverty and inferior schooling for their children. If these youngsters were growing up just to replace their parents as the next generation of low-paid manual workers, the present situation could go on forever. But this is not how things happen.

Children of immigrants do not grow up to be low-paid foreign workers but U.S. citizens, with English as their primary language and American-style aspirations. In my study with Rubén G. Rumbaut of more than 5,200 second-generation children in the Miami and San Diego school systems, we found that 99 percent spoke fluent English and that by age 17 less than a third maintained any fluency in their parents' tongues. Two-thirds of these youths had aspirations for a college degree and a professional-level occupation. The proportion aspiring to a postgraduate education varied significantly by nationality, but even among the most impoverished groups the figures were high.

The trouble is that poor schools, tough neighborhoods, and the lack of role models to which their parents' poverty condemns them make these lofty aspirations an unreachable dream for many. Among Mexican parents, the largest group in our survey as well as in the total immigrant population, just 2.6 percent had a college education. Even after controlling for their paltry human capital, Mexican immigrants' incomes are significantly lower than those of workers with comparable education and work experience. Similar conditions were found among other sizable immigrant groups such as Haitians, Laotians, Nicaraguans, and Cambodians. Children born to these immigrants are caught between the pitiful jobs held by their parents and an American future blocked by a lack of resources and suitable training. Add to this the effects of race discrimination—because the majority of today's second generation is nonwhite by present U.S. standards—and the stage is set for serious trouble.

The future of children growing up under these conditions is not entirely unknown, for there are several telling precedents. Journalistic and scholarly writings concerning the nearly five million young inner-city Americans who are not only unemployed but unemployable—and the more than 300,000 young men of color who crowd the American prison system—commonly neglect to mention that this underclass population did not materialize out of thin air but is the human aftermath of earlier waves of labor migration. The forebears of today's urban underclass were the southern-black and Puerto Rican migrants who moved to the industrializing cities of the Northeast and Midwest in the mid-twentieth century in search of unskilled factory employment. They too willingly performed the poorly paid menial jobs of the time and were, for that reason, preferred by industrial employers. Yet when their

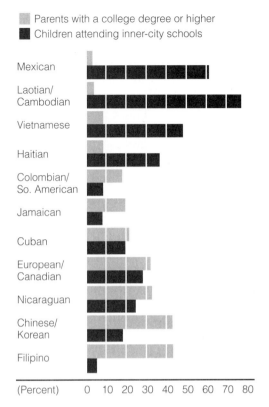

■ Parents with a college degree or higher
■ Children attending inner-city schools

(Percent)   0   10   20   30   40   50   60   70   80

CHART 2 Relative Parental Education and Children's Inner-city School
Attendance Among U.S. Immigrants, 1992

SOURCE: *CILS data*

children and grandchildren grew up, they found the road into the American middle class blocked by poverty, lack of training, and discrimination. The entrapment of this redundant population in American inner cities is the direct source of the urban underclass and the nightmarish world of drugs, gangs, and violence that these cities battle every day.

Children of poor immigrants are encountering similar and even more difficult conditions of blocked opportunity and external discrimination. In the postindustrial era, the American labor market has come to resemble a metaphoric hourglass, with job opportunities concentrated at the top (in professional and technical fields requiring an advanced education) and at the bottom (in low-paid menial services and agriculture). New migrants respond by crowding into the bottom of the hourglass, but their children, imbued with American-style aspirations, resist accepting the same jobs. This means that they must bridge in the course of a single generation the gap between their parents' low education and the college-level training required to access well-paid non-menial jobs. Those who fail, and there are likely to be many, are just a step short of the same labor-market redundancy that has trapped descendants of earlier black and Puerto Rican migrants.

Assimilation under these conditions does not lead upward into the U.S. middle class but downward into poverty and permanent disadvantage. This outcome is not the fault of immigrant parents or their children but of the objective conditions with which they must cope. All immigrants are imbued with a strong success drive—otherwise they wouldn't have made the uncertain journey to a new land—and all have high ambitions for their children. But family values and a strong work ethic do not compensate for the social conditions that these children face.

Parents' educational expectations are quite high, even higher than their children's. Expectations vary significantly by nationality, but among all groups, 50 percent or more of parents believe that their offspring will attain a college degree. Yet the resources required to achieve this lofty goal—parental education, family income, quality of schools attended—often are not there. The differences found among immigrant nationalities are illustrated in charts 1 and 2, which show the wide disparities in parents' income and education and in their children's attendance at poor inner-city school. Groups that comprise the largest and fastest-growing components of contemporary immigration, primarily Mexicans, have the lowest human-capital endowments and incomes, and their children end up attending mostly inner-city schools.

Effects of these disparities do not take long to manifest themselves in the form of school achievement and the probability of dropping out of school. Parental education and occupation are consistently strong predictors of children's school achievement. Each additional point in parental socioeconomic status (a composite of parents' education, occupation, and home ownership) increases math-test scores by 8 percentile points and reading by 9 points in early adolescence (after controlling for other variables). Living in a family with both parents present also increases performance significantly and reduces the chances of leaving school. Growing up in an intact family and attending a suburban school in early adolescence cuts down the probability of dropping out by high school by a net 11 percent, or approximately half the average dropout rate (again controlling for other variables).

Differences in academic outcomes are illustrated in chart 3, which presents math-test scores and school-inactivity rates of immigrants' children, again broken down by nationality. While the correlation is not perfect, the groups with the lowest family incomes and educational endowments—and highest probability of attending inner-city schools—also tend to produce the most disadvantaged children, both in terms of test scores and the probability of achieving a high-school diploma.

At San Diego's Hoover High, there's a group that calls itself the Crazy Brown Ladies. They wear heavy makeup, or "ghetto paint," and reserve derision for classmates striving for grades ("schoolgirls" is the Ladies' label for these lesser beings). Petite Guatemalan-born Iris de la Puente never joined the Ladies, but neither did she make it through high school. The daughter of a gardener

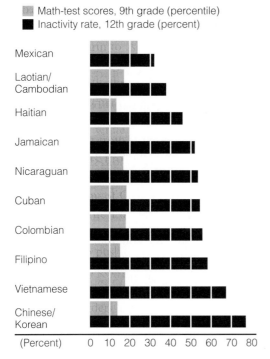

Math-test scores, 9th grade (percentile)
Inactivity rate, 12th grade (percent)

CHART 3 Relative Math-Test Scores and School Inactivity Among U.S. Immigrants' Children, 1996

SOURCE: *CILS data*

and a seamstress, she has lived alone with her mother for several years, since her father was deported and did not return. Mrs. De la Puente repeatedly exhorted Iris to stay in school, but her message was empty. The pressure of work kept the mother away from home for many hours, and her own modest education and lack of English fluency did not give her a clue how to help Iris. By ninth grade, the girl's grade-point average had fallen to a C and she was just hanging in there, hoping for a high-school diploma. When junior year rolled around, it was all over. "Going to college would be nice, but it was clear that it was not for me," Iris said. Getting a job, no matter how poorly paid, became the only option. As far as the immigrant second generation is concerned, it simply is not true that "where there's a will, there's a way." No matter how ambitious parents and children are, no matter how strong their family values and dreams of making it in America, the realities of poverty, discrimination, and poor schools become impassable barriers for many. Like Iris de la Puente, these youths find that the dream of a college education is just that. The same children growing up in inner cities encounter a ready alternative to education in the drug gangs and street culture that already saturate their environment. The emergence of a "rainbow underclass" that includes the offspring of many of today's immigrants is an ominous but distinct possibility.

The short-term economic benefits of immigration are easy to understand and equally easy to appropriate by the urban firms, ranches, and farms that employ this labor, ensuring their profitability. Absent heroic social supports, the long-term consequences are borne by children growing up under conditions of severe disadvantage and by society at large. If the United States wants to keep indulging its addiction to cheap foreign workers, it had better do so with full awareness of what comes next. For immigrants and their children are people, not just labor, and they cannot be dismissed so easily when their work is done. The aftermath of immigration depends on what happens to these children. The prospects for many, given the obstacles at hand, appear dim.

DISCUSSION QUESTIONS

1. What does Portes identify as factors that limit the educational and life options of children of low-wage immigrants?

2. Why does Portes argue that social factors are more important than values and aspirations for social mobility?

29

Are Asian Americans Becoming "White"?

BY MIN ZHOU

"I never asked to be white. I am not literally white. That is, I do not have white skin or white ancestors. I have yellow skin and yellow ancestors, hundreds of generations of them. But like so many other Asian Americans of the second generation, I find myself now the bearer of a strange new status: white, by acclimation. Thus it is that I have been described as an 'honorary white,' by other whites, and as a 'banana' by other Asians . . . to the extent that I have moved away from the periphery and toward the center of American life, I have become white inside."
—Eric Liu, The Accidental Asian (p. 34)

Are Asian Americans becoming "white?" For many public officials the answer must be yes, because they classify Asian-origin Americans with European-origin Americans for equal opportunity programs. But this classification is premature and based on false premises. Although Asian Americans as a group have attained the career and financial success equated with being white, and although many have moved next to or have even married whites, they still remain culturally distinct and suspect in a white society.

At issue is how to define Asian American and white. The term "Asian American" was coined by the late historian and activist Yuji Ichioka during the ethnic consciousness movements of the late 1960s. To adopt this identity was to reject the western-imposed label of "Oriental." Today, "Asian American" is an umbrella category that includes both U.S. citizens and immigrants whose ancestors came from Asia, east of Iran. Although widely used in public discussions, most Asian-origin Americans are ambivalent about this label, reflecting the difficulty of being American and still keeping some ethnic identity: Is one, for example, Asian American or Japanese American?

From *Contexts* 3(1): 29–37. Copyright © 2004 by the American Sociological Association.

Similarly, "white" is an arbitrary label having more to do with privilege than biology. In the United States, groups initially considered nonwhite, such as the Irish and Jews, have attained "white" membership by acquiring status and wealth. It is hardly surprising, then, that nonwhites would aspire to becoming "white" as a mark of and a tool for material success. However, becoming white can mean distancing oneself from "people of color" or disowning one's ethnicity. Pan-ethnic identities—Asian American, African American, Hispanic American—are one way the politically vocal in any group try to stem defections. But these group identities may restrain individual members' aspirations for personal advancement.

VARIETIES OF ASIAN AMERICANS

Privately, few Americans of Asian ancestry would spontaneously identify themselves as Asian, and fewer still as Asian American. They instead link their identities to specific countries of origin, such as China, Japan, Korea, the Philippines, India or Vietnam. In a study of Vietnamese youth in San Diego, for example, 53 percent identified themselves as Vietnamese, 32 percent as Vietnamese American, and only 14 percent as Asian American. But they did not take these labels lightly; nearly 60 percent of these youth considered their chosen identity as very important to them.

Some Americans of Asian ancestry have family histories in the United States longer than many Americans of Eastern or Southern European origin. However, Asian-origin Americans became numerous only after 1970, rising from 1.4 million to 11.9 million (4 percent of the total U.S. population), in 2000. Before 1970, the Asian-origin population was largely made up of Japanese, Chinese and Filipinos. Now, Americans of Chinese and Filipino ancestries are the largest subgroups (at 2.8 million and 2.4 million, respectively), followed by Indians, Koreans, Vietnamese and Japanese (at more than one million). Some 20 other national-origin groups, such as Cambodians, Pakistanis, Laotians, Thai, Indonesians and Bangladeshis, were officially counted in government statistics only after 1980; together they amounted to more than two million Americans in 2000.

The sevenfold growth of the Asian-origin population in the span of 30-odd years is primarily due to accelerated immigration following the Hart-Celler Act of 1965, which ended the national origins quota system, and the historic resettlement of Southeast Asian refugees after the Vietnam War. Currently, about 60 percent of the Asian-origin population is foreign-born (the first generation), another 28 percent are U.S.-born of foreign-born parents (the second generation), and just 12 percent were born to U.S.-born parents (the third generation and beyond).

Unlike earlier immigrants from Asia or Europe, who were mostly low-skilled laborers looking for work, today's immigrants from Asia have more varied backgrounds and come for many reasons, such as to join their families, to invest their money in the U.S. economy, to fill the demand for highly skilled labor, or to escape war, political or religious persecution and economic hardship. For example, Chinese, Taiwanese, Indian, and Filipino Americans tend to be over-represented among scientists, engineers, physicians and other skilled professionals, but less educated, low-skilled workers are more common among Vietnamese, Cambodian, Laotian, and Hmong Americans, most of whom entered the United States as refugees. While middle-class immigrants are able to start their American lives with high-paying professional careers and comfortable suburban lives, low-skilled immigrants and refugees often have to endure low-paying menial jobs and live in inner-city ghettos.

Asian Americans tend to settle in large metropolitan areas and concentrate in the West. California is home to 35 percent of all Asian Americans. But recently, other states such as Texas, Minnesota and Wisconsin, which historically received few Asian immigrants, have become destinations for Asian American settlement. Traditional ethnic enclaves, such as Chinatown, Little Tokyo, Manilatown, Koreatown, Little Phnom Penh, and Thaitown, persist or have emerged in gateway cities, helping new arrivals to cope with cultural and linguistic difficulties. However, affluent and highly-skilled immigrants tend to bypass inner-city enclaves and settle in suburbs upon arrival, belying the stereotype of the "unacculturated" immigrant. Today, more than half of the Asian-origin population is spreading out in suburbs surrounding traditional gateway cities, as well as in new urban centers of Asian settlement across the country.

Differences in national origins, timing of immigration, affluence and settlement patterns profoundly inhibit the formation of a pan-ethnic identity. Recent arrivals are less likely than those born or raised in the United States to identify as Asian American. They are also so busy settling in that they have little time to think about being Asian or Asian American, or, for that matter, white. Their diverse origins include drastic differences in languages and dialects, religions, cuisines and customs. Many national groups also bring to America their histories of conflict (such as the Japanese colonization of Korea and Taiwan, Japanese attacks on China, and the Chinese invasion of Vietnam).

Immigrants who are predominantly middle-class professionals, such as the Taiwanese and Indians, or predominantly small business owners, such as the Koreans, share few of the same concerns and priorities as those who are predominantly uneducated, low-skilled refugees, such as Cambodians and Hmong. Finally, Asian-origin people living in San Francisco or Los Angeles among many other Asians and self-conscious Asian Americans develop a stronger ethnic identity than those living in predominantly Latin Miami or predominantly European Minneapolis. A politician might get away with calling Asians "Oriental" in Miami but get into big trouble in San Francisco. All of these differences create obstacles to fostering a cohesive pan-Asian solidarity. As Yen Le Espiritu shows, pan-Asianism is primarily a political ideology of U.S.-born, American-educated, middle-class Asians rather than of Asian immigrants, who are conscious of their national origins and overburdened with their daily struggles for survival.

UNDERNEATH THE MODEL MINORITY:
"WHITE" OR "OTHER"

The celebrated "model minority" image of Asian Americans appeared in the mid-1960s, at the peak of the civil rights and the ethnic consciousness movements, but before the rising waves of immigration and refugee influx from Asia. Two articles in 1966—"Success Story, Japanese-American Style," by William Petersen in the *New York Times Magazine,* and "Success of One Minority Group in U.S.," by the *US News & World Report* staff—marked a significant departure from how Asian immigrants and their descendants had been traditionally depicted in the media. Both articles congratulated Japanese and Chinese Americans on their persistence in overcoming extreme hardships and discrimination to achieve success, unmatched even by U.S.-born whites, with "their own almost totally unaided effort" and "no help from anyone else." (The implicit contrast to other minorities was clear.) The press attributed their winning wealth and respect in American society to hard work, family solidarity, discipline, delayed gratification, non-confrontation and eschewing welfare.

This "model minority" image remains largely unchanged even in the face of new and diverse waves of immigration. The 2000 U.S. Census shows that Asian Americans continue to score remarkable economic and educational achievements. Their median household income in 1999 was more than $55,000—the highest of all racial groups, including whites—and their poverty rate was under 11 percent, the lowest of all racial groups. Moreover, 44 percent of all Asian Americans over 25 years of age had at least a bachelor's degree, 18 percentage points more than any other racial group. Strikingly, young Asian Americans, including both the children of foreign-born physicians, scientists, and professionals and those of uneducated and penniless refugees, repeatedly appear as high school valedictorians and academic decathlon winners. They also enroll in the freshman classes of prestigious universities in disproportionately large numbers. In 1998, Asian Americans, just 4 percent of the nation's population, made up more than 20 percent of the undergraduates at universities such as Berkeley, Stanford, MIT and Cal Tech. Although some ethnic groups, such as Cambodians, Lao, and Hmong, still trail behind other East and South Asians in most indicators of achievement, they too show significant signs of upward mobility. Many in the media have dubbed Asian Americans the "new Jews." Like the second-generation Jews of the past, today's children of Asian immigrants are climbing up the ladder by way of extraordinary educational achievement.

One consequence of the model-minority stereotypes is that it reinforces the myth that the United States is devoid of racism and accords equal opportunity to all, fostering the view that those who lag behind do so because of their own poor choices and inferior culture. Celebrating "model minorities" can help impede other racial minorities' demands for social justice by pitting minority groups against each other. It can also pit Asian Americans against whites. On the surface, Asian Americans seem to be on their way to becoming white, just like the offspring of earlier European immigrants. But the model-minority image implicitly casts Asian Americans as different from whites. By placing Asian Americans above whites, this image still sets them apart from other Americans, white or nonwhite, in the public mind.

There are two other less obvious effects. The model-minority stereotype holds Asian Americans to higher standards, distinguishing them from average Americans. "What's wrong with being a model minority?" a black student once asked, in a class I taught on race, "I'd rather be in the model minority than in the downtrodden minority that nobody respects." Whether people are in a model minority or a downtrodden minority, they are still judged by standards different from average Americans. Also, the model-minority stereotype places particular expectations on members of the group so labeled, channeling them to specific avenues of success, such as science and engineering. This, in turn, makes it harder for Asian Americans to pursue careers outside these designated fields. Falling into this trap, a Chinese immigrant father gets upset when his son tells him he has changed his major from engineering to English. Disregarding his son's talent for creative writing, such a father rationalizes his concern, "You have a 90 percent chance of getting a decent job with an engineering degree, but what chance would you have of earning income as a writer?" This thinking represents more than typical parental concern; it constitutes the self-fulfilling prophecy of a stereotype.

The celebration of Asian Americans rests on the perception that their success is unexpectedly high. The truth is that unusually many of them, particularly among the Chinese, Indians and Koreans, arrive as middle-class or upper middle-class immigrants. This makes it easier for them and their children to succeed and regain their middle-class status in their new homeland. The financial resources that these immigrants bring also subsidize ethnic businesses and services, such as private after-school programs. These, in turn, enable even the less fortunate members of the groups to move ahead more quickly than they would have otherwise.

NOT SO MUCH BEING "WHITE"
AS BEING AMERICAN

Most Asian Americans seem to accept that "white" is mainstream, average and normal, and they look to whites as a frame of reference for attaining higher social position. Similarly, researchers often use non-Hispanic whites as the standard against which other groups are compared, even though there is great diversity among whites, too. Like most immigrants to the United States, Asian immigrants tend to believe in the American Dream and measure their achievements materially. As a Chinese immigrant said to me in an interview, "I hope to accomplish nothing but three things: to own a home, to be my own boss, and to send my children to the Ivy League." Those with sufficient education, job skills and money manage to move into white middle-class suburban neighborhoods immediately upon arrival, while others work intensively to accumulate enough savings to move their families up and out of inner-city ethnic enclaves. Consequently, many children of Asian ancestry have lived their entire childhood in white communities, made friends with mostly white peers, and grown up speaking only English. In fact, Asian Americans are the most acculturated non-European group in the United States. By the second generation, most have lost fluency in their parents' native languages (see "English-Only Triumphs, but the Costs are High," *Contexts*, Spring 2002). David Lopez finds that in Los Angeles, more than three-quarters of second-generation Asian Americans (as opposed to one-quarter of second-generation Mexicans) speak only English at home. Asian Americans also intermarry extensively with whites and with members of other minority groups. Jennifer Lee and Frank Bean find that more than one-quarter of married Asian Americans have a partner of a different racial background, and 87 percent of those marry whites; they also find that 12 percent of all Asian Americans claim a multiracial background, compared to 2 percent of whites and 4 percent of blacks.

Even though U.S.-born or U.S.-raised Asian Americans are relatively acculturated and often intermarry with whites, they may be more ambivalent about becoming white than their immigrant parents. Many only cynically agree that "white" is synonymous with "American." A Vietnamese high school student in New Orleans told me in an interview, "An American is white. You often hear people say, hey, so-and-so is dating an 'American.' You know she's dating a white boy. If he were black, then people would say he's black." But while they recognize whites as a frame of reference, some reject the idea of becoming white themselves: "It's not so much being white as being American," commented a Korean-American student in my class on the new second generation. This aversion to becoming white is particularly common among second-generation college students who have taken ethnic studies courses, and among Asian-American community activists. However, most of the second generation continues to strive for the privileged status associated with whiteness, just like their parents. For example, most U.S.-born or U.S.-raised Chinese-American youth end up studying engineering, medicine, or law in college, believing that these areas of study guarantee a middle-class life.

Second-generation Asian Americans are also more conscious of the disadvantages associated with being nonwhite than their parents, who as immigrants tend to be optimistic about overcoming the disadvantages of this status. As a Chinese-American woman points out from her own experience, "The truth is, no matter how American you think you are or try to be, if you have almond-shaped eyes, straight black hair, and a yellow complexion, you are a foreigner by default. . . . You can certainly be as good as or even better than whites, but you will never become accepted as white." This remark echoes a commonly-held frustration among second-generation, U.S.-born Asians who detest being treated as immigrants or foreigners. Their experience suggests that whitening has more to do with the beliefs of white America, than with the actual situation of Asian Americans. Speaking perfect English, adopting mainstream cultural values, and even intermarrying members of the dominant group may help reduce this "otherness" for particular individuals, but it has little effect on the group as a whole. New stereotypes can emerge and un-whiten Asian Americans, no matter how "successful" and "assimilated" they have become. For example, Congressman David Wu once was invited by the Asian-American employees of the U.S. Department of Energy to give a speech in celebration of Asian-American Heritage Month. Yet, he and his Asian-American staff were not allowed into the department building, even after pre-senting their congressional Identification, and were repeatedly asked about their citizenship and country of origin. They were told that this was standard procedure for the Department of Energy and that a congressional ID card was not a reliable document. The next day, a congressman of Italian descent was allowed to enter the same building with his congressional ID, no questions asked.

The stereotype of the "honorary white" or model minority goes hand-in-hand with that of the "forever foreigner." Today, globalization and U.S. –Asia relations, combined with continually high rates of immigration, affect how Asian Americans are perceived in American society. Many historical stereo-types, such as the "yellow peril" and "Fu Manchu" still exist in contemporary American life, as revealed in such highly publicized incidents as the murder of Vincent Chin, a Chinese American mistaken for Japanese and beaten to death by a disgruntled while auto worker in the 1980s; the trial of Wen Ho Lee, a nuclear scientist suspected of spying for the Chinese government in the mid-1990s; the 1996 presidential campaign finance scandal, which implicated Asian Americans in funneling foreign contributions to the Clinton campaign; and most recently, in 2001, the Abercrombie & Fitch t-shirts that depicted Asian cartoon characters in stereotypically negative ways, with slanted eyes, thick glasses and heavy Asian accents. Ironically, the ambivalent, conditional nature of their acceptance by whites prompts many Asian Americans to organize pan-ethnically to fight back—which consequently heightens their racial dis-tinctiveness. So becoming white or not is beside the point. The bottom line is: Americans of Asian ancestry still have to constantly prove that they truly are loyal Americans.

REFERENCES

Liu, Eric. 1988. *The Accidental Asian*. New York: Random House.

DISCUSSION QUESTIONS

1. What barriers to a pan-ethnic identification among recent Asian immi-grants does Zhou identify?

2. What does Zhou argue are the sociological explanations for the economic success of the so-called "model minority"?

30

Wealth and Racial Stratification

BY MELVIN L. OLIVER AND THOMAS M. SHAPIRO

Income is what the average American family uses to reproduce daily existence in the form of shelter, food, clothing, and other necessities. In contrast, *wealth* is a storehouse of resources, it's what families own and use to produce income. Wealth signifies a control of financial resources that, when combined with income, provides the means and the opportunity to secure the "good life" in whatever form is needed—education, business, training, justice, health, material comfort, and so on. In this sense, wealth is a special form of money not usually used to purchase milk and shoes or other life necessities; rather it is used to create opportunities, secure a desired stature and standard of living, and pass class status along to one's children.

Wealth has been a neglected dimension of social science's concern with the economic and social status of Americans in general and racial minorities in particular. . . . During the past decade, sociologists and economists have begun to pay more attention to the issue of wealth. The growing concentration of wealth at the top, and the growing racial wealth gap, have become important public-policy issues that undergird many political debates but, unfortunately, not many policy discussions. . . .

Understanding racial inequality, with respect to the distribution of power, economic resources, and life chances, is a prime concern of the social sciences. Most empirical research on racial inequality has focused on the economic dimension, which is not surprising considering the centrality of this component for life chances and well-being in an industrial society. . . .

Income is a tidy and valuable gauge of the state of present economic inequality. Indeed, a strong case can be made that reducing racial discrimination in the labor market has resulted in increasing the income of racial minorities and, thus, narrowing the hourly wage gap between minorities and Whites. The command of resources that wealth entails, however, is more encompassing than income or education, and closer in meaning and theoretical significance to the traditional connotation of economic well-being and access to life chances. . . .

As important is the fact that wealth taps not only contemporary resources, but also material assets that have historic origins and future implications. Private wealth thus captures inequality that is the product of the past, often passed down from generation to generation. Conceptualizing racial inequality through wealth revolutionizes the concept of the nature and magnitude of inequality, and of whether it is decreasing or increasing. Although most recent analyses have concluded that contemporary class-based factors are most important in understanding the sources of continuing racial inequality, a focus on wealth sheds light on both the historical and the contemporary impacts not only of class but also of race. Income is an important indicator of racial inequality; wealth allows an examination of racial stratification. . . .

HISTORICAL TRENDS AND CONTEXT OF WEALTH DISTRIBUTION IN THE UNITED STATES

Wealth inequality is today, and always has been, more extreme than income inequality. Wealth inequality is more lopsided in the United States than in Europe. Recent trends in asset ownership do not alleviate inequality concerns or issues. In general, inequality in asset ownership in the United States between the bottom and top of the distribution domain has been growing. Wealth inequality was at a 60-year high in 1989, with the top 1 percent of U.S. citizens controlling 39 percent of total U.S. household wealth. The richest 1 percent owned 48 percent of the total. These themes have been amply described in the work of Wolff (1994, 1996a, 1996b). Household wealth inequality increased sharply between 1983 and 1989. There was a modest attenuation in 1992, but the level of wealth concentration was still greater in 1992 than in 1983.[1]

---

[1] Editors' note: This gap has increased since.

Black Wealth/White Wealth (Oliver and Shapiro, 1995a) decomposed the results of a regression analysis to give Blacks and Whites the same level of income, human capital, demographic, family, and other characteristics. . . . Taking the average Black household and endowing it with the same income, age, occupational, educational, and other attributes as the average White household still leaves a $25,794 racial gap in mean net financial assets. These residual gaps should not be cast wholly to racial dynamics; nonetheless, the . . . analyses offer a powerful argument to directly link race in the American experience to the wealth-creation process.

As important is the finding that more than two-thirds of Blacks have no net financial assets, compared to less than one-third of Whites. This near absence of assets has extreme consequences for the economic and social well-being of the Black community, and of the ability of families to plan for future social mobility. If the average Black household were to lose an income stream, the family would not be able to support itself without access to public support. At their current levels of net financial assets, nearly 80 percent of Black families would not be able to survive at poverty-level consumption for three months. Comparable figures for Whites—although large in their own right—are one-half that of Blacks. Thinking about the social welfare of children, these figures take on more urgency. Nine out of ten Black children live in households that have less than three months of poverty-level net financial assets; nearly two-thirds live in households with zero or negative net financial assets (Oliver and Shapiro, 1989, 1990, 1995a, 1995b). . . .

Because home ownership plays such a large role in the wealth portfolios of American families, it is a prime source of the differences between Black and White net worth. Home ownership rates for Blacks are 20 percent lower than rates for Whites; hence, Blacks possess less of this important source of equity. Discrimination in the process of securing home ownership plays a significant role in how assets are generated and accumulated. The reality of residential segregation also plays an important role in the way home ownership figures in the wealth portfolio of Blacks. Because Blacks live, for the most part, in segregated areas, the value of their homes is less, demand for them is less, and thus their equity is less (Oliver and Shapiro, 1995b; Massey and Denton, 1994). . . .

Similar findings on gross differences between Hispanics and Whites also have been uncovered (Eller and Fraser, 1995; Flippen and Tienda, 1997; O'Toole, 1998; Grant, 2000). Hispanics have slightly higher, but not statistically different, net worth figures than Blacks, based on the 1993 Survey of Income and Program Participation (SIPP); however, these findings are not sufficiently nuanced to capture the diversity of the Hispanic population. Data from the Los Angeles Survey of Urban Inequality show substantial differences in assets and net financial assets between recent immigrants who are primarily from Mexico and Central America and U.S.-born Hispanics (Grant, 2000).

Likewise, place of birth and regional differences among Hispanic groups also complicate a straightforward interpretation of this national-level finding. For example, Cuban Americans, we would hypothesize, have net worth figures comparable to Whites because of their dominance in an ethnic economy in which they own small and medium-sized businesses (Portes and Rumbaut, 1990). They have a far different set of economic life chances than Blacks and other recent Hispanic immigrants by way of their more significant wealth accumulation. For recent Hispanic immigrants, these figures suggest real vulnerability for the economic security of their households and children. . . .

The case of Asians is quite similar to that for Hispanics, in that it is necessary to be mindful of their diversity, in terms of both national origin and immigrant status. Changes in immigration rules have favored those who bring assets into the country over those without assets; as a consequence, recent immigrants, from Korea, for example, are primarily individuals and families with assets, and once they arrive, they convert these assets into other asset-producing activities—e.g., small businesses. . . . Data from Los Angeles again underscore the importance of immigrant status and place of birth. U.S.-born Asians have both net worth and net financial assets approaching those of White Los Angelenos; foreign-born Asians, however, report lower wealth than U.S.-born Asians but higher wealth than all other ethnic and racial groups (Grant, 2000).

American Indians form a unique case when it comes to assets. They are asset rich but control little of these assets. Most Indian assets are held in tribal or individual Indian trust (Office of Trust Responsibilities, 1995). Thus, any accounting of the assets of individual Indian households is nearly impossible to calculate, given their small population and these "hidden" assets.

The dearth of studies of wealth in the United States has hampered efforts to develop both wealth theory and information. For more than 100 years, the prime sources concerning wealth status came from estate tax records, biographies of the super rich, various listings of the wealthiest, and like sources. In other words, something was known about those who possessed abundant amounts of wealth, but virtually nothing was known about the wealth status of average American families. . . .

UNDERSTANDING THE TRENDS

. . . First, racial differences are important in terms of wealth. In the case of Blacks, for whom there is the most clear-cut data, it is obvious that racial factors are implicit in these findings. Second, national origin and immigration status may explain a great deal of the differences among Hispanic and Asian groups; but this is a key area of research that needs to be explored further. Third, American Indians pose a different set of challenges for understanding wealth accumulation because the Indian community has significant wealth in terms of land and assets, but those assets are under federal government control. The public policies needed to address economic inequality for American Indians would require a legal solution.

The analytical power derived from examining racial stratifications through the lens of wealth is most obvious in the case of Blacks. Blacks and Whites face different structures of investment opportunity, which have been affected historically and contemporaneously by both class and race. Three concepts can be used to provide a sociologically grounded approach to understanding racial differences in wealth accumulation. . . .

The first concept, "racialization of state policy," refers to how state policy has impaired the ability of many Blacks to accumulate wealth—and discouraged them from doing so—from the beginning of slavery throughout American history. From the first codified decision to enslave Blacks, to the local ordinances that barred Blacks from certain occupations, to the welfare-state policies of the recent past that discouraged wealth accumulation, the state has erected major barriers to Black economic self-sufficiency. In particular, state policy has structured the context within which it has been possible to acquire land, build community, and generate wealth. . . .

One of the key policies of the federal government that encourages home ownership is the deductibility of mortgage interest on homes (a companion is the deferral of capital gains on the sale of principal residences). The state subsidizes home ownership by allowing the marginal tax rate to be deducted from a family's income taxes. Housing policies that encourage ownership are a part of the "hidden" welfare state that cost the federal government about $94 billion in fiscal expenditures (Howard, 1997). Home ownership may very well be sound social policy, but it is an uneven process that clearly benefits some groups over others (Jackman and Jackman, 1980; Ong and Grigsby, 1988). . . .

The second concept, the "economic detour," helps explain the relatively low level of entrepreneurship among, and the small scale of the businesses owned by, Black Americans. Although Blacks have traditionally sought out opportunities for self-employment, they have faced an environment, especially from the postbellum period to the middle of the twentieth century, in which they were restricted by laws from participation in business as free economic agents (Butler, 1991). These policies had a devastating impact on the ability of Blacks to build and maintain successful enterprises—the kind that anchor communities and spur economic development. Not only were Blacks limited to a restricted Black market, to which others also had easy access, but they were unable to tap the more lucrative and expansive mainstream White market. When businesses were developed that competed in size and scope with White businesses, intimidation and, in many cases, violence were used to curtail their expansion or destroy them altogether. The lack of major assets and indigenous community economic development has thus played a crucial role in limiting the wealth-accumulating ability of Blacks.

The past certainly casts a long shadow on the economic status of Blacks, but discrimination is not limited to the past. Recent studies of housing discrimination (Yinger, 1995, 1998), consumer markets (Yinger 1998), employment practices (Darity and Mason, 1998), and mortgage lending (Ladd, 1998) indicate pervasive and persistent discrimination—individual and institutional—in the last years of the twentieth century; in fact, the "economic detour" concept still operates in the most important way typical American families accumulate assets—home equity.

The third concept, the "sedimentation of racial inequality" is synthetic in nature. The idea is that, in pivotal ways, the cumulative effects of the past have ostensibly cemented Blacks to the bottom of society's economic hierarchy. A history of low wages (Leiberson, 1980), poor schooling and segregation affected not one generation of Blacks but practically all Blacks well into the twentieth century. The best indicator of this is wealth—or lack thereof. Wealth is one indicator of material disparity that captures the historical legacy of low wages, personal and organizational discrimination, and institutionalized racism. The low level of wealth accumulation evidenced by current generations of Blacks best represents the economic status of Blacks in the American social structure. In contrast, Whites in general—but well-off Whites in particular—had far greater structured opportunities to amass assets and use their secure financial status to pass their wealth and its benefits from generation to generation. What is often not acknowledged is that the same social system that fosters the accumulation of private wealth for many Whites denies it to Blacks, thus forging an intimate connection between White wealth accumulation and Black poverty. Just as Blacks have had "cumulative disadvantages," many Whites have had "cumulative advantages." Because wealth builds over a lifetime and is then passed along to kin, it is an essential indicator of Black economic well-being.

An understanding of the trends focuses attention on how past racial inequality in policy and practices translates into current racial stratification in the form of vastly different wealth resources for Black and White families, even among those with roughly equal accomplishments. Social injustice is not just an artifact of the past; contemporary institutional discrimination contributes to generating and maintaining the racial wealth gap.

REFERENCES

Butler, J. 1991. *Entrepreneurship and Self-Help Among Black Americans: A Reconsideration of Race and Economics.* Albany, N.Y.: State University of New York Press.

Darity, W., Jr., and P. Mason. 1998. Evidence on discrimination in employment: Codes of color, codes of gender. *Journal of Economic Perspectives* 12(2):63-90.

Eller, T., and W. Fraser. 1995. Asset Ownership of Households: 1993. *U.S. Bureau of the Census. Current Population Reports, P70-47.* Washington, D.C.: U.S. Government Printing Office.

Flippen, C., and M. Tienda. 1997. Racial and Ethnic Differences in Wealth Among the Elderly. Paper presented at the 1997 Annual Meeting of the Population Association of America, Washington, D.C.

Grant, D. 2000. A demographic portrait of Los Angeles, 1970-1990. In *Prismatic Metropolis: Analyzing Inequality in Los Angeles,* L. Bobo, M. Oliver, J. Johnson Jr., and A. Valenzuela, eds. New York: Russell Sage Foundation.

Howard, C. 1997. *The Hidden Welfare State: Tax Expenditures and Social Policy in the United States.* Princeton: Princeton University Press.

Jackman, M., and R. Jackman. 1980. Racial inequalities in home ownership. *Social Forces* 58:1221-1233.

Ladd, H. 1998. Evidence on discrimination in mortgage lending. *Journal of Economic Perspectives* 12(2):41-62.

Leiberson, S. 1980. *A Piece of the Pie.* Berkeley: University of California Press.

Massey, D., and N. Denton. 1994. *American Apartheid: Segregation and the Making of the Underclass.* Cambridge: Harvard University Press.

O'Toole, B. 1998. Family net asset levels in the greater Boston region. Paper presented at the Greater Boston Social Survey Community Conference, John F. Kennedy Library, Boston, Mass., November.

Office of Trust Responsibilities. 1995. *Annual Report of Indian Lands.* Washington, D.C.: U.S. Department of the Interior.

Oliver, M., and T. Shapiro. 1989. Race and wealth, *Review of Black Political Economy* 17:5-25.

———1990. Wealth of a nation: At least one-third of households are asset poor. *American Journal of Economics and Sociology* 49:129-151.

———1995a. *Black Wealth/White Wealth: A New Perspective on Racial Inequality.* New York: Routledge.

———1995b. Them that's got shall get. In *Research in Politics and Society,* M. Oliver R. Ratcliff, and T. Shapiro, eds. Greenwich, Conn.: JAI Press Vol. 5.

Ong, P. and E. Grigsby. 1988. Race and life cycle effects on home ownership in Los Angeles, 1970 to 1980. *Urban Affairs Quarterly* 23:601-615.

Portes, A., and Rumbaut, R. 1990. *Immigrant America.* Berkeley: University of California Press.

Wolff, E. 1994. Trends in household wealth in the United States, 1962-1983 and 1983-1989. *Review of Income and Wealth* 40:143-174.

———1996a. *Top Heavy: A Study of Increasing Inequality of Wealth in America.* Updated and expanded edition. New York: Free Press.

———1996b. International comparisons of wealth inequality. *Review of Income and Wealth* 42:433-451.

Yinger, J. 1995. *Closed Doors, Opportunities Lost: The Continuing Costs of Housing Discrimination.* New York: Russell Sage Foundation.

———1998. Evidence on discrimination in consumer markets. *Journal of Economic Perspectives* 12(2):23-40.

DISCUSSION QUESTIONS

1. What advantages does an analysis of wealth add to the study of racial stratification?

2. Why do Oliver and Shapiro think that differences in wealth between Black, Asian, Hispanics, Native Americans, and Whites are related to a history of past practices?

31

The Making
of Groveland

BY MARY PATTILLO-McCOY

The Great Migration—the movement of blacks from the South to
urban centers in the North from the start of World War I through the
... 1960s—has had a profound impact on the racial makeup of northern
(and western) receiving cities. The historical paintings of the African American
artist Jacob Lawrence depict the southern train stations crowded with families
looking for a better life. Black newspapers published in the North fueled the
migration fever with announcements that there were jobs to be had....

In many northern cities like Chicago, this growing population was rele-
gated to geographic space that did not increase with the same rapidity. With
no new housing being built from the depression era through World War II,
the Black Belt became dangerously overcrowded. But with the ending of the
war, blacks and whites alike started looking for new homes, changing the racial
geography of the city. The Black Belt grew, and Black Metropolis began to
claim more territory as its own. Between 1950 and 1970, block after block,
neighborhood after neighborhood dramatically changed from white to
African American occupancy. Groveland lay in the path of expansion of
Chicago's South Side black community. One Groveland resident put it simply:
"There'd be about seven or eight of us [blacks] and about thirty of them
[whites]. But, you know, you'd see a white family would move out and a black

From *Black Picket Fences: Privilege and Peril among the Black Middle Class* (Chicago, IL:
University of Chicago Press, 1990), pp. 31–38. Reprinted by permission of University of
Chicago Press.

one moved in, and soon it was more and more of us." Census statistics bear out this story. In 1960, there were only six blacks in all of Groveland out of a total population of 12,710. By 1970, just ten years later, blacks made up over 80 percent of the neighborhood's population. By 1990, the neighborhoods was 98 percent black. The neighborhoods that surround Groveland went through similar transformations, some a bit sooner and some later.

Many Chicago neighborhoods experienced some kind of racial or ethnic change during this unstable period, and each employed specific ways of addressing such changes, including violence. Yet in stark contrast to the fire-bombings, mob attacks, and nightly protests mounted in other Chicago areas, racial changes in Groveland proceeded with little turbulence. Some of Groveland's current residents were youngsters when their families moved them into the neighborhood. They remember being chased by groups of white youth, or being the targets of glass milk bottles hurled by their white neighbors, but these were isolated incidents. There were no organized efforts in Groveland to impede the geographic advances of African Americans. In fact, the only organized efforts on the part of Groveland residents were for positive ends. The white priest of the local Catholic church, along with a concerned black parent who lived across the street from the church, created an interracial baseball league to foster friendships between the new and the old residents. One white resident, who later left the neighborhood to enter the priesthood, recalled the role of the Catholic church in easing tensions: "I remember when the neighborhood, as they say, changed. Monsignor Welch said, 'The people who are coming are God's people, and if anything negative happens you will have to answer to me.' Mom and I stayed.". . .

Groveland did not violently erupt during this transition because the African American newcomers shared many qualities with the former residents of the neighborhood. The predominantly single-family housing and "suburban-style" living in Groveland and the surrounding neighborhoods attracted black families who were able to purchase homes rather than rent. The percentages of families who own their homes in Groveland has not dropped below 70 percent since 1950. In fact, the *only* neighborhood characteristic that dramatically changed during this transition period was the race of its residents. Between 1960 and 1970, the median years of schooling *increased* from 12.1 years to 12.4 years. During the same time, unemployment jumped slightly, from just over 2 percent to 4 percent, but the latter figure included unemployment for both males and females, whereas the former counted only male unemployment. Finally, the percentage of people living below the poverty level *fell* from 6 percent in 1960 to 5 percent in 1970. In many ways, the new black residents were slightly better off economically and educationally than the whites who were leaving. . . .

Many of the first black residents to move to Groveland in the 1960s are still there in the 1990s. According to the 1990 census, over 40 percent of Grovelandites moved into their current residences before 1970. The neighborhood's median age of thirty-six reflects this sizable contingent of older adults, many of whom have retired. This stability has translated into enduring institutions such as strong churches and schools, an active and productive business association and business district, and a responsive neighborhood park. Also, residents have close family and friendship ties. For example, Anna Morris moved to Groveland with her parents and sisters when she was a young girl in the early 1960s. Now approaching forty, Mrs. Morris is raising her own family in Groveland. With such a lengthy family investment in the neighborhood, she easily described her circle of family and friends in both network and geographic terms. "You'd never know it, but it's like one big family around here," she began after a game of volleyball at the Groveland Park gym. "You'd never know all the people that's related." Filling in the details, Anna Morris reported that Diedra, Lucky, and Lance—all regulars at the gym—were brothers and sister. And Spider, who worked at the field house, had shared a locker in high school with Mrs. Morris's own sister, Julie. "'Cause see I live on this block," she explained as she positioned her left hand as a marker of the street where she lived. "And my mother lives on this block." She held her right hand parallel to her left, showing that she and her mother lived only one block apart. Even her husband's family lives near: "And then Moe's mama lives on this block," she continued, crossing her left hand over her right to indicate her mother-in-law's street, one block away. As Mrs. Morris's family demonstrates, the social fabric in Groveland is thick, nurtured by high rates of residential stability and the growth of extended families.

Groveland is not, however, an island of longtime primary relationships, quiet streets, and committed institutions. A congestion of large auto-body shops and retail stores at the north end and a commercial park at the south end surround the interior of brick homes and green grasses. These physical boundaries are formidable, but not impermeable. . . .

Also, Groveland is not the same neighborhood it was when the process of racial change was nearly complete in 1970. The changes in the economy that have had a profound impact on urban black communities have also affected Groveland, albeit somewhat indirectly. Deindustrialization, the movement of manufacturing jobs to the suburban ring, and the general stagnation of the economy during the 1970s have left urban black neighborhoods with few job options for those with limited skills. The neighborhoods hardest hit by economic changes have lost business; housing has deteriorated and institutions cannot survive, leading to the rise in various social problems. Groveland has been somewhat buffered from the immediate impact of these structural changes. In 1980, almost two-thirds of Groveland's population were employed in white-collar occupations. This group was more likely to benefit, rather than suffer, from economic restructuring that favored technical and professional experience. Still, the neighborhood's professionals and managers live beside residents in lower service positions and manufacturing jobs; the two groups have had to deal with economic changes together.

The deteriorating economic and social conditions in the poorest areas prompt the residents of those areas to continue their attempts to escape, usually into neighborhoods like Groveland. While the racial transformation of Groveland occurred rapidly, class changes in the neighborhood are developing more gradually. They do not, however, go unnoticed by residents. The continuous movement of families out of nearby poor neighborhoods and into Groveland has brought economic heterogeneity to Groveland and neighborhoods like it around the periphery of Chicago's core ghetto. . . .

Twenty-eight-year-old Tommy Smith grew up in the northernmost section of Groveland, closest to the expanding poverty core. His description of the recent changes on his block suggests an unsure future for the neighborhood.

> There may be five houses on the block—no six—six houses on the block that once upon a time was full with residents that grew up and started at the same time as me. They either died or moved away. Out of that, one of those houses is knocked down, one is vacant, one is burned down, and one is empty on a Section 8 listing, and one is, two are empty. . . . And, the people who did move in two of the vacant houses were low-income-housed people. And in that, again, they just did not take care of their property. And did not care much about the individuals surrounding them. And, no, I did not know them very well.

Although most of the streets in Groveland are not experiencing such visible decline, the interrelated issues of population succession, property maintenance, and low-income newcomers are often raised among residents. Most academic examinations of neighborhood change privilege processes of *racial* change, but black neighborhoods can experience class-based changes that similarly threaten their stability. The containment of black urban (and even suburban) communities as a result of large discriminatory forces ensures that black middle-class neighborhoods are the most vulnerable to processes of class-based destabilization. Groveland parents, like Tommy Smith, devise strategies to manage growing heterogeneity. Other families have picked up and moved out of Groveland, often headed for Chicago's southern suburbs. These movers are replicating the patterns of out-migration and creating new black communities on the suburban periphery of the established Black Belt.

The heterogeneity that some residents are fleeing is apparent when comparing the income distribution in Groveland to the distribution in Beltway, a white middle-class neighborhood in Chicago. Even though the median family income in Groveland is slightly higher than the median family income in Beltway, the poverty rate in Groveland is *four times* the poverty rate in Beltway. The concurrence of these facts is possible because in Groveland, relatively high-earning black families live alongside poor families, whereas Beltway's income distribution contains a clustering of families in the center of the distribution. Groveland's hollowed-out distribution, compared to Beltway's bell-shaped curve, illustrates a polarization of the black class structure. At the neighborhood level, the higher poverty rate and more disparate income distribution translate into a shaky balance of residents. Most Grovelandites (poor and middle class) share the desire to keep the neighborhood safe, clean, and thriving, but a smaller proportion than in Beltway have the financial resources to contribute to these goals. The higher poverty rate also places a certain strain on public and social services, most especially the schools, a strain with which the white neighborhood is less familiar. These differences underlie many of the pressures that black middle-class families face that similar white families do not.

Overall, then, the Great Migration, the growth of the Black Belt, constraints on the exodus of middle-class blacks, and economic restructuring have all contributed to the present makeup of Groveland—a racially homogeneous, economically diverse neighborhood.

REFERENCES

Grant, David, Melvin Oliver, and Angela James. 1996. "African Americans: Social and Economic Bifurcation." pp. 379–409 in *Ethnic Los Angeles.* Roger Waldinger and Mehdi Bozorgmehr, eds. New York: Russell Sage Foundation.

Wilson, William Julius. 1978. *The Declining Significance of Race: Blacks and Changing American Institutions.* Chicago: University of Chicago Press.

DISCUSSION QUESTIONS

1. Describe the Black residents who initially moved into the formerly predominantly White area of Groveland during the 1960s.

2. How did this community change through the 1970s and 1980s?

3. How does Pattillo-McCoy think that Black middle-class neighborhoods differ from White middle-class neighborhoods?

32

A Tale of Two Classes

The Socio-Economic Divide among Black Americans Under 35

BY CELESTE M. WATKINS

For one segment of the generation of African Americans under 35, the current era is the best of times. Credentialed with college as well as graduate- and professional-school degrees, skilled in the use of the latest technological wizardry, possessed of the "right" jobs with the "right" potential, and supported by competitive salaries, they have benefited enormously from the expanded opportunity American society now offers more and more African Americans—and they readily express the self-confidence such success breeds.

But what is the apt characterization of the current era for the other segment of African Americans under 35?

For this group—who are not credentialed with higher education degrees, and thus, who are not a part of the heady swirl of the white-collar world; who may not have even high school diplomas, or, if they do, have neither the education nor the training that would make those diplomas meaningful; who are consigned now to the low-wage service jobs of the economy, if they have a job at all—the right words or phrases are far more difficult to select.

For the latter group, is the current era a new beginning, a climb out of the material and psychological swamp of continual un- and underemployment? The sharp, quick decline in the black unemployment rate in the late 1990s, powered by many of the so-called hard-core unemployed taking the low-wage

From *The State of Black America 2001*, ed. by Lee Daniels (New York: National Urban League, 2001), pp. 67–81. Reprinted by permission.

service jobs the booming economy offered, showed that most of those in this profoundly disparaged cohort know all about the value of work, want to work, and will whenever they are given a chance.

Will they now, in the period when the air is thick with job cutbacks among American companies and foreboding about a recession, still be given a chance?

Or will the American economy return to its old "tradition" of not just double-digit black unemployment, but high double-digit black unemployment?

And if that happens, what will Black America do?

This issue—the future of those in their generation who have had relatively little opportunity—is the most critical question under-35 African Americans face. For the ways in which blacks of different socioeconomic groups seek to deal with and help each other will, in large measure, determine the future economic, social, and political health of Black America itself. . . .

It is not, of course, as if class divisions are something new to Black America. What are new, however, are their scope and the forces behind it. When William Julius Wilson argued this very point in his groundbreaking 1978 book *The Declining Significance of Race*, asserting that the life chances of individual African Americans have more to do with their economic class positions than their day-to-day encounters with racism, he faced a wave of criticisms.[1] . . .

Years later, Wilson's argument is less controversial. In fact, it is taken as a widely agreed upon thesis by scholars and non-scholars alike. Social scientists point out the widening class gap within the African-American community as income inequality increases across all American households. They highlight the implications of such a gap in terms of varying access to education, jobs, and the networks that can lead to socioeconomic mobility. Others, particularly blacks themselves, notice the gap as well, often in the diversity of class presentation in public spaces and within African-American familial networks. . . .

While the condition of poorer blacks has occupied social scientists for decades, increased attention is now being paid to the experiences of working-class, middle-class, and upper-class blacks. These more precise ways of defining the experiences of African-Americans have led to at least three conclusions. First, as Wilson argued, the opportunities and experiences of African Americans heavily depend on their class status. Second, within the black community, the gap is indeed widening between the "haves" and the "have-nots." Third, such a widening gap has major implications for the interclass relations of blacks as well as for the broader policy agenda of the African-American community.

THE DEVELOPMENT OF THE DIVIDE: THE EVOLUTION OF CLASS IN BLACK AMERICA

. . . Segregation encouraged the interaction of classes within the black community and served to offset some of the effects of the systematic economic and political disempowerment of blacks. Black business owners, doctors and other professionals serviced predominately, and often exclusively, black clientele.

Poorer blacks often lived in the same or in close-by neighborhoods to blacks that were more affluent. While the income of many blacks was derived almost solely from black clientele, others challenged the hegemonic arrangements that kept them out of the industrial economy, eventually opening access for blacks to many of the jobs that provided economic stability and potential socioeconomic mobility. By the mid-1900s that challenge included increasingly successful efforts to wrest their share of local political power out of the hands of predominately white political machines. . . .

Following World War II, large structural changes in the economy had major implications for the American class structure, affecting the African-American community significantly. In fact, the largest changes in black mobility occurred in the 1950s and 1960s. Increases in African-American employment in the government and private sector, sparked by the passage of the Fair Employment Practices Act and the economic expansion after the war, sharply increased the size of the African-American working and middle classes. By the 1970s, some suburban communities opened to African Americans, allowing the more affluent members of predominately African-American neighborhoods to move. In addition, increased economic opportunity through the gains of the Civil Rights Movement and affirmative action programs gave many African Americans increased opportunities to solidify their positions in the American middle class.

At the same time, however, economic shifts and public policies in the 1970s and 1980s intensified the stressful economic situation of the inner-city communities where many lower and some working-class blacks remained. Hit hard by economic recessions and inflation, poverty rates for the nation's central cities rose from 12.7 percent to 19 percent between 1969 and 1985. . . .

Throughout the 1990s, scholars and journalists paid increased attention to those at the bottom of the economic ladder in their writings. Terms such as "the underclass" and "ghetto poor" were used to describe those inhabiting the core of America's central cities who, for the most part, had been shut out of the technological and economic advances of the past decades.[2] This group had evolved as a result of current and historic discrimination, the dynamics of the low-wage labor market, the decentralization of businesses, middle- and working-class outmigration, un- and underemployment, and limited education and skills. These macro-structural changes were linked to the concentration of poverty in inner cities as well as the rise in black female-headed families, out-of-wedlock births, and crime. Expressing long-term spells of poverty and welfare receipt, this group was a small segment of the low-income African-American families residing in inner cities, yet received major attention by scholars and policy makers alike.

The result of all these forces and developments is a more variegated class structure within Black America, with each stratum facing similar but often distinct challenges. In 1998, 26 percent of African Americans (9.1 million) were living in poverty, compared with 8 percent of non-Hispanic whites (15.8 million).[3] The condition of the black lower class and ghetto poor persists, with scholars fine-tuning their understanding of and policy prescriptions for this group, composed of not only the chronically un- and underemployed but also the working poor. . . .

Scholars and journalists have also raised concerns about the actual economic and political power of the black middle class and have highlighted that while the black-white income gap my be decreasing, the wealth gap remains sizable.[4] In addition, while African Americans have gained political power in central cities, the surrounding, predominantly-white suburban areas have continued to attract both businesses and residents, challenging the influence that these localities can wield. In 1999, while the median income of African-American households was $27,910, the highest ever recorded, it still significantly trailed the $44,366 median income of white non-Hispanic households.[5]

Yet, one cannot deny that, at the same time, African Americans have gained significant ground on both economic and political fronts. . . . More and more blacks have become a part of the nation's political life through public service in elected and appointed positions. In 1999, about 47 percent of African-American householders were homeowners.[6] . . . In fact, about 6.1 percent of black households had annual incomes over $100,000 in 1999.[7]

THE WIDENING CLASS GAP: ISSUES AND COMPLICATIONS

. . . Is a widening class gap among African Americans necessarily problematic? Some would argue that because income inequality is rising across American households, the gap should not be a concern. In fact, some would argue that such inequality is simply further evidence that the trends for blacks are becoming more in line with the trends for the rest of society. Perhaps such a gap will encourage those African Americans on the bottom of the economic ladder to strive for mobility as they see more and more blacks taking their places on the higher rungs. . . .

While these points are well taken, a widening class gap within Black America presents similar problems as rising income inequality crosses all American households. While class mobility may be a beneficial and favorable goal, when that mobility occurs only for those already on the higher rungs of the economic ladder, intractable inequality rises. Inequality of outcomes (in this case, income) on the surface may not be a problem. However, in American society, too often inequality of outcomes leads to and reinforces inequality of opportunities, creating not a class system but a caste-like one. As some climb to the top, others will be left on the bottom with very few opportunities for upward mobility and a future so bleak that deprivation, particularly in a land of wealth, could potentially lead to desperation.

African Americans under 35 must forcefully come to grips with this issue. The reason is that although the problem is not new, they have come of age in an American society in which they have opportunities to compose their racial identities at a time of largely implicit rather than explicit racial tension. While racism is still very much a part of this generation's consciousness and overall life experiences, its often subtle manifestations in the members' daily experiences has enabled some to create political consciousness that may not hinge entirely on racial solidarity. Many note the class gap and wrestle with it in their daily encounters with African Americans higher and lower on the socioeconomic ladder. Many even think about it when engaging in dialogue about the future leadership needs of Black America. How does one represent the interests of a social group whose issues and objectives seem increasingly diverse and at times, in opposition to each other? There is a lack of clarity on how best to respond to the class gap, both on a broader community level and in the day-to-day interactions of blacks across class lines. Two complications are central.

First, many African Americans want to live in socioeconomically diverse black neighborhoods. But their desire is often sapped by the instability that pervades neighborhoods with significant numbers of poor residents. The resulting internal conflict frequently ends with those better-off opting for less economically diverse communities that wholeheartedly encourage upward mobility.[8] Social scientists Katherine Newman, Elijah Anderson, and Mary Pattillo-McCoy all highlight these tensions in neighborhoods whose members represents not only varying class positions but also different ways of thinking about the appropriate values and norms for their communities. At the same time, those with limited opportunities for advancement are often frustrated by the perceived lack of ongoing support and resources afforded by more successful members of the community. . . .

Second, because of their history in this country, African Americans have a shared desire for economic, social, and political equality with the rest of America. However, some blacks are closer than others to achieving this goal, raising the question of whether the means by which this goal is to be pursued should be the same throughout the black community. . . . Largely through racial integration, blacks, particularly those under 35, are faced with the fact that their experiences—professional, educational and cultural—are often vastly different from others sharing their racial, but not socioeconomic, class backgrounds. In fact, depending on their backgrounds, some blacks would argue that in certain aspects, they have more in common with some of their white counterparts than with blacks of a different socioeconomic class. . . .

BRIDGING THE DIVIDE: STRATEGIES AND DIRECTIVES FOR THE UNDER 35 GENERATION

It is imperative to respond to the socioeconomic divide on both intra- and interracial levels. The former must take place largely through the interpersonal interactions of people of diverse socioeconomic groups. The latter must take place through a broader political agenda that is sensitive to the socioeconomic diversity within Black America.

. . . Increased class-based tension is certainly an understandable—indeed, unavoidable—by-product of rising income inequality within the black community. In his book *Harlemworld: Doing Race and Class in Contemporary Black America,* anthropologist John L. Jackson argues that it is problematic, however, to think of Black America as two overtly discrete worlds—one lower class, one middle class— that rarely, if ever, interact in meaningful ways. While black suburban flight has encouraged geographic distance between blacks of different classes, there are often opportunities through familial ties, churches and other institutions, and in shared public spaces, for blacks to interact across class lines, mitigating some of this social distance.[9]

In order to respond to potential tensions, at least three actions are necessary, particularly from the generation of 25–35 year old African Americans poised to become leaders in their communities. First, there must be a cross-pollination of resources through familial ties, churches and other community organizations, and opportunities created by individuals. Research suggests that often what encourages mobility, and the behaviors necessary for mobility, are social networks that are extensive and diverse. . . .

Second, there must be a fundamental acceptance among black people of their own diversity and, to borrow from Jackson, of different ways of "doing blackness" within the African-American community. Questioning racial authenticity due to political views, socioeconomic status, or presentation of self is futile at best. . . .

Third, it is imperative for African Americans to be involved in debates that shape the political and economic future of this country. Allowing a diversity of views encourages creative, achievable, and effective approaches to responding to what ultimately perpetuates much of the class gap in America—inequality of opportunity. . . . Power in America is largely gained through economic resources obtained through education, ownership, and networks. Many blacks have obtained, or are on their way to obtaining, this power on an individual level. . . . The ability to leverage power allows one to sit at the table when decisions facing the nation are made. As a large heterogeneity of social and political views of African Americans come to the forefront, it is possible to garner broad support for some initiatives among African Americans yet respect and appreciate those times when there is little opportunity to gather support simply on the basis of racial solidarity. . . .

TOWARD A BROADER POLITICAL AGENDA

First, it is important to remember that the class gap within Black America is not just the black community's problem. It is America's problem. Just as affluent African Americans should be encouraged to consider their responsibilities toward less affluent blacks, it is also essential to challenge those forces in American society that created and sustain widening income inequality in

America. After all, those large reservoirs of low-wage workers who are poorly educated were not created by black elites. These reservoirs are the product of decades of systematic racial subjugation—and that is a problem for which America as a whole bears responsibility.

Second, policies designed to increase and sustain labor force participation can greatly assist those on the bottom third of the socioeconomic class structure. . . . Policies that encourage us to make work pay through the Earned Income Tax Credit, living wages, and family friendly employment benefits assist all low-income families and greatly assist families in the bottom third of the black economic class structure. . . .

Third, we must support policies that allow working- and middle-class Americans to secure assets and build wealth, which would greatly aid African Americans in these socioeconomic groups. Policy conversations often focus on the needs of lower-class and middle-class Americans. Working-class Americans then, find themselves attracted to demands for education and training, affordable health care, and community development attractive to low-income families as well as home ownership policies, entrepreneurship opportunities, and wealth building incentives attractive to the middle class. These kinds of policy initiatives have major implications for the economic stability and mobility of black households and many of them, because they appeal across racial lines, have great potential for garnering widespread political support.

Fourth, the transmission of wealth intergenerationally positions African Americans to solidify and improve upon their economic standing and to reduce wealth inequality within the American socioeconomic class structure. . . . Through initiatives designed to educate households on how to not only build but pass on wealth, African-Americans could solidify the economic gains made in the past several years and position themselves to eliminate the wealth gap between themselves and their white counterparts.

NOTES

1. Wilson, William Julius, 1978. *The Declining Significance of Race: Blacks & Changing American Institutions.* Chicago: University of Chicago Press.

2. Wilson, William Julius. 1987. *The Truly Disadvantaged: The Inner City, the Underclass, & Public Policy.* Chicago: University of Chicago Press. Jencks, Christopher & Paul Peterson. 1991. *The Urban Underclass.* Washington, DC: Brookings Institution. Gans, Herbert. 1995. *The War Against the Poor: The Underclass & Anti-Poverty Policy.* New York: Basic Books.

3. The poverty rate of blacks has declined from a high of 38 percent in 1969. U.S. Census Bureau. 2000. *The Black Population in the United States,* March 1999. Washington, DC: On-line data.

4. Oliver, Melvin L. & Thomas M. Shapiro. 1995. *Black Wealth/White Wealth: A New Perspective on Racial Inequality.* New York: Routledge. Conley, Dalton. 1999. *Being Black, Living in the Red: Race, Wealth, & Social Policy in America.* Berkeley: University of California Press.

5. U.S. Census Bureau. 2002. *Money Income in the United States: 1999.* Current Population Reports, P60-209. U.S. Government Printing Office. Washington, DC.

6. U.S. Census Bureau. 2001. *Report on Residential Vacancies & Homeownership.* Press Release. Washington, DC.

7. U.S. Census Bureau. 2000. *Money Income in the United States: 1999.* Current Population Reports, P60-209. U.S. Government Printing Office. Washington, DC.

8. Anderson, Elijah. 1990. *Streetwise: Race, Class, & Change in an Urban Community.* Chicago: University of Chicago Press. Anderson, Elijah. 1999. *Code of the Street: Decency, Violence, & the Moral Life of the Inner City.* New York: W.W. Norton. Newman, Katherine.1999. *No Shame in My Game.* New York: Knopf & Russell Sage. Pattillo-McCoy, Mary. 1999. *Black Picket Fences: Privilege and Peril Among the Black Middle Class.* Chicago: University of Chicago Press.

9. Jackson, John L. 2001. *Harlemworld: Doing Race & Class in Contemporary Black America.* Chicago: University of Chicago Press.

DISCUSSION QUESTIONS

1. What generational differences does Watkins see as she analyzes social class among Black Americans?

2. What policies and practices can help bridge the gap between the prospects of people of color under 35 who are from different social classes?

33

Experiencing Success

Structuring the Perception of Opportunities for West Indians

BY VILNA F. BASHI BOBB AND AVERIL Y. CLARKE

The question of social mobility is a critical one for West Indian migrants in the United States and for the lives of their children. . . .

We want to look at what first- and second-generation West Indians believe about the possibilities and opportunities for "getting ahead" in American society. Research to date has focused on "objective" indicators of socioeconomic status attainment, but none has assessed what members of the different generations believe about this success or lack thereof in their own social experience.[1] Although writings on West Indians have shown that social experience is important in explaining first- and second-generation perceptions of the social structure of economic opportunity, few works have emphasized social experience directly, and even fewer have compared the social experience of the first generation with that of the second. Here we use respondents' own words to report what the immigrant and second generations understand success to be. . . .

This study is based on two sets of interviews with West Indian immigrants and U.S.-born children of immigrants from the West Indies. The data on the immigrant generation's perspectives on social mobility come from Vilna Bashi's

1996 interviews with forty-four black immigrants from St. Vincent and the Grenadines and Trinidad and Tobago who migrated to New York City between 1930 and 1980 (Bashi 1997). From 1998 to 1999, Averil Clarke (2001) interviewed fifty-five college-educated women of African American and West Indian ancestry to study how fertility and nuptiality decisions affected the ability to earn a college degree. For the present [article], Clarke used data only for those women whose parents emigrated from the West Indies to New York, and supplemented their stories with fifteen additional interviews of native-born persons, all between the ages of fifteen and forty-five, whose parents migrated from these same islands to New York City. . . .

EDUCATION AND SOCIAL MOBILITY
FOR WEST INDIANS

The first-generation immigrants interviewed for this study strongly believed that education would generate returns in the labor market; they were also committed to the idea of getting an education themselves. Many came to the United States expressly to gain an education, some exclusively for education and others in order to work and obtain schooling simultaneously.[2] A well-known and common scheme—for women at least—was to enter the country under a visa that allowed them to start work as babysitters or housekeepers (employers in these niches typically filed the necessary papers under exceptions to the usual immigration laws), and then to go to school to learn the skills that would permit a career change (Bashi 1998). . . .

The second generation was just as committed to the belief in education as a tool necessary for advancement. Almost every second-generation respondent directly or indirectly mentioned education as key factor in what they had achieved (or failed to achieve) to date. Roy[3] told us that his professional success was due to "the ability to learn and understand what I do." . . . Jonathan maintained that "some of the values" that he possessed were "due to [his parents] being immigrants." He continued, "My mom and my dad always explained to me that you had to work to get whatever you wanted. Education was one of the ways of getting something in this country. So they came here and worked hard and I believe they instilled those things into me."

Although second-generation West Indians agreed with their parents that education was the key to success or failure in the economic sector, some perceived limitations in the value of an education and felt that a singular focus on education was not likely to suffice for economic advancement. For example, although Nordrick did not attribute his career success to his record of educational achievement, he added that, when he discovered the fact that his educational credentials surpassed those of his white coworkers, he realized he was being discriminated against in terms of salary and promotions. Kyle explained that the major obstacles to his success were the structure of society and racial barriers. Consider his comments on how a degree from a black institution did not carry the same weight as one from a majority white institution when applying to graduate school.

> Like going to my university. The way this world in America views that is that I can have the same grades as a person in a University of Penn, and just because he or she went to Penn and I went to Virginia State, I would get overlooked with a quickness. I feel like to be coerced to go to a pre-dominantly white school or some other school because of a reputation—you know, I feel the person should speak for himself not the reputation of the school.

. . . Much like their parents, second-generation West Indians continue to hail the value of an education for gaining economic success, and they often attribute failure to inadequate, misdirected, or inappropriate education. Yet the second-generation and long-resident immigrants also perceive that there are limits to the returns to education, and more than a few of them would be suspicious of claims that education is the only or even the main ingredient to success in their field. . . . These perceptions emerge from the U.S.-born generation's understanding of how the racial structure in the United States negatively affects black people who seek to achieve higher social or economic status.

SOCIAL EXPERIENCE AS A FACTOR
IN UNDERSTANDING THE ROLE
OF EDUCATION

What are the social experiences that underpin the attitudes we have described? First-generation immigrants bring with them extensive experience in the economic and education systems of "Third World" immigrant-sending countries. This includes but is not limited to the idea that "education was a privilege, not a right, in the West Indies," as Randall, a first-generation immigrant, put it. During colonialism the West Indian education system was modeled after the British education system, which, albeit in neocolonial fashion, still exists today. In the West Indies it is not the case that all students can advance as far as they like. At various points one must take examinations in order to move to the next level, and there are only a very limited number of slots available for advancement. Thus, not only must you take qualifying examinations, but it is your placement among all exam-takers that determines whether you will win a coveted slot that allows you to further your education.

Sufficient experience in this context generates investment in the idea that lack of educational opportunity limits one's economic mobility (since it literally does so). . . .

Immigrant West Indians encourage their second-generation children to adopt the West Indian immigrant worldview in an environment where, despite severe inequities, education appears to be free and available to all. However, in the United States, education does not always translate into attractive social positions. In this context, black immigrant parents' attempt to pass on their intense attachment to education is a "harder sell" to black children who know the "raw deal" of the black experience in this different racial structure. Education is the culturally accepted means for achieving social mobility, and our second-generation respondents (only two-thirds of whom held college degrees) generally agreed with this but saw race as a mediating factor. Moving our lens from the first to the second generation, we move from a generation that wholly believes in education as a means for mobility to one that does not. What needs to be explained is not why some young black people believe in the power of education, but why it is that many young blacks do not. We emphasize that this difference between the generations is due to structural limitations that restrict blacks in the United States and the fact that these limitations differ from those that affect blacks who live in other nations. . . .

It is evident that West Indians have used geographic mobility to take advantage of opportunities for socioeconomic mobility for more than a century. "Strategic flexibility" is the term Carnegie (1987) has coined to describe the West Indian philosophy of organizing one's life in order to be ready to take advantage of whatever opportunities for work and migration may arise. . . . Emigration has long been a primary means of economic advancement, from the emancipation in 1834 of black slaves in the islands until today. . . .

Our first-generation informants corroborated the existence of and adherence to the "strategic flexibility" philosophy, particularly as it relates to seeking education for job advancement. For example, one informant explained that when he first came to New York, he "did a little of everything," starting with jobs in carpentry and then moving on to spraypainting. "You've gotta have more than one skill," he said. . . . Randall told a similar story when asked if his work history consistently used one of the skills he had honed. "No, I did a couple of other jobs, I work other little places. Listen, let me tell you, any place that was offering me five or ten dollars more than what I was doing, I'm gone. Oh yes, that's the way." To take advantage of opportunities for advancement, some people moved, not only from one job to another, but across state borders. . . .

Among other things, strategic flexibility depends upon an understanding of the world in which personal accommodation to economic and social systems is part and parcel of economic mobility. Struggling to gain an education becomes just one kind of adjustment or accommodation that one must make. Furthermore, education is bound up with the migration act itself. . . .

With regard to socialization in the education system of the United States, second-generation immigrants experience a system in which all education is not equal (e.g., "black" schools are valued less than "white" schools), and this inequality affects their belief in the value of an education. . . . It is well known that primary education systems in the United States differ in quality by regional status within and across counties. The children of West Indian immigrants living in urban areas (as most do) are likely to attend a school with a predominantly black student body and inadequate funding and resources. . . .

The American school system is not always helpful in educating young blacks, despite their high aspirations and hard work.

In sum, although both generations see education as the primary factor in social advancement, their differing social experiences lead them to interpret the importance of education somewhat differently. For the second generation, a lack of experience in the postcolonial Third World education system and economy has a dampening influence on their perceptions about the rewards associated with education. The investment in strategic flexibility driven by Third World–style limits to social and economic opportunity is another quality possessed by the first generation but lacking in the second.

While these experiences are enough to cause differences in the generations' perspectives on the rewards to education, we also believe that racial socialization has its own independent effect. The way in which black West Indian immigrants as a group are able to distance themselves economically and psychologically from racism in the United States leaves intact their beliefs about the rewards of education. In contrast, the second generation, as well as those who immigrated at young ages or have remained in the United States for long periods of time, becomes immersed in the American racial hierarchy and has no such distance. These West Indians have a less optimistic understanding of the meaning of blackness for their chances for advancement opportunities in the economic arena.

RACISM AND SOCIAL ADVANCEMENT FOR WEST INDIANS

... West Indians in the immigrant generation are shocked or surprised by racism in the United States. "We don't have that back home," as one person explained and many others expressed in similar words. West Indian migrants come to the United States with an idea of race developed in their Caribbean homelands. Immigrants described the West Indian racial structure as akin to a class system, where access to sources of human capital—particularly education—makes a significant difference in one's ability to achieve social mobility (Gopaul-McNichol 1993; Foner 1985; Dominguez 1975). ...

Evidence suggests that there is an ideal typical trajectory for the adaptation of West Indian immigrants to racism in the United States (Bashi 1996). When they arrive, West Indians believe that, because they are committed to hard work and social advancement, and because they are foreigners, racist behavior will not be directed toward them. They soon learn that racism is directed toward all black people (i.e., racism is not a set of social behaviors applied by white Americans just toward black Americans, as many immigrants initially believe). Once they learn that racism indeed applies to them, West Indians develop coping strategies to ignore, avoid, or overlook it. Then they learn a second lesson—that living with racism is a demoralizing process, and that racism cannot be wholly ignored or avoided.

Immigrant responses to racism in the United States indicate an understanding of how racism has demoralized African Americans, though only one immigrant reported being demoralized herself. Although they saw themselves as targets of racism, West Indians still reported that they did not let racism bother them or get in the way of achieving their dreams.

> I lived in St. Vincent, where there is a good enclave of so-called white people, and we worked together but we didn't see black and white, we saw people. We worked together, socialized together. . . . So when coming to America, and they were calling you "black" . . . You work in the hospital and they would spit on you and tell you don't touch them and tell you to take your black so-and-so away from them . . . After a while you learn to rise above it, like water off the duck's back.

Of course, racism did bother some immigrants enough that it became one reason for deciding to leave New York City and return home. Still, they did not seem to think that racism was a problem that hindered mobility. . . .

Second-generation respondents unanimously stated that racism should be an issue of great concern in our society. They readily recited incidents and situations that they believed were indicative of racism and its deleterious consequences for blacks. As Nordrick said in a matter-of-fact tone, "Racism is still a big problem. It's still nowhere near equal." . . . Nordrick stated that he was affected by racism early in his professional life.

> During my eight years of working construction management, just seeing numerous white bosses bring their children, nieces, sons, and daughters in, bring them into a position above me, making more money than me, and they had no formal education and I had been working there for years. There was basically no question for me to ask. So eventually I caught on and knew that I had to move on.

. . . While the U.S.-born and U.S.-raised respondents were unanimous in their perceptions of the problem of racism, there was more variance in the degrees to which they felt that racism affected them personally. . . . Carla declared that minorities face racism as they attempt to climb the rungs of the job ladder. However, when she spoke about racism's effect on her own life, Carla said that she herself had not been its victim.

In the same vein, Christopher, who mentioned racism in the military, said, "Me, no, I haven't been personally affected by racism. I rise above the top when it comes to that. I know how to deal with it, I guess. I don't let it really get to me.". . .

CONCLUSION

Both first- and second-generation West Indians believe that education is key to achieving social mobility. In a racialized society, however, race and ethnicity mediate the impact of education. Both generations experience racism, but they perceive its personal and societal impact differently. . . . Despite having experienced racism in the United States, immigrants believed that racism need not be an obstacle to mobility, but that it is something readily ignored or avoided. We believe that these differences can be explained largely by one's ability to invoke West Indian ethnic status, which is related to the degree that the social experience common to a foreign-born, foreign-raised West Indian is incorporated into one's socialization.

The first generation has the maximum opportunity to invoke West Indian ethnic identity—they have the social experiences of a foreign reference point, a network of coethnics with which to live and work, and a foreign accent—all of which translate into social distance from a racially hierarchical social system. The second generation is less strategically flexible, particularly in those areas most influenced by the racial system. Social experience variables, rather than immigrant status or cultural ethnicity, best explain differences between the generations. What characterizes the first generation is not just their foreign birth but that they have moved from a black-majority society to the white-majority United States and successfully remained, reaching employment and educational goals that most of those who stayed behind can only dream of. What characterizes the second generation is not just their American birth but that they have difficulty marshaling their West Indianness in a society that racializes black people with little regard to ethnicity. As Linda, a second-generation respondent, told us,

> We are affected by race 'cause a lot of people just see us as black Americans.
> We're treated, as far as I know, the same ways black Americans are treated
> because we're clumped in that category. But I think our own perception
> is different than that of black American because our parents had such a
> drive and a strive that other Americans did not have. We come from a
> totally different perspective of work ethic and a different perspective of,
> you know, their education. . . . So I think we're affected because we are
> clumped in those categories, but I think that our perception is different.

Since subsequent American-born generations will have less and less ability to invoke West Indian ethnicity, they may end up exposed to the oppressive aspects of American racial structure. In contrast, ethnic West Indians may be relieved of racism's worst effects if West Indian ethnicity continues to be recognized as a marker of difference. Thus, West Indians of both the first and the second generations may gain from investing in the cultural difference stereotype.

NOTES

1. By *social experience,* we mean immersion in a social context of sufficient duration to socialize a person or group so that they are fully familiar with the way that social context operates.

2. Over 95 percent of the first-generation respondents arrived in New York City well before 1980. They were here, then, before the 1986 crackdown on employers who hired foreigners without proper working papers. They were also here before New York City's near bankruptcy and fiscal crisis in 1975–76, which brought about restrictions in City University of New York admission

policies for bachelor's degree programs. Thus, it was relatively easy for these immigrants to combine both work and schooling under the immigration, employment, and college admissions climates they faced in New York City between 1965 and 1976.

3. We invented pseudonyms to refer to our respondents. None of their real names are used in this article.

REFERENCES

Bashi, Vilna. 1996. "'We Don't Have That Back Home': West Indian Immigrant Perspectives on American Racism." Paper presented at the annual meeting of the American Sociological Association, New York, August.

———.1997. "Survival of the Knitted: The Social Networks of West Indian Immigrants." Ph.D. diss., University of Wisconsin, Madison.

———.1998. "Racist Ideology in Immigration Policy and its 'Effect' on Black Immigrant Social Networks." Paper presented at the annual meeting of the Social Science History Association, Chicago, November.

Carnegie, Charles V. 1987. "A Social Psychology of Caribbean Migration: Strategic Flexibility in the West Indies." In Barry Levine (ed.) *The Caribbean Exodus*. New York: Praeger.

Clarke, Averil. 2001. "Black Women, Silver Spoons: How African American Women Manage Social Mobility, Fertility, and Nuptiality Decisions." Ph.D. diss., University of Pennsylvania.

Dominguez, Virginia. 1975. *From Neighbor to Stranger: The Dilemma of Caribbean Peoples in the United States*. Occasional Papers #5. New Haven: Yale University, Antilles Research Program.

Foner, Nancy. 1978. *Jamaica Farewell: Jamaican Migrants in London*. Berkeley: University of California Press.

———.1985. "Race and Color: Jamaican Migrants in London and New York City." *International Migration Review* 19:708–727.

Gopaul-McNichol, Sharon-Ann. 1993. *Working with West Indian Families*. New York: The Guilford Press.

DISCUSSION QUESTIONS

1. Why do Bobb and Clarke think that social experiences are important in modifying one's values regarding education?

2. Define "strategic flexibility" and describe how immigrants use it to achieve social mobility.

3. How do members of the second generation differ from their immigrant parents in how they are prepared to be successful in the United States?

InfoTrac College Edition: Bonus Reading

http://infotrac.thomsonlearning.com

You can use the InfoTrac College Edition feature to find an additional reading pertinent to this section. You can also search on the author's name or key-words to locate the following reading:

Brandon, Peter David. 2002. "The Living Situations of Children in Immigrant Families in the United States." *International Immigration Review* 26 (Summer): 416–437.

Brandon uses the Current Population Survey, done annually by the U.S. Census Bureau, to look at the living arrangements of children, that is whether children live with married parents or in single-parent households. He com-pares the living arrangements of immigrant children to U.S.-born White chil-dren with U.S.-born parents, finding that, except for foreign-born Black youth and some Hispanic children, foreign-born children are more likely to live with married parents than are U.S.-born White children. This pattern shifts by the third generation, when the percentage of children living with single parents increases. Brandon uses a framework of *segmented assimilation* to describe this pattern.

STUDENT EXERCISES

1. Immigration is transforming many communities in the United States. Who are the immigrants in your community? Where do they live? What work do they do? Did or do their children attend school with you? After you have made your own assessment, review newspaper and visual media cover-age to understand popular images of these immigrants. Are they presented as a positive influence in the community or as a negative one? How do media images communicate ideas about immigration?

2. Voting is an important right of citizenship, yet a very low percentage of those eligible actually exercise the franchise. Voting among young people has been particularly low. Interview ten students from some of your other classes; ask them whether or not they vote, and ask their reasons for participating or not in this act of citizenship. Using your interview results, explain why you think voter turnout among youth is low.

PART VI

INTRODUCTION

Institutional Segregation and Inequality

The United States has made progress toward reducing inequality by changing key laws that excluded people from entering the country, becoming citizens and voting, and accessing work opportunities and education. However, even after the passage of key Civil Rights legislation and important Supreme Court decisions we still have **institutionalized racism,** that is, power and privileges based on race are still a part of most areas of social life. The history of discrimination, both how race was a factor in the labor market, in education, and in other spheres of social life and the legacy of the ideas that allegedly justified those arrangements, still plague us. Institutionalized racism persists in the patterns of segregation in most areas of society. People of different races are socially and spatially separated from each other. Segregation means that people's daily lives are structurally very different; such differences are linked to opportunities. Differences in power and resources also mean variation in a group's or individual's abilities to secure housing, provide for their families, advance educationally, protect themselves from exploitation, as well as generally enjoy the fruits of society.

In its 1954 *Brown v. The Board of Education of Topeka, Kansas* decision, the Supreme Court declared segregation unconstitutional. Brown included four public education cases woven together that contested the 1898 *Plessy v. Ferguson* Supreme Court decision. In that case, the Court had recognized segregation as constitutional because it assumed that communities could provide separate but equal facilities in schools, public transportation, and other spheres of life. In the South, legalized segregation—separate facilities sanctioned by *de jure segregation*—shaped very different lives for people based on race. The system that resulted was referred to as Jim Crow. In the North, people were also socially and spatially separated, but rather than the separation being legally enforced, the system was made up of common practices, or *de facto segregation* (resulting from economic or social factors rather than law).

Both systems caused harm, just as our new post-Civil Rights segregation is detrimental to the lives of both disadvantaged and advantaged persons. At this time, when the nation includes a great deal of racial and ethnic diversity, we could share experiences and shape a vision that promotes equality. Instead, the shape of our institutions means that struggle and hardship characterize many people's lives, while more privileged groups know little about those lives other than what the media project. Part VI explores the segregated nature of many contemporary social institutions and what that means for various segments of the population. Here scholars document discriminatory treatment on the part of private corporations and different levels of government, so that employment options, social policies, and educational opportunities all reflect patterns of exclusion or differences in treatment. Segregation is also supported by the actions of private individuals, whose choices about residence and educational options reflect their own racial privileges.

Racial oppression in the United States was originally grounded in economic exploitation, so scholars study employment patterns to see how, and the degree to which, race still shapes access to work. In Section A, Cedric Herring ("Is Job Discrimination Dead?") finds strong documentation for practices that are continuing to limit employment options for Black Americans in working- and middle-class jobs. The U.S. political economy is shifting from industrial jobs to service jobs, a process called *economic restructuring*. This restructuring is shaping the work opportunities of different groups, with some groups becoming more disenfranchised while others still find decent jobs. Whether one can find a good job is, as the popular adage goes, often a matter of "who you know." And, as Deirdre A. Royster shows ("Race and The Invisible Hand: How White Networks Exclude Black Men from Blue-Collar Jobs"), racism works like an "invisible hand," shaping the networks and connections that Black men have, leaving them less likely than White men to find employment, even in working-class jobs. Her research provides a rich account of the processes by which discrimination operates against Black men, as well as other groups who do not have access to White networks.

Patterns of segregation and inequality are also found in the work worlds for new immigrants. Pierrette Hondagneu-Sotelo ("Families on the Frontier") examines how immigrant women, especially those from Latin America, have found a racialized and gendered niche in private homes, hotels, nursing homes, and hospitals. Many women send their wages back home to support families still in their home nations. The new transnational families, in which family members live in two countries, are different from those of the past. Hondagneu-Sotelo notes that, earlier, men would cross borders for industrial and agricultural work to support their families in home countries, but now it is women who cross borders.

Young people in the inner city have particular trouble finding work in this new postindustrial economy for a range of reasons. Katherine Newman ("Getting a Job in the Inner City") looks at how determined young people who need work negotiate urban labor markets. Their segregated environments shape very different options for them than for those for young people growing up in suburban communities.

The economic rewards promised by civil rights legislation and improved educational attainment have not materialized for all people of color, including new immigrant groups. The articles on employment help explain why communities of color are increasingly stratified, with some people doing well and others struggling to escape poverty and its complications. Other patterns of segregation and inequality in government or state policy directly influence family life and health for racial and ethnic groups. Section B, "Families, Communities, and Welfare," opens with a look at environmental racism. The quality of our communities is fundamental to well-being. Raquel Pinderhughes ("The Impact of Race on Environmental Quality") explores how people of color are more at risk for exposure to environmental hazards because they tend to be concentrated in impoverished areas and also lack the power to challenge major polluters. Only recently have activists recognized this fact and built a broad coalition of diverse groups working for environmental justice.

We see that rather than protecting people of color, social policies can cause harm to communities because people in the community find it difficult to influence either public opinion or politicians. We find this trend to be the case with minority families, many of whom are poor and vulnerable. Linda Burnham ("Welfare Reform, Family Hardship, and Women of Color") looks at the impact of the Personal Responsibility and Work Opportunities Reconciliation Act (PRWORA) of 1996, or what is known as "welfare reform." She finds that this legislation has *not* provided women with a path out of poverty. Racial minority and immigrant women frequently face the hardships of homelessness and hunger without a safety net. The policies that lead to their plight are often the result of negative stereotypes—what Patricia Hill Collins (2000) calls controlling images, ones that separate racial "minority" women from "majority" women. The stereotypes mean that social policies are based on myths rather than the social contexts of poor women's lives, and, as a result, cause more poverty and hardship. As people's material conditions decline, it becomes harder for them to secure the power to change their lives. These families have been historically neglected by government agencies, and current trends indicate that many agencies intervene in their lives in negative ways—a burden that more privileged families seldom face.

Today we do not have laws that explicitly treat people differently, so it is often hard to identify racial disparities in treatment by the state. But it is essential that we do, so as to understand the new construction of racism. Dorothy Roberts ("Child Welfare as a Racial Justice Issue") argues that the fact that children of color are more likely to be in foster care is evidence of the racial inequality within state policies. As the clients in the child welfare systems around the nation changed from a majority of White children to a majority of children of color, the remedy also changed—from providing services and keeping children at home to removing them from homes and placing them in foster care. Negative ideas about people of color still found in the media and other segments of the society foster the type of mistreatment that Roberts addresses. Beth Ritchie examines this practice in "The Social Construction of the Immoral Black Mother," in which she shows that we are not addressing the real needs of these women because the policies relevant to them are shaped by myths and misinformation. When she interviewed Black women in low-income communities, she found that they faced barriers to mothering that social policies neither recognized nor helped them address.

The communities where people live are shaped by the racial and ethnic makeup of a given area. Peggy Levitt ("Salsa and Ketchup: Transnational Migrants Straddle Two Worlds") shows how families negotiate their place in communities when their ethnic identity makes them part of two worlds—the community in which they live and their community of origin. Her study of Dominican and Gujarati immigrants (the latter a state on the west coast of India) demonstrates how immigrant families create community in a new context, even while maintaining strong ties to their initial homelands. This work also describes the adaptations that families have to make in the context of racial and ethnic stratification.

Patterns of segregation shape what people experience and what people know about the world. In 1968, Congress passed fair housing legislation to address discrimination in housing options. As we enter the twenty-first century, however, our housing is still not integrated. In Section C of Part VI, Douglas Massey ("Residential Segregation and Neighborhood Conditions in U.S. Metropolitan Areas") argues that residential segregation means a concentration of poverty in urban areas with consequences for educational options and other life chances.

As part of an effort to provide opportunities in the face of residential segregation, efforts have been made to desegregate and integrate public schools. Now those trends are reversing. Erica Frankenberg and Chungmei Lee ("Race in American Public Schools: Rapidly Resegregating School Districts") address the resegregation in public schools that is a consequence of both residential segregation and the ending of court-ordered desegregation plans. Not only are majority Black and Latino schools likely to receive fewer resources, but the social isolation thus engendered is also a factor in the development of individuals and groups.

What do these patterns of segregation of neighborhoods and schools mean for young people coming of age? Students struggle to survive in these isolated settings. If they move on to college or to majority White educational settings, they have to combat the negative stereotypes held by White students who have had little interaction with people of color. Research tells us that many students do not get to college. Nancy Lopez ("Race–Gender Experiences and Schooling") explores the gender gap in educational attainment among second-generation Caribbean youth in New York City. Like Bobb and Clarke's article "Experiencing Success," (Part V, Section C) Lopez is interested in how the social construction of race and gender shapes the experiences of Caribbean youth, pushing the young men into the criminal justice system and the women toward education.

Heidi Lasley Barajas and Jennifer Pierce ("The Significance of Race and Gender in School Success among Latinas and Latinos in College") also explore gender differences via interviews with Latina and Latino college students. Gendered experiences play a role in the strategies they employ to adjust to college and the challenge of mobility. While sports (with its focus on the hard work of the individual) frame the view of the men, the women, who remember periods of exclusion, employ more collective relationships—getting and giving support to others. We see in this section that segregation limits opportunities for young people of color, particularly urban youth, while those who do succeed academically have to scale social barriers. Understanding how people cope in educational institutions can help us shape institutions that do not demand as much sacrifice by some groups.

As we saw in Part V, the denial of justice to people of color began with the Constitution. After decades of struggle there has been some progress in securing civil rights, but mistreatment of people of color is common. There is racial profiling on the highways, city streets, airports, and other public places. Police and other law enforcement can be a hostile presence in minority communities. In our final section in the examination of segregation and inequality, Section D, we look at how social control—the practices employed to push individuals and groups to conform to the rules of the society—is often in conflict with social justice. In a democratic society, ideally all individuals should be treated fairly by the laws, law enforcement officials, and the courts. However, race still plays a role in the treatment of individuals, and segregation in isolated communities provides fertile ground for more harsh treatment.

We have touched upon how images of Black and Latino youth as troublemakers can influence their attitudes toward education. In Section D, we look at the issue of police and law enforcement. Eduardo Bonilla-Silva ("Keeping Them in Their Place") uses empirical studies to argue that Black people have a unique relationship with law enforcement. Because of their segregation, surveillance and differential treatment is invisible to many White Americans, who are often taught through the media to fear minority men as potential criminals. As Lani Guinier and Gerald Torres ("Watching the Canary") note, these practices lead to racial profiling, which makes moving about in public spaces hazardous for Black and Latino youth, regardless of their social class. We must look carefully at such practices, particularly the increasing criminalization of behavior and the incarceration of so many people of color. As a society we are granting more power to the criminal justice system, even though this is a poor substitute for other social reforms such as education. As noted above, the separation of the different groups of people means that only a few actually see and question these new practices.

Christina Swarns ("The Uneven Scales of Capital Justice") discusses how race and social class play a role in capital criminal cases, particular where the death penalty is involved. The U.S. Supreme Court declared the death penalty unconstitutional in 1972, but it has been reinstated in many states across the nation since 1976. Swarm examines the biases that still exist at various levels of the criminal justice system. Her article explores the implications of many of the factors addressed in the Guinier and Torres article. She explores why unchecked power influences who is likely to be prosecuted, found guilty, and sentenced to death. In addition to identifying the bias in the system, we can see how patterns of segregation originating in some areas reverberate in the courts and prisons.

Ideologies as well as spatial borders, are still used to separate us. Fear of the "other" has jeopardized citizenship rights and social justice for racial groups at different points in history and up to the present. After the bombing of Pearl Harbor by the Japanese and the U.S. entrance into World War II, President Franklin Roosevelt issued Executive Order 9066 that called for the internment of Japanese immigrants and Japanese Americans living on the West Coast for security reasons. Roger Daniels ("Detaining Minority Citizens, Then and Now") uses that history as a vantage point from which to examine recent trends, including the experiences of many Arab Americans, who after the tragedies of September 11, 2001, found themselves racialized and jeopardized. Understanding the history and the implications of policies that are supposed to make *some of us* feel safe is important, since what we need is to preserve social justice for all citizens.

Perhaps the challenge for this new century is to dismantle both the current racial segregation and the ideologies that target people of color as problems. Such injustices keep us from working together to make this nation a land of equality and social justice for all. In our final part, Part VII, we will look at the work that members of excluded groups, such as African Americans, American Indians, and Asian Americans did to change the legal opportunity structure. Here we might find lessons on how to work to change the racial landscape.

REFERENCES

Collins, Patricia Hill. 2000. *Black Feminist Thought: Knowledge, Consciousness, and Empowerment.*, 2nd ed. New York: Routledge.

34

Is Job Discrimination Dead?

BY CEDRIC HERRING

n November 1996, Texaco settled a case for $176 million with African-American employees who charged that the company systematically denied them promotions. Texaco originally vowed to fight the charges. When irrefutable evidence surfaced, however, Texaco changed its position. *The New York Times* released a tape recording of several Texaco executives referring to black employees as "niggers" and "black jelly beans" who would stay stuck at the bottom of the bag. Texaco also ultimately acknowledged that they used two promotion lists—a public one that included the names of blacks and a secret one that excluded all black employee names. The $176 million settlement was at the time the largest amount ever awarded in a discrimination suit.

Much has changed in American race relations over the past 50 years. In the old days, job discrimination against African Americans was clear, pervasive and undeniable. There were "white jobs" for which blacks need not apply, and there were "Negro jobs" in which no self-respecting white person would be found. No laws prohibited racial discrimination in employment. Indeed, in several states laws required separation of blacks and whites in virtually every public realm. Not only was racial discrimination the reality of the day, but also many whites supported the idea that job discrimination against blacks was appropriate. In 1944, 55 percent of whites admitted to interviewers that they thought whites should receive preference over blacks in access to jobs, compared with only 3 percent who offered such opinions in 1972.

Many blatant forms of racism have disappeared. Civil rights laws make overt and covert acts of discrimination illegal. Also, fewer Americans admit to traditional racist beliefs than ever before. Such changes have inspired many scholars and social commentators to herald the "end of racism" and to declare that we have created a color-blind society. They point to declines in prejudice, growth in the proportion of blacks who hold positions of responsibility, a closing of the earnings gap between young blacks and young whites and other evidence of "racial progress."

However, racial discrimination in employment is still widespread; it has just gone underground and become more sophisticated. Many citizens, especially whites who have never experienced such treatment, find it hard to believe that such discriminatory behavior by employers exists. Indeed, 75 percent of whites in a 1994 survey said that whites were likely to lose a job to a less-qualified black. Nevertheless, clear and convincing evidence of discriminatory patterns against black job seekers exists.

In addition to the landmark Texaco case, other corporate giants have made the dishonor roll in recent years. In 2000, a court ordered Ford Motor Company to pay $9 million to victims of sexual and racial harassment. Ford also agreed to pay $3.8 million to settle another suit with the U.S. Labor Department involving discrimination in hiring women and minorities at seven of the company's plants. Similarly in 1999, Boeing agreed to pay $82 million to end racially based pay disparities at its plants. In April 2000, Amtrak paid $16 million to settle a race discrimination lawsuit that alleged Amtrak had discriminated against black employees in hiring, promotion, discipline and training. And in November 2000, the Coca-Cola Company settled a federal lawsuit brought by black employees for more than $190 million. These employees accused Coca-Cola of erecting a corporate hierarchy in which black employees were clustered at the bottom of the pay scale, averaging $26,000 a year less than white workers.

The list of companies engaged in discrimination against black workers is long and includes many pillars of American industry, not just marginal or maverick firms. Yet when incidents of discrimination come into public view, many of us are still mystified and hard-pressed for explanations. This is so, in part, because discrimination has become so illegitimate that companies expend millions of dollars to conceal it. They have managed to discriminate without using the blatant racism of the old days. While still common, job discrimination against blacks has become more elusive and less apparent.

HOW COMMON?

Most whites think that discriminatory acts are rare and sensationalized by a few high-profile cases and that the nation is well on its way to becoming a color-blind society. According to a 2001 Gallup survey, nearly 7 in 10 whites (69 percent) said that blacks are treated "the same as whites" in their local communities.

The numbers, however, tell a different story. Annually, the federal government receives about 80,000 complaints of employment discrimination, and another 60,000 cases are filed with state and local fair employment practices commissions. One recent study found that about 60 percent of blacks reported racial barriers in their workplace in the last year, and a 1997 Gallup survey found that one in five reported workplace discrimination in the previous month.

The results of "social audits" suggest that the actual frequency of job discrimination against blacks is even higher than blacks themselves realize. Audit studies test for discrimination by sending white and minority "job seekers" with comparable resumes and skills to the same hiring firms to apply for the same job. The differential treatment they receive provides a measure of discrimination. These audits consistently find that employers are less likely to interview or offer jobs to minority applicants. For example, studies by the Fair Employment Practices Commission of Washington, D.C., found that blacks face discrimination in one out of every five job interviews and that they are denied job offers 20 percent of the time. A similar study by the Urban Institute matched equally qualified white and black testers who applied for the same jobs in Chicago. About 38 percent of the time, white applicants advanced further in the hiring process than equally qualified blacks. Similarly, a General Accounting Office audit study uncovered significant discrimination against black and Latino testers. In comparison with whites, black and Latino candidates with equal credentials received 25 percent fewer job interviews and 34 percent fewer job offers.

These audit studies suggest that present-day discrimination is more sophisticated than in the old days. For example, discriminating employers do not explicitly deny jobs to blacks; rather, they use the different phases of the hiring process to discriminate in ways that are difficult to detect. In particular, when comparable resumes of black and white testers are sent to firms, discriminatory firms systematically call whites first and repeatedly until they exhaust their list of white applicants before they approach their black prospects. They offer whites jobs on the spot but tell blacks that they will give them a call back in a few weeks. These mechanisms mean that white applicants go through the hiring process before any qualified blacks are even considered.

Discriminatory employers also offer higher salaries and higher-status positions to white applicants. For example, audit studies have documented that discriminatory employment agencies often note race in the files of black applicants and steer them away from desirable and lucrative positions. A Fair Employment Practices Commission study found that these agencies, which control much of the applicant flow into white-collar jobs, discriminate against black applicants more than 60 percent of the time.

Surprisingly, many employers are willing to detail (in confidence to researchers) how they discriminate against black job seekers. Some admit refusing to consider any black applicants. Many others admit to engaging in recruitment practices that artificially reduce the number of black applicants who know about and apply for entry-level jobs in their firms. One effective way is to avoid ads in mainstream newspapers. In one Chicago study, more than 40 percent of the employers from firms within the city did not advertise their entry-level job openings in mainstream newspapers. Instead, they advertised job vacancies in neighborhood or ethnic newspapers that targeted particular groups, mainly Hispanics or white East European immigrants. For the employer who wants to avoid blacks, this strategy can be quite effective when employment ads are written in languages other than English, or when the circulation of such newspapers is through channels that usually do not reach many blacks.

Employers described recruiting young workers largely from Catholic schools or schools in white areas. Besides avoiding public schools, these employers also avoided recruiting from job-training, welfare and state employment service programs. Consequently, some job-training programs have had unanticipated negative effects on the incomes and employment prospects of their African-American enrollees. For instance, research on the effect of such training programs on the earnings and employability of black inner-city residents found that those who participated in various job-training programs earned less per month and had higher unemployment rates than their counterparts who had not participated in such programs.

WHO SUFFERS?

Generally, no black person is immune from discriminatory treatment. A few factors make some even more vulnerable to discrimination than others. In particular, research has shown that African Americans with dark complexions are likelier to report discrimination—one-half do—than those with lighter complexions. Job discrimination is also associated with education in a peculiar fashion: Those blacks with more education report more discrimination. For example, in a Los Angeles study, more than 80 percent of black workers with college degrees and more than 90 percent of those with graduate-level educations reported facing workplace discrimination. Black immigrants are more likely than nonimmigrants to report discrimination experiences, residents of smaller communities report more than those of larger ones, and younger African Americans report more than older ones. Rates of job discrimination are lower among those who are married than among those who are not wed. Research also shows that some employment characteristics also appear to make a difference: African Americans who are hired through personal contacts report discrimination less often, as do those who work in the manufacturing sector and those who work for larger firms.

Discrimination exacts a financial cost. African Americans interviewed in the General Social Survey in 1991 who reported discrimination in the prior year earned $6,200 less than those who reported none. (In addition, blacks earn $3,800 less than whites because of differences in educational attainment, occupation, age and other factors.) A one-time survey cannot determine whether experiences of discrimination lead to low income or whether low income leads to feeling discriminated against. Multivariate research based on data from the Census Bureau, which controls for education and other wage-related factors, shows that the white-black wage gap (i.e., "the cost of being black") has continued to be more than 10 percent—about the same as in the mid-1970s. Moreover, research looking at the effects of discrimination over the life course suggests a cumulative effect of discrimination on wages such that the earnings gap between young blacks and whites becomes greater as both groups age.

HOW CAN THERE BE DISCRIMINATION?

Many economists who study employment suggest that job discrimination against blacks cannot (long) exist in a rational market economy because jobs are allocated based on ability and earnings maximization. Discrimination, they argue, cannot play a major role in the rational employer's efforts to hire the most productive worker at the lowest price. If employers bypass productive workers to satisfy their racism, competitors will hire these workers at lower-than-market wages and offer their goods and services at lower prices, undercutting discriminatory employers. When presented with evidence that discrimination does occur, many economists point to discriminators' market monopoly: Some firms, they argue, are shielded from competition and that allows them to act on their "taste for discrimination." These economists, however, do not explain why employers would prefer to discriminate in the first place. Other economists suggest that employers may rationally rely on "statistical discrimination." Lacking sufficient information about would-be employees, employers use presumed "average" productivity characteristics of the groups to which the potential employees belong to predict who will make the best workers. In other words, stereotypes about black workers (on average) being worse than whites make it "justifiable" for employers to bypass qualified black individuals. In these ways, those economists who acknowledge racial discrimination explain it as a "rational" response to imperfect information and imperfect markets.

In contrast, most sociologists point to prejudice and group conflict over scarce resources as reasons for job discrimination. For example, racial groups create and preserve their identities and advantages by reserving opportunities for their own members. Racially based labor queues and differential terms of employment allow members to allocate work according to criteria that have little to do with productivity or earnings maximization. Those who discriminate against blacks often use negative stereotypes to rationalize their behavior after the fact, which, in turn, reinforces racism, negative stereotypes and caricatures of blacks.

In particular, labor market segregation theory suggests that the U.S. labor market is divided into two fundamentally different sectors: (1) the primary sector and (2) the secondary sector. The primary sector is composed of jobs that offer job security, work rules that define job responsibilities and duties, upward mobility, and higher incomes and earnings. These jobs allow incumbents to accumulate skills that lead to progressively more responsibility and higher pay. In contrast, secondary sector jobs tend to be low-paying, dead-end jobs with few benefits, arbitrary work rules and pay structures that are not related to job tenure. Workers in such jobs have less motivation to develop attachments to their firms or to perform their jobs well. Thus, it is mostly workers who cannot gain employment in the primary sector who work in the secondary sector. Race discrimination—sometimes by employers but at times by restrictive unions and professional associations fearful that the inclusion of blacks may drive down their overall wages or prestige—plays a role in determining who gets access to jobs in the primary sector. As a consequence, African Americans are locked out of jobs in the primary labor market, where they would receive higher pay and better treatment, and they tend to be crowded into the secondary sector. And these disparities compound over time as primary sector workers enhance their skills and advance while secondary sector workers stay mired in dead-end jobs.

An alternative sociological explanation of African-American disadvantage in the U.S. labor market is what can be referred to as "structural discrimination." In this view, African Americans are denied access to good jobs through practices that appear to be race-neutral but that work to the detriment of African Americans. Examples of such seemingly race-neutral practices include seniority rules, employers' plant location decisions, policy makers' public transit decisions, funding of public education, economic recessions and immigration and trade policies.

In the seniority rules example, if blacks are hired later than whites because they are later in the employers' employment queue (for whatever reason), operating strictly by traditional seniority rules will ensure greater job security and higher pay to whites than to African Americans. Such rules virtually guarantee that blacks, who were the last hired, will be the "first fired" and the worst paid. The more general point is that employers do not have to be prejudiced in implementing their seniority rules for the rules to have the effects of structural discrimination on African Americans. Unequal outcomes are built into the rules themselves.

These same dynamics apply when (1) companies decide to locate away from urban areas with high concentrations of black residents; (2) policy makers decide to build public transit that provides easy access from the suburbs to central city job sites but not from the inner city to central city job sites or to suburban job sites; (3) public education is funded through local property tax revenues that may be lower in inner-city communities where property values are depressed and higher in suburban areas where property values are higher and where tax revenues are supplemented by corporations that have fled the inner city; (4) policy makers attempt to blunt the effects of inflation and high interest rates by allowing unemployment rates to climb, especially when they climb more rapidly in African-American communities; and (5) policy makers negotiate immigration and trade agreements that may lead to lower employer costs but may also lead to a reduction in the number of jobs available to African Americans in the industries affected by such agreements. Again, in none of these cases do decision makers need to be racially prejudiced for their decisions to have disproportionately negative effects on the job prospects or life chances of African Americans.

WHAT CAN BE DONE?

Employment discrimination, overt or covert, is against the law, yet it clearly happens. Discrimination still damages the lives of African Americans. Therefore, policies designed to reduce discrimination should be strengthened and expanded rather than reduced or eliminated, as has recently occurred. Light must be shed on the practice, and heat must be applied to those who engage in it. Some modest steps can be taken to reduce the incidence and costs of racial discrimination:

Conduct More Social Audits of Employers in Various Industries of Varying Sizes and Locations

In 2000, the courts upheld the right of testers (working with the Legal Assistance Foundation of Chicago) to sue discriminatory employers. Expanded use of evidence from social audits in lawsuits against discriminatory employers provides more information about discriminatory processes, arms black applicants more effectively and provides greater deterrence to would-be discriminators who do not want to be exposed. Even when prevention is not successful, documentation from social audits makes it easier to prosecute illegal discrimination. As in the Texaco case, it has often been through exposure and successful litigation that discriminatory employers mended their ways.

Restrict Government Funding to and Public Contracts with Firms that Have Records of Repeated Discrimination Against Black Applicants and Black Employees

The government needs to ensure that discriminatory employers do not use taxpayer money to carry out their unfair treatment of African Americans. Firms that continue discriminating against blacks should have their funding and their reputations linked to their performance. Also, as lawsuits over this issue proliferate, defense of such practices becomes an expensive proposition. Again, those found guilty of such activities should have to rely on their own resources and not receive additional allocations from the state. Such monetary deterrence may act as a reminder that racial discrimination is costly.

Redouble Affirmative Action Efforts

Affirmative action consists of activities undertaken specifically to identify, recruit, promote or retain qualified members of disadvantaged minority groups to overcome the results of past discrimination and to deter discriminatory practices in the present. It presumes that simply removing existing impediments is not sufficient for changing the relative positions of various groups. In addition, it is based on the premise that to truly affect unequal distribution of life chances, employers must take specific steps to remedy the consequences of discrimination.

Speak Out When Episodes of Discrimination Occur

It is fairly clear that much discrimination against African Americans goes unreported because it occurs behind closed doors and in surreptitious ways. Often, it is only when some (white) insider provides irrefutable evidence that such incidents come to light. It is incumbent upon white Americans to do their part to help stamp out this malignancy.

Now that racial discrimination in employment is illegal, stamping it out should be eminently easier to accomplish. The irony is that because job discrimination against blacks has been driven underground, many people are willing to declare victory and thereby let this scourge continue to flourish in its camouflaged state. If we truly want to move toward a color-blind society, however, we must punish such hurtful discriminatory behaviors when they occur, and we should reward efforts by employers who seek to diversify their workforce by eliminating racial discrimination. This is precisely what happened in the landmark Texaco case, as well as the recent Coca-Cola settlement. In both cases, job discrimination against African Americans was driven above ground, made costly to those who practiced it and offset by policies that attempted to level the playing field.

DISCUSSION QUESTIONS

1. Why does Herring think that the 1964 Civil Rights Act has not ended employment discrimination?
2. What is an audit study? How is this tool used to investigate discriminatory practices?
3. What is the impact of new discriminatory practices on minorities?

35

Race and the Invisible Hand

How White Networks Exclude Black Men from Blue Collar Jobs

BY DEIRDRE A. ROYSTER

I n the late 1970s, [African American sociologist Williams Julius Wilson] published an extremely influential book, *The Declining Significance of Race*. In this book, Wilson argued that race was becoming less and less important in predicting the economic possibilities for well-educated African Americans. In other words, the black-led Civil Rights movement had been successful in removing many of the barriers that made it difficult, if not impossible, for well-trained blacks to gain access to appropriate educational and occupational opportunities. Wilson argued that this new pattern of much greater (but not perfect) access was unprecedented in the racial stratification system in the United States and that it would result in significant and lasting gains for African American families with significant educational attainment.

Recent research on the black middle class has only partially supported Wilson's optimistic prognosis. While blacks did experience significant educational and occupational gains during the 1970s, their upward trajectory appears to have tapered off in the 1980s and 1990s. Moreover, some blacks have found themselves tracked into minority-oriented community relations positions within the professional and managerial occupational sphere. Even more troubling are

From Deirdre A. Royster, *Race and the Invisible Hand: How White Networks Exclude Black Men from Blue-Collar Jobs* (Berkeley, CA: University of California Press, 2003), pp. 18–23, 58–59, 180–189. Copyright © 2003 by The Regents of the University of California. Reprinted by permission.

data indicating that the proportion of blacks who attend and graduate from college appears to be shrinking, with the inevitable result that fewer blacks will have the credentials and skills necessary to get the better jobs in the growing technical and professional occupational categories. Despite real concerns about the stability of the black middle class and some glitches in the workings of the professional labor market, no one doubts that a substantial portion of the black population now enjoys access to middle-class opportunities and amenities— including decent homes, educational facilities, public services, and most importantly, jobs—commensurate with their substantial education and job experience, or in economic terms, their endowment of human capital.

While other scholars were investigating his theories about the black middle class, Wilson became distressed about the pessimistic prospects of blacks who were both poorly educated and increasingly isolated in urban ghettos with high rates of poverty and unemployment. His main concern was that changing labor demands that increase opportunities for highly skilled workers have the potential of making unskilled black labor obsolete. According to Wilson's next two books, *The Truly Disadvantaged* and *When Work Disappears,* this group's inability to gain access to mobility-enhancing educational opportunities is exacerbated by the further problems of a deficiency of useful employment contacts, lack of reliable transportation, crowded and substandard housing options, a growing sense of frustration, and an image among urban employers that blacks are undesirable workers, not to mention the loss of manufacturing and other blue-collar jobs. These factors, and a host of others, contribute to the extraordinarily difficult and unique problems faced by the poorest inner-city blacks in attempting to advance economically. Wilson and hundreds of other scholars—even those who disagree with certain aspects of his thesis—argue that this group needs special assistance in order to overcome the obstacles they face.

If Wilson intended the Declining Significance of Race thesis (and its Underclass corollary) to apply mainly to well educated blacks and ghetto residents, then Wilson only explained the life chances of at most 30 to 40 percent of the black population.[1] The rest of the black population neither resides in socially and geographically isolated ghettoes, nor holds significant human capital, in the form of college degrees or professional work experience. Looking at five-year cohorts beginning at the turn of the century, Mare found that the cohort born between 1946 and 1950 reached a record high when 13 percent of its members managed to earn bachelor's degrees. Recent cohorts born during the Civil Rights era (1960s) have not reached the 13 percent high mark set by the first cohort to benefit from Civil Rights era victories. As a result, today the total percentage of African Americans age 25 and over who have four or more years of college is just under 14 percent.[2] According to demographer Reynolds Farley, while college attendance rates for white males (age 18–24) have rebounded from dips in the 1970s back to about 40 percent, black male rates have remained constant at about 30 percent since the 1960s. Figures like these suggest that Civil Rights era "victories" have not resulted in increasing percentages of blacks gaining access to college training. Instead, most blacks today attempt to establish careers with only modest educational credentials, just as earlier cohorts did. Thus the vast majority of blacks are neither extremely poor nor particularly well educated; most blacks would be considered lower-middle- or working-class and modestly educated. That is, most blacks (75 percent) lack bachelor's degrees but hold high school diplomas or GEDs; most blacks (92 percent) are working rather than unemployed; and most (79 percent) work at jobs that are lower-white- or blue-collar rather than professional.[3] Given that modestly educated blacks make up the bulk of the black population, it is surprising that more attention has not been devoted to explicating the factors that influence their life chances.

Wilson's focus on the extremes within the black population, though understandable, points to a troubling underspecification in his thesis: it is unclear whether Wilson sees individuals with modest educational credentials—high school diplomas, GEDs, associate's degrees, or some college or other post-secondary training, but not the bachelor's degree—as cobeneficiaries of civil rights victories alongside more affluent blacks. The logic of his thesis implies that as long as they do not reside in socially and geographically isolated communities filled with poor and unemployed residents, from which industrial jobs have departed, then modestly educated blacks, like highly educated blacks, ought to do about as well as their white counterparts. . . .

Because he argues that *past* racial discrimination created the ghetto poor, or underclass, while macro-economic changes—and not current racial discrimination—explain their current economic plight, Wilson's perspective implies that white attempts to exclude blacks are probably of little significance today. In addition, Wilson offers a geographic, rather than racial, explanation for whites' labor market advantages when he argues that because most poor whites live outside urban centers, they do not suffer the same sort of structural dislocation or labor obsolescence as black ghetto residents. If Wilson's reasoning holds, there is no reason to expect parity among the poorest blacks and whites in the United States without significant government intervention. Despite a conspicuous silence regarding the prospects for parity among modestly educated blacks and whites, Wilson's corpus of research and theory offers the most race and class integrative market approach available. First, Wilson specifies how supply and demand mechanisms work differently for blacks depending on their class status. Specifically, he argues that there is now a permanent and thriving pool (labor supply) of educationally competitive middle-class blacks, while simultaneously arguing that changes in the job structure in inner cities have disrupted the employment opportunities (labor demand) for poorer blacks. Second, Wilson argues that contemporary racial disparity results, by and large, from the structural difficulties faced by poor blacks rather than racial privileges enjoyed by (or racial discrimination practiced by) poor or more affluent whites. One of the questions guiding this study is whether the life chances of modestly educated whites and blacks are becoming more similar, as with blacks and whites who are well-educated, or more divergent, as with blacks and whites on the bottom. . . .

The fifty young men interviewed for this study may have been, in some ways, atypical. For example, none of them had dropped out of school, and all were extremely polite and articulate. I suspect that these men were among the easiest to contact because of high residential stability and well-maintained friendship networks. Their phone numbers had remained the same or they had kept in touch with friends since graduating from high school two to three years earlier. Because of these factors, I may have tapped into a sample of men who were more likely to be success stories than most would have been. Researchers call this sampling dilemma creaming, because the sample may reflect those who were most likely to rise to the top or be seen as the cream of the crop, rather than those of average or mixed potential. In this study, however, it may have been an advantage to have "creamed," since I wanted to compare black and white men with as much potential for success as possible. Moreover, men who are personable and who have stable residences and friendship networks might be most able to tap into institutional and personal contacts in their job searches—one of my main research queries. . . .

While I don't think there were idiosyncratic differences among the black and white men I studied or between the men I found and their same-race peers that I didn't find, some of the positive attributes of my sample suggest that my findings may generalize only to young men who generally play by the rules. Of course, it isn't all that clear what proportion of young working-class males (black or white) try to play by the rules—maybe the vast majority try to do so. Nor is it clear by what full set of criteria my subjects did, in fact, play by the rules. My sample includes men who had brushes with the law as well as some who might be considered "goody two-shoes." In other words I'm not sure that the specific men with whom I spoke are atypical among working-class men, but I am willing to acknowledge that they may be. Perhaps what is most important to remember about the sample, who seemed to me to be pretty ordinary, "All-American" men, is my contention that this set of men *ought* to have similar levels of success in the blue-collar labor market—if, that is, we have finally reached a time when race doesn't matter. . . .

One narrative, the achievement ideology, asserts that formal training, demonstrated ability, and appropriate personal traits will assure employment access and career mobility. The second narrative, the contacts ideology, emphasizes personal ties and affiliations as a mechanism for employment referrals, access, and mobility. . . . The achievement ideology has persistently dominated American understanding of occupational success, even though everyone, it seems, is willing to admit that "who you know" is at least as important as "what you know" in gaining access to opportunities in American society. All of the men in this study, for example, said that contacts were very important in establishing young men like themselves in careers. One offered a more nuanced explanation: "It's not [just] who you know, it's how they know you." That is, it is not simply knowing the right people that matters; it is sharing the right sort of bonds with the right people that influences what those people would be willing to do to assist you.

The black and white men in this study had more achievements in common than contacts. They were trained in the same school, in many of the same trades, and by the same instructors. They had formal access to the same job listing services and work-study programs. Instructors and students alike agreed, and records confirm, that in terms of vocational skills and performance, the blacks and whites in this sample were among the stronger students. . . .

Black Glendale graduates trail behind their white peers. They are employed less often in the skilled trades, especially within the fields for which they have been trained; they earn less per hour; they hold lower status positions; they receive fewer promotions; and they experience more and longer periods of unemployment. No set of educational, skill, performance, or personal characteristics unique to either the black or white students differentiates them in ways that would explain the unequal outcomes. . . . Only their racial status and the way it situates them in racially exclusive networks during the school-work transition process adequately explain their divergent paths from seemingly equal beginnings.

In this study of one variant of the "who you know" versus "what you know" conundrum, it is manifestly and perpetually evident that racial dynamics are a key arbiter of employment outcome. Yet challenging the power of the achievement ideology in American society requires a careful exposition of how factors such as race throw a wrench into the presumption of meritocracy. In addition, the contacts ideology must be uniquely construed to take into account the significance of racially determined patterns of affiliation within a class, in this case the working class.

RACE, AN ARBITER
OF EMPLOYMENT NETWORKS

Researchers have long argued that black males lack access to the types of personal contacts that white males appear to have in abundance.[4] I would argue that it's more than not having the right contacts. In terms of social networks, black men are at a disadvantage in terms of configuration, content, and operation, a disadvantage that is exacerbated in sectors with long traditions of racial exclusion, such as the blue-collar trades. Even when blacks and whites have access to some of the same connections, as in this study, care must be taken to examine exactly what transpires. For example, I noted that black and white males were assisted differently by the same white male teachers. If I had only asked students whether they considered their shop teachers contacts on which they could rely, equal numbers of black and white males would have answered affirmatively. But this would have told us nothing about *how* white male teachers *chose to know and help* their black male students. The teachers chose to verbally encourage black students, while providing more active assistance to white students. I discovered a munificent flow of various forms of assistance, including vacancy information, referrals, direct job recruitment, formal and informal training, vouching behaviors, and leniency in supervision. For white students, this practice, which repeated neighborhood and community patterns within school walls, served to convert institutional ties (as teachers) into personal ones (as friends) that are intended to and do endure well beyond high school. . . .

Even without teachers consciously discriminating, significant employment information and assistance remained racially privatized within this public school context. In that white male teachers provided a parallel or shadow transition system for white students that was not equally available to black students, segregated networks still governed the school-work transition at Glendale even though classrooms had long been desegregated.

The implications for black men are devastating. Despite having unprecedented access to the same preparatory institution as their white peers, black males could not effectively use the institutional connection to establish successful trade entry. Moreover, segregation in multiple social arenas, beyond schools, all but precluded the possibility of network overlaps among working-class black and white men. As a result, black men sought employment using a truncated, resource-impoverished network consisting of strong ties to other blacks (family, friends, and school officials) who like themselves lacked efficacious ties to employment.

Beyond school, matters were even worse. Without being aware of it, white males' descriptions of their experiences revealed a pattern of intergenerational intraracial assistance networks among young and older white men that assured even the worst young troublemaker a solid place within the blue-collar fold. The white men I studied were not in any way rugged individualists; rather they survived and thrived in rich, racially exclusive networks.

For the white men, neighborhood taverns, restaurants, and bars served as informal job placement centers where busboys were recruited to union apprentice programs, pizza delivery boys learned to be refrigeration specialists, and dishwashers studying drafting could work alongside master electricians then switch back to drafting if they wished. I learned of opportunities that kept coming, even when young men weren't particularly deserving. One young man had been able to hold onto his job after verbally abusing his boss. Another got a job installing burglar alarms after meeting the vice president of the company at a cookout—without ever having to reveal his prison record, which included a conviction for burglary.

Again and again, the white men I spoke with described opportunities that had landed in their laps, not as the result of outstanding achievements or personal characteristics, but rather as the result of the assistance of older white neighbors, brothers, family friends, teachers, uncles, fathers, and sometimes mothers, aunts, and girlfriends (and their families), all of whom overlooked the men's flaws. It never seemed to matter that the men were not A students, that they occasionally got into legal trouble, that they lied about work experiences from time to time, or that they engaged in horseplay on the job. All of this was expected, brushed off as typical "boys will be boys" behavior, and it was sometimes the source of laughter at the dining room table. In other words, there were no significant costs for white men associated with being young and inexperienced, somewhat immature, and undisciplined.

The sympathetic pleasure I felt at hearing stories of easy survival among working-class white men in an era of deindustrialization was only offset by the depressing stories I heard from the twenty-five black men. Their early employment experiences were dismal in comparison, providing a stark and disturbing contrast. Whereas white men can be thought of as the second-chance kids, black men's opportunities were so fragile that most could not have recovered from even the relatively insignificant mishaps that white men reported in passing.

Black men were rarely able to stay in the trades they studied, and they were far less likely than white men to start in one trade and later switch to a different one, landing on their feet. Once out of the skilled trade sphere, they sank to the low-skill service sector, usually retail or food services. The black men had numerous experiences of discrimination at the hands of older white male supervisors, who did not offer to help them and frequently denigrated them, using familiar racial epithets. The young black men I spoke with also had to be careful when using older black social contacts. More than one man indicated to me that, when being interviewed by a white person, the wisest course of action is to behave as if you don't know anyone who works at the plant, even if a current worker told you about the opening. These young black men, who had been on the labor market between two and three years, were becoming discouraged. While they had not yet left the labor force altogether, many (with the help of parents) had invested time and resources in training programs or college courses that they and their families hoped would open up new opportunities in or beyond the blue-collar skilled labor market. Many of the men had begun to lose the skills they had learned in high school; others, particularly those who'd had a spell or two of unemployment, showed signs of depression.

My systematic examination of the experiences of these fifty matched young men leads me to conclude that the blue-collar labor market does not function as a market in the classic sense. No pool of workers presents itself, offering sets of skills and work values that determine who gets matched with the most and least desirable opportunities. Rather, older men who recruit, hire, and fire young workers choose those with whom they are comfortable or familiar. Visible hands trump the "invisible hand"—and norms of racial exclusivity passed down from generation to generation in American cities continue to inhibit black men's entry into the better skilled jobs in the blue-collar sector.

Claims of meritocratic sorting in the blue-collar sector are simply false; equally false are claims that young black men are inadequately educated, inherently hostile, or too uninterested in hard work or skill mastery to be desirable workers. These sorts of claims seek to locate working-class black men's employment difficulties in the men's alleged deficits—bad attitudes, shiftlessness, poor skills—rather than in the structures and procedures of worker selection that are typically under the direct control of older white men whose preferences, by custom, do not reflect meritocratic criteria.

Few, if any, political pressures, laws, or policies provide sufficient incentives or sanctions to prevent such employers from arbitrarily excluding black workers or hiring them only for menial jobs for which they are vastly overqualified. Moreover, in recent years, affirmative action policies that required that government contracts occasionally be awarded to black-owned firms or white-owned firms that consistently hire black workers have come under attack—eroding the paltry incentives for inclusion set forth during the Civil

Rights era, nearly forty years ago. Indeed, there is far less pressure today than in the past for white-owned firms to hire black working men. And given persistent patterns of segregation—equivalent to an American apartheid, according to leading sociologists—there remain few incentives for white men to adopt young black men into informal, neighborhood-generated networks. As a result, occupational apartheid reigns in the sector that has always held the greatest potential for upward mobility, or just basic security, for modestly educated Americans.

IDEOLOGY AND THE DEFENSE OF RACIALIZED EMPLOYMENT NETWORKS

The public perception of the causes of black men's labor difficulties—namely, that the men themselves are to be blamed—contrasts with my findings. And my research is consistent with that of hundreds of social scientists who have demonstrated state-supported and informal patterns of racial exclusion in housing, education, labor markets, and even investment opportunities. Racism continues to limit the life chances of modestly educated black men....

WHITE PRIVILEGE, BLACK ACCOMMODATION

How, then, do black males, if they wish to earn a living in the surviving trades, negotiate training and employment opportunities in which networks of gatekeepers remain committed to maintaining white privilege? The present research suggests that the options are few: either accommodation to the parameters of a racialized system or failure in establishing a successful trade career. The interviews revealed that forms of black accommodation begin early, as when young men avoided training in trades of interest because they were known to hold little promise for integration and advancement. For those who made such discoveries later, accommodation took the form of disengaging from specific trades, such as electrical construction, and pursuing whatever jobs became available. For some, the disengaging process involved the claim that they were never really committed to the original trade field, but I suspect that such claims merely served to soften the blow of almost inevitable career failure. For the determined, accommodation required suppressing anger at racially motivated insults and biased employment decisions in the majority-white trade settings. If this strategy wore thin, two difficult accommodations remained. The first involved finding a work setting—not necessarily within one's trade—in which the workplace culture was, if not actively receptive to black inclusion, at least neutral. The second involved finding ways to work in white-dominated fields without having to work beside whites.

A word needs to be said about a particularly troubling accommodative behavior adopted by the black men: not actively and persistently pursuing offers of assistance. It is not clear to what extent the black men were fully cognizant of the extent and potency of whites' informal networks or of the cultural norms governing their operation. But, while it is evident that the older white men who were network gatekeepers did not extend the same access and support to the black men, the black men may also have been less proactive in pursuing older white men who might have assisted them.

Generally, the white men in the study appear to have more actively followed up on offers of assistance. And although their careers developed much more smoothly than those of the black men, they were certainly not without the difficulties of not being hired, workplace dissatisfaction, competing vocational interests, and unemployment. Nevertheless, they returned, sometimes repeatedly, to contacts for further assistance. Certainly, demonstrating proactivity toward a typically racially exclusive white network would be especially problematic for black men.

Undoubtedly, black men's exclusion from white personal settings where easy informal contact is facilitated, like neighborhoods and family, contributes to black men's reluctance to pursue whites for assistance. In addition, black men's lack of personal familiarity with normative expectations among whites probably hampers their efforts to imitate their white peers' more forward network behaviors. Furthermore, any efforts by blacks to engage in such behaviors might not be similarly regarded as appropriate, and might instead be interpreted as aggressive, "uppity," or indicative of a feeling of entitlement. Finally, black men's early experiences of racial exclusion, bias, and hostility in the school and the workplace inform not only their assessment of employment prospects, but also their actual employment strategies. Given these complicated contingencies, perhaps the somewhat hesitant responses of black men are, on the whole, not unreasonable.

CONCLUSION

Black men have paid a great price for exclusion from blue-collar trades and the networks that supply those trades, but they have not paid it alone. The pain of black men's unemployment and underemployment spreads across black communities in a ripple effect. Less able to contribute financially to the care of children and parents, or to combine resources with black women or assist other men with work entry and "learning the ropes" on the job, black men withdraw from the support structures that they need and that they are needed to support emotionally as well as economically. The enduring power of segregated networks in the blue-collar trades is as responsible as segregated neighborhoods for the existence of extremely poor and isolated black communities and of the disproportionately black and male prison population—in fact, more so. While many black families live in stable communities that are mostly, if not entirely, black, the inability to find remunerative jobs that do not require expensive college training makes living decently anywhere extremely difficult. And the loss of manufacturing jobs cannot account for black men's underemployment in the remaining blue-collar fields—especially construction, auto mechanics, plumbing, computer repair, and carpentry. . . .

My findings demonstrate that, without governmental initiatives that provide strong incentives for inclusion, white tradesmen will have no reason to open their networks to men of color. As a result, the work trajectories of white and black men who start out on an equal footing will continue to diverge into skilled and unskilled work paths because of business-as-usual patterns of exclusion. Although there are few precedents for intervening in the private sector, there are strong precedents for intervening in the public sector, where the tax dollars of majority and minority citizens must not be redistributed in ways that condone customs of exclusion. . . .

Without the government taking a lead, the young black men I studied—who played by the rules—are unlikely to ever reach their potential as skilled workers or to take their places as blue-collar entrepreneurs, as so many of their white peers are poised to do. This tragedy could have been averted. My hope is that it will be averted in the next generation.

NOTES

1. Haywood Horton, Beverlyn Lundy, Cedric Herring, and Melvin E. Thomas, "Lost in the Storm: The Sociology of the Black Working Class, 1850 to 1990," *American Sociological Review* 65, no. 1 (2000): 128–137.

2. Robert Mare, "Changes in Educational Attainment and School Enrollment," in *State of the Union,* ed. Reynolds Farley (New York: Russell Sage Foundation, 1995); Nancy Folbre, *The Field Guide to the U.S. Economy* (New York: New Press, 1999).

3. Ibid.

4. Richard Freeman and Harry J. Holzer, *The Black Youth Employment Crisis* (Chicago: University of Chicago Press, 1986); William Julius Wilson, *The Truly Disadvantaged* (Chicago: University of Chicago Press, 1987); Paul Osterman, *Getting Started: The Youth Labor Market* (Cambridge: Massachusetts Institute of Technology Press, 1980).

DISCUSSION QUESTIONS

1. Why is economic advancement difficult for inner-city Black men with modest levels of education?

2. What are the differences in the way networks operated for Black men and White men in Royster's study? What are the implications of those outcomes for their futures?

3. Rather than just let market forces work as they will, why does Royster think government intervention is important?

36

Families on the Frontier

From Braceros in the Fields to Braceras in the Home

BY PIERRETTE HONDAGNEU-SOTELO

Why are thousands of Central American and Mexican immigrant women living and working in California and other parts of the United States while their children and other family members remain in their countries of origin? . . . I argue that U.S. labor demand, immigration restrictions, and cultural transformations have encouraged the emergence of new transnational family forms among Central American and Mexican immigrant women. Postindustrial economies bring with them a labor demand for immigrant workers that is differently gendered from that typical of industrial or industrializing societies. In all postindustrial nations, we see an increase in demand for jobs dedicated to social reproduction, jobs typically coded as "women's jobs." In many of these countries, such jobs are filled by immigrant women from developing nations. Many of these women, because of occupational constraints—and, in some cases, specific restrictionist contract labor policies—must live and work apart from their families.

My discussion focuses on private paid domestic work, a job that in California is nearly always performed by Central American and Mexican immigrant women. Not formally negotiated labor contracts, but rather informal occupational constraints, as well as legal status, mandate the long-term

From *Latinos: Remaking America*, ed. by Marcelo Suarez-Orozco and Mariela M. Paez
(Berkeley, CA: University of California Press, 2002), pp. 259–266. Copyright © 2002 by
The Regents of the University of California. Reprinted by permission.

spatial and temporal separation of these women from their families and children. For many Central American and Mexican women who work in the United States, new international divisions of social reproductive labor have brought about transnational family forms and new meanings of family and motherhood. In this respect, the United States has entered a new era of dependency on braceras. Consequently, many Mexican, Salvadoran, and Guatemalan immigrant families look quite different from the images suggested by Latino familism.

This [article] is informed by an occupational study I conducted of over two hundred Mexican and Central American women who do paid domestic work in private homes in Los Angeles (Hondagneu-Sotelo 2001). Here, I focus not on the work but on the migration and family arrangements conditioned by the way paid domestic work is organized today in the United States. I begin by noting the ways in which demand for Mexican—and increasingly Central American—immigrant labor shifted in the twentieth century from a gendered labor demand favoring men to one characterized by robust labor demand for women in a diversity of jobs, including those devoted to commodified social reproduction. Commodified social reproduction refers to the purchase of all kinds of services needed for daily human upkeep, such as cleaning and caring work. The way these jobs are organized often mandates transnational family forms. . . .

I . . . note the parallels between family migration patterns prompted by the Bracero Program and long-term male sojourning, when many women sought to follow their husbands to the United States, and the situation today, when many children and youths are apparently traveling north unaccompanied by adults, in hopes of being reunited with their mothers. In the earlier era, men were recruited and wives struggled to migrate; in a minority of cases, Mexican immigrant husbands working in the United States brought their wives against the latters' will. Today, women are recruited for work, and increasingly, their children migrate north some ten to fifteen years after their mothers. Just as Mexican immigrant husbands and wives did not necessarily agree on migration strategies in the earlier era, we see conflicts among today's immigrant mothers in the United States and the children with whom they are being reunited. In this regard, we might suggest that the contention of family power in migration has shifted from gender to generation. . . .

GENDERED LABOR DEMAND
AND SOCIAL REPRODUCTION

Throughout the United States, a plethora of occupations today increasingly rely on the work performed by Latina and Asian immigrant women. Among these are jobs in downgraded manufacturing, jobs in retail, and a broad spectrum of service jobs in hotels, restaurants, hospitals, convalescent homes, office buildings, and private residences. In some cases, such as in the janitorial industry and in light manufacturing, jobs have been re-gendered and re-racialized so that jobs previously held by U.S.-born white or black men are now increasingly held by Latina immigrant women. Jobs in nursing and paid domestic work have long been regarded as "women's jobs," seen as natural outgrowths of essential notions of women as care providers. In the late twentieth-century United States, however, these jobs have entered the global marketplace, and immigrant women from developing nations around the globe are increasingly represented in them. In major metropolitan centers around the country, Filipina and Indian immigrant women make up a sizable proportion of HMO nursing staffs—a result due in no small part to deliberate recruitment efforts. Caribbean, Mexican, and Central American women increasingly predominate in low-wage service jobs, including paid domestic work.

This diverse gendered labor demand is quite a departure from patterns that prevailed in the western region of the United States only a few decades ago. The relatively dramatic transition from the explicit demand for Mexican and Asian immigrant *male* workers to demand that today includes women has its roots in a changing political economy. From the late nineteenth century until 1964, the period during which various contract labor programs were in place, the economies of the Southwest and the West relied on primary extractive industries. As is well known, Mexican, Chinese, Japanese, and Filipino immigrant workers, primarily men, were recruited for jobs in agriculture, mining, and railroads. These migrant workers were recruited and incorporated in ways that mandated their long-term separation from their families of origin.

As the twentieth century turned into the twenty-first, the United States was once again a nation of immigration. This time, however, immigrant labor is not involved in primary, extractive industry. Agribusiness continues to be a financial leader in the state of California, relying primarily on Mexican immigrant labor and increasingly on indigenous workers from Mexico, but only a fraction of Mexican immigrant workers are employed in agriculture. Labor demand is now extremely heterogeneous and is structurally embedded in the economy of California (Cornelius 1998). In the current period, which some commentators have termed "postindustrial," business and financial services, computer and other high-technology firms, and trade and retail prevail alongside manufacturing, construction, hotels, restaurants, and agriculture as the principal sources of demand for immigrant labor in the western region of the United States.

As the demand for immigrant women's labor has increased, more and more Mexican and (especially) Central American women have left their families and young children behind to seek employment in the United States. Women who work in the United States in order to maintain their families in their countries of origin constitute members of new transnational families, and because these arrangements are choices that the women make in the context of very limited options, they resemble apartheid-like exclusions. These women work in one nation-state but raise their children in another. Strikingly, no formalized temporary contract labor program mandates these separations. Rather, this pattern is related to the contemporary arrangements of social reproduction in the United States.

WHY THE EXPANSION IN PAID
DOMESTIC WORK?

Who could have foreseen that as the twentieth century turned into the twenty-first, paid domestic work would become a growth occupation? Only a few decades ago, observers confidently predicted that this job would soon become obsolete, replaced by such labor-saving household devices as automatic dishwashers, disposable diapers, and microwave ovens and by consumer goods and services purchased outside the home, such as fast food and dry cleaning (Coser 1974). Instead, paid domestic work has expanded. Why?

The exponential growth in paid domestic work is due in large part to the increased employment of women, especially married women with children, to the underdeveloped nature of child care centers in the United States, and to patterns of U.S. income inequality and global inequalities. National and global trends have fueled this growing demand for paid domestic services. Increasing global competition and new communications technologies have led to work speedups in all sorts of jobs, and the much bemoaned "time bind" has hit professionals and managers particularly hard (Hochschild 1997). Meanwhile, normative middle-class ideals of child rearing have been elaborated (consider the proliferation of soccer, music lessons, and tutors). At the other end of the age spectrum, greater longevity among the elderly has prompted new demands for care work.

Several commentators, most notably Saskia Sassen, have commented on the expansion of jobs in personal services in the late twentieth century. Sassen located this trend in the rise of new "global cities," cities that serve as business and managerial command points in a new system of intricately connected nodes of global corporations. Unlike New York City, Los Angeles is not home to a slew of Fortune 500 companies, but in the 1990s it exhibited remarkable economic dynamism. Entrepreneurial endeavors proliferated and continued to drive the creation of jobs in business services, such as insurance, real estate, public relations, and so on. These industries, together with the high-tech and entertainment industries in Los Angeles, spawned many high-income managerial and professional jobs, and the occupants of these high-income positions require many personal services that are performed by low-wage immigrant workers. Sassen provides the quintessentially "New York" examples of dog walkers and cooks who prepare gourmet take-out food for penthouse dwellers. The Los Angeles counterparts might include gardeners and car valets, jobs filled primarily by Mexican and Central American immigrant men, and nannies and house cleaners, jobs filled by Mexican and Central American immigrant women. In fact, the numbers of domestic workers in private homes counted by the Bureau of the Census doubled from 1980 to 1990 (Waldinger 1996).

I favor an analysis that does not speak in terms of "personal services," which seems to imply services that are somehow private, individual rather than social, and are superfluous to the way society is organized. A feminist concept that was originally introduced to valorize the nonremunerated household work of women, *social reproduction* or alternately, *reproductive labor,* might be more usefully employed. Replacing *personal services* with *social reproduction* shifts the focus by underlining the objective of the work, the societal functions, and the impact on immigrant workers and their own families.

Social reproduction consists of those activities that are necessary to maintain human life, daily and intergenerationally. This includes how we take care of ourselves, our children and elderly, and our homes. Social reproduction encompasses the purchasing and preparation of food, shelter, and clothing; the routine daily upkeep of these, such as cooking, cleaning and laundering; the emotional care and support of children and adults; and the maintenance of family and community ties. The way a society organizes social reproduction has far-reaching consequences not only for individuals and families but also for macrohistorical processes (Laslett and Brenner 1989).

Many components of social reproduction have become commodified and outsourced in all kinds of new ways. Today, for example, not only can you purchase fast-food meals, but you can also purchase, through the Internet, the home delivery of customized lists of grocery items. Whereas mothers were once available to buy and wrap Christmas presents, pick up dry cleaning, shop for groceries and wait around for the plumber, today new businesses have sprung up to meet these demands—for a fee.

In this new milieu, private paid domestic work is just one example of the commodification of social reproduction. Of course, domestic workers and servants of all kinds have been cleaning and cooking for others and caring for other people's children for centuries, but there is today an increasing proliferation of these services among various class sectors and a new flexibility in how these services are purchased.

GLOBAL TRENDS
IN PAID DOMESTIC WORK

Just as paid domestic work has expanded in the United States, so too it appears to have grown in many other postindustrial societies, in the "newly industrialized countries" (NICs) of Asia, in the oil-rich nations of the Middle East, in Canada, and in parts of Europe. In paid domestic work around the globe, Caribbean, Mexican, Central American, Peruvian, Sri Lankan, Indonesian, Eastern European, and Filipina women—the latter in disproportionately large numbers—predominate. Worldwide, paid domestic work continues its long legacy as a racialized and gendered occupation, but today, divisions of nation and citizenship are increasingly salient.

The inequality of nations is a key factor in the globalization of contemporary paid domestic work. This has led to three outcomes: (1) Around the globe, paid domestic work is increasingly performed by women who leave their own nations, their communities, and often their families of origin to do the work. (2) The occupation draws not only women from the poor socioeconomic classes, but also women who hail from nations that colonialism has made much poorer than those countries where they go to do domestic work. This explains why it is not unusual to find college-educated women from the middle class working in other countries as private domestic workers. (3) Largely because of the long, uninterrupted schedules of service required, domestic workers are not allowed to migrate as members of families.

Nations that "import" domestic workers from other countries do so using vastly different methods. Some countries have developed highly regulated, government-operated, contract labor programs that have institutionalized both the recruitment and the bonded servitude of migrant domestic workers. Canada and Hong Kong provide paradigmatic examples of this approach. Since 1981 the Canadian federal government has formally recruited thousands of women to work as live-in nannies/housekeepers for Canadian families. Most of these women came from Third World countries in the 1990s (the majority came from the Philippines, in the 1980s from the Caribbean), and once in Canada, they must remain in live-in domestic service for two years, until they obtain their landed immigrant status, the equivalent of the U.S. "green card." This reflects, as Bakan and Stasiulis (1997) have noted, a type of indentured servitude and a decline in the citizenship rights of foreign domestic workers, one that coincides with the racialization of the occupation. When Canadians recruited white British women for domestic work in the 1940s, they did so under far less controlling mechanisms than those applied to Caribbean and Filipina domestic workers. Today, foreign domestic workers in Canada may not quit their jobs or collectively organize to improve the conditions under which they work.

Similarly, since 1973 Hong Kong has relied on the formal recruitment of domestic workers, mostly Filipinas, to work on a full-time, live-in basis for Chinese families. Of the 150,000 foreign domestic workers in Hong Kong in 1995, 130,000 hailed from the Philippines, and smaller numbers were drawn from Thailand, Indonesia, India, Sri Lanka, and Nepal (Constable 1997, p. 3). Just as it is now rare to find African American women employed in private domestic work in Los Angeles, so too have Chinese women vanished from the occupation in Hong Kong. As Nicole Constable reveals in her detailed study, Filipina domestic workers in Hong Kong are controlled and disciplined by official employment agencies, employers, and strict government policies. Filipinas and other foreign domestic workers recruited to Hong Kong find themselves working primarily in live-in jobs and bound by two-year contracts that stipulate lists of job rules, regulations for bodily display and discipline (no lipstick, nail polish, or long hair, submission to pregnancy tests, etc.), task timetables, and the policing of personal privacy. Taiwan has adopted a similarly formal and restrictive government policy to regulate the incorporation of Filipina domestic workers (Lan 2000).

In this global context, the United States remains distinctive, because it takes more of a laissez-faire approach to the incorporation of immigrant women into paid domestic work. No formal government system or policy exists to legally contract foreign domestic workers in the United States. Although in the past, private employers in the United States were able to "sponsor" individual immigrant women who were working as domestics for their "green cards" using labor certification (sometimes these employers personally recruited them while vacationing or working in foreign countries), this route is unusual in Los Angeles today. Obtaining legal status through labor certification requires documentation that there is a shortage of labor to perform a particular, specialized occupation. In Los Angeles and in many parts of the country today, a shortage of domestic workers is increasingly difficult to prove. And it is apparently unnecessary, because the significant demand for domestic workers in the United States is largely filled not through formal channels of foreign recruitment but through informal recruitment from the growing number of Caribbean and Latina immigrant women who are *already* legally or illegally living in the United States. The Immigration and Naturalization Service, the federal agency charged with enforcement of illegal-migration laws, has historically served the interests of domestic employers and winked at the employment of undocumented immigrant women in private homes.

As we compare the hyperregulated employment systems in Hong Kong and Canada with the more laissez-faire system for domestic work in the United States, we find that although the methods of recruitment and hiring and the roles of the state in these processes are quite different, the consequences are similar. Both systems require the incorporation as workers of migrant women who can be separated from their families.

The requirements of live-in domestic jobs, in particular, virtually mandate this. Many immigrant women who work in live-in jobs find that they must be "on call" during all waking hours and often throughout the night, so there is no clear line between working and nonworking hours. The line between job space and private space is similarly blurred, and rules and regulations may extend around the clock. Some employers restrict the ability of their live-in employees to receive phone calls, entertain friends, attend evening ESL classes, or see boyfriends during the workweek. Other employers do not impose these sorts of restrictions, but because their homes are located in remote hillsides, suburban enclaves, or gated communities, live-in nannies/housekeepers are effectively restricted from participating in anything resembling social life, family life of their own, or public culture.

These domestic workers—the Filipinas working in Hong Kong or Taiwan, the Caribbean women working on the East Coast, and the Central American and Mexican immigrant women working in California constitute the new "braceras." They are literally "pairs of arms," disembodied and dislocated from their families and communities of origin, and yet they are not temporary sojourners.

REFERENCES

Bakan, Abigail B., and Daiva Stasiulis. 1997. "Foreign Domestic Worker Policy in Canada and the Social Boundaries of Modern Citizenship," In Abigail B. Bakan and Daiva Stasiulis (eds.), *Not One of the Family: Foreign Domestic Workers in Canada,* Toronto: University of Toronto Press, pp. 29–52.

Constable, Nicole. 1997. *Maid to Order in Hong Kong: Stories of Filipina Workers.* Ithaca and London: Cornell University Press.

Cornelius, Wayne. 1998. "The Structural Embeddedness of Demand for Mexican Immigrant Labor: New Evidence from California." In Marcelo M. Suárez-Orozco (ed.), *Crossings: Mexican Immigration in Interdisciplinary Perspectives.* Cambridge, MA: Harvard University, David Rockefeller Center for Latin American Studies, pp. 113–44.

Coser, Lewis. 1974. "Servants: The Obsolescence of an Occupational Role." *Social Forces* 52:31–40.

Hochschild, Arlie. 1997. *The Time Bind: When Work Becomes Home and Home Becomes Work.* New York: Metropolitan Books, Henry Holt.

Hondagneu-Sotelo, Pierrette. 1994. *Doméstica: Immigrant Workers and Their Employers.* Berkeley: University of California Press.

Lan, Pei-chia. 2000. "Global Divisions, Local Identities: Filipina Migrant Domestic Workers and Taiwanese Employers." Dissertation, Northwestern University.

Laslett, Barbara, and Johanna Brenner. 1989. "Gender and Social Reproduction: Historical Perspectives," *Annual Review of Sociology* 15:381–404.

Waldinger, Roger, and Mehdi Bozorgmehr. 1996. "The Making of a Multicultural Metropolis." In Roger Waldinger and Mehdi Bozorgmehr (eds.), *Ethnic Los Angeles.* New York: Russell Sage Foundation, pp. 3–37.

DISCUSSION QUESTIONS

1. What does Hondagneu-Sotelo mean by the term transnational families?

2. Define social reproductive work and explain why Hondagneu-Sotelo thinks this work is performed by immigrant women.

3. Social reproductive work is part of the industrial societies of the United States, Canada, and Hong Kong, but what are the different national strategies for securing workers?

37

Getting a Job
in the Inner City

BY KATHERINE S. NEWMAN

I f you drive around the suburban neighborhoods of Long Island or Westchester County, you cannot miss the bright orange "Help Wanted" signs hanging in the windows of fast food restaurants. Teenagers who have the time to work can walk into most of these shops and land a job before they finish filling out the application form. In fact, labor scarcity (for these entry-level jobs) is a problem for employers in these highly competitive businesses. Therefore, though it cuts into their profits, suburban and small-town employers in the more affluent parts of the country are forced to raise wages and redouble their efforts to recruit new employees, often turning to the retiree labor force when the supply of willing youths has run out.

From the vantage point of central Harlem, this "seller's market" sounds like a news bulletin from another planet. Jobs, even lousy jobs, are in such short supply that inner-city teenagers are all but barred from the market, crowded out by adults who are desperate to find work. Burger Barn managers rarely display those orange signs; some have never, in the entire history of their restaurants, advertised for employees. They can depend upon a steady flow of willing applicants coming in the door—and they can be very choosy about whom they sign up. In fact, my research shows that among central Harlem's fast food establishments, the ratio of applicants to available jobs is 14:1....

From Katherine S. Newman, *No Shame in My Game: The Working Poor in the Inner City* (New York: Random House, 1999), pp. 62–70. Copyright © 1999 by Russell Sage Foundation. Used by permission of Alfred A. Knopf, a division of Random House, Inc.

Long lines of job-seekers depress the wages of those lucky enough to pass through the initial barriers and find a job. Hamburger flippers in central Harlem generally do not break the minimum wage. Longtime workers, like Kyesha Smith, do not see much of a financial reward for their loyalty. After five years on the job, she was earning $5 an hour, only 60 cents more than the minimum wage at the time. . . .

Why do people seek low-wage jobs in places like Burger Barn? How do they go about the task in labor markets that are saturated with willing workers? What separates the success stories, the applicants who actually get jobs, from those who are rejected from these entry-level openings? These are questions that require answers if we are to have a clear picture of how the job market operates in poverty-stricken neighborhoods like Harlem.

WHY WORK?

. . . Jessica has worked at a fast food restaurant in the middle of Harlem since she was seventeen, her first private-sector job following several summers as a city employee in youth program. During her junior and senior year, Jessie commuted forty-five minutes each way to school, put in a full day at school, and then donned her work uniform for an eight-hour afternoon/evening shift. Exhausted by the regimen, she took a brief break from work toward the end of her senior year, but returned when she graduated from high school. Now, at the age of twenty-one, she is a veteran fast food employee with an unbroken work record of about three years.

Jessica had several motivations for joining the workforce when she was a teenager, principal among them the desire to be independent of her mother and provide for her own material needs. No less important, however, was her desire to escape the pressures of street violence and what appeared to be a fast track to nowhere among her peers. For in Jessica's neighborhood, many a young person never sees the other side of age twenty. . . .

Street violence, drive-by shootings, and other sources of terror are obstacles that Jessica and other working-poor people in her community have to navigate around. But Jessica knows that troubles of this kind strike more often among young people who have nothing to do but spend time on the street. Going to work was, for her, a deliberate act of disengagement from such a future.

William, who has worked in the same Burger Barn as Jessica, had the same motivation. A short, stocky African-American, who was "a fat, pudgy kid" in his teen years, Will was often the butt of jokes and the object of bullying in the neighborhood. . . .

As he crested into his teenage years, he wanted some way to occupy his time that would keep him clear of the tensions cropping up in his South Bronx housing project. Lots of boys his age were getting into drugs, but Will says, "Fortunately I was never really into that type of thing." After a stint with a summer youth corps job, he found his "own thing": working for Burger Barn. . . .

For Stephanie, the trouble wasn't just in the streets, it was in her house. When she was in her teens, Stephanie's mother began taking in boarders in their apartment, young men and their girlfriends who did not always get along with her. The home scene was tense and occasionally violent, with knives flashing. The worse it got, the more Stephanie turned her attention to her job. She focused on what her earnings could do to rescue her from this unholy home life. Because she had her own salary, she was able to put her foot down and insist that her mother get rid of the troublemakers. . . .

There are many "push factors" that prod Harlem youth to look for work. Yet there are many positive inducements as well. Even as young teens, Jessica, William, and Stephanie were anxious to pay their own way, to free their families from the obligation to take care of all their needs. In this, they are typical of the two hundred Burger workers I tracked, the majority of whom began their work lives when they were thirteen to fifteen years old. This early experience in the labor force usually involves bagging groceries or working off the books in a local bodega, a menial job under the watchful eye of an adult who, more often than not, was a friend of the family or a relative who happened to have a shop. . . .

While middle-class parents would feel they had abrogated a parental responsibility if they demanded that their kids handle these basic costs (not to mention the frills), many poor parents consider the "demand" perfectly normal. Whether American-born or recent immigrants, these parents often began working at an early age themselves and consequently believe that a "good" kid should not be goofing off in his or her free time—summers, after school, and vacations—but should be bringing in some cash to the family.

Burger Barn earnings will not stretch to cover a poor family's larger items like food and shelter, and in this respect entry-level jobs do not underwrite any real independence. They do make it possible for kids . . . to participate in youth culture. Many writers have dismissed teenage workers on these grounds, complaining that their sole (read "trivial") motivation for working (and neglecting school) is to satisfy childish desires for "gold chains and designer sneakers."[1] Jessie and Will do want to look good and be cool. But most of their wages are spent providing for basic expenses. When she was still in high school, Jessica paid for her own books, school transportation, lunches, and basic clothing expenses. Now that she has graduated, she has assumed even more of the cost of keeping herself. Her mother takes care of the roof over their heads, but Jessie is responsible for the rest, as well as for a consistent contribution toward the expense of running a home with other dependent children in it. . . .

Pressures build early for the older children in these communities to take jobs, no matter what the wages, in order to help their parents make ends meet. Ana Gonzales is a case in point. Having reached twenty-one years old, she had been working since she was fifteen. Originally from Ecuador, Ana followed her parents, who emigrated a number of years before and presently work in a factory in New Jersey. Ana completed her education in her home country and got a clerical job. She emigrated at eighteen, joining two younger brothers and a twelve-year-old sister already in New York. Ana has ambitions for going back to college, but for the moment she works full-time in a fast food restaurant in Harlem, as does her sixteen-year-old brother. Her sister is responsible for cooking and caring for their five-year-old-brother, a responsibility Ana assumes when she is not at work or attending her English as a Second Language class.

The Gonzales family is typical of the immigrant households that participate in the low-wage economy of Harlem and Washington Heights, and it bears a strong resemblance to the Puerto Rican families in other parts of the city.[2] Parents work, adolescent children work, and only the youngest of the children are able to invest themselves in U.S. schooling. Indeed, it is often the littlest who is deputized to master the English language on behalf of the whole family. Children as young as five or six are designated as interpreters responsible for negotiations between parents and landlords, parents and teachers, parents and the whole English-speaking world beyond the barrio.

The social structure of these households is one that relies upon the contributions of multiple earners for cash earnings, child care, and housework. Parents with limited language skills (and often illegal status) are rarely in a position to support their children without substantial contributions from the children themselves. . . .

Older workers, especially women with children to support, have other motives for entering the low-wage labor market. Like most parents, they have financial obligations: rent, clothes, food, and all the associated burdens of raising kids. Among the single mothers working at Burger Barn, however, the options for better-paying jobs are few and the desire to avoid welfare is powerful. This is particularly true for women who had children when they were in their teens and dropped out of school to take care of them.

Latoya . . . had her first child when she was sixteen. She was married at seventeen and then had another. But the marriage was shaky; her husband was abusive and is in jail now. Latoya learned about being vulnerable, and has made sure she will never become dependent again. She lives with Jason, her common-law husband, a man who is a skilled carpenter, and they have a child between them. Jason makes a good living, a lot more money than Latoya can earn on her own. Now that she has three kids, plus Jason's daughter by his first marriage, she has occasionally been tempted to quit work and just look after them. After all, it is hard to take care of four kids, even with Jason's help, and work a full-time job at the same time. She barely has the energy to crawl into bed at night, and crumbles at the thought of the overnight shifts she is obliged to take.

But Latoya's experience with her first husband taught her that no man is worth the sacrifice of her independence.

> *This was my first real job. . . . I take it seriously, you know. . . . It means a lot to me. It give you—what's the word I'm lookin' for? Security blanket. 'Cuz, a lot of married women, like when I was married to my husband, when he left, the burden was left on me. . . .*

For Latoya, as for many other working mothers, working is an insurance policy against dependence on men who may not be around for the long haul.

NOTES

1. Academics, too, often focus upon adolescents' expanding interests in acquiring consumer goods as a primary motivation for employment; see Ellen Greenberger and L. Steinberg, *When Teenagers Work* (Basic Books, 1986), and Laurence Steinberg, *Beyond the Classroom* (Simon & Schuster, 1996). This is due, in part, to the changing composition of the teenage workforce, once largely made up of youth from lower classes. See Joseph F. Kett, *Rites of Passage* (Basic Books, 1977). The teenage workforce today, which has grown enormously since the 1950s, contains just as many solidly middle-class young people, whose total earnings, in most cases, may be spent on luxuries.

2. See Mercer Sullivan, *"Getting Paid": Youth, Crime, and Work in the Inner City* (Cornell University Press, 1989), for an account of Puerto Rican families in Sunset Park, Brooklyn.

DISCUSSION QUESTIONS

1. What motivates the young people Newman interviewed in Harlem to seek work?

2. Why is there a great deal of competition for these low-wage jobs in the inner-city? What skills seem to be important in securing employment?

3. What do the young people Newman interviewed do with their earnings?

38

The Impact of Race on Environmental Quality

An Empirical and Theoretical Discussion

BY RAQUEL PINDERHUGHES

INTRODUCTION

Many toxic substances threaten the health and well being of people in the United States. Human exposure to toxins occurs through various pathways—pesticides in the food, heavy metals, synthetic chemicals, plastic residues, pesticides and other chemical products in the water, and toxic chemicals in the air and soil. When individuals are exposed to environmental hazards the substances often behave synergistically, creating more dangerous effects than those expected by exposure to the hazards individually. Cancer, heart disease, diseases of the respiratory system, neurological damage, birth defects and genetic mutations, miscarriage, lowered sperm count, and sterility are some of the adverse health effects associated with exposure to environmental hazards.

The toxic pollution problem is composed of several interrelated parts which are involved in the process of production, use and disposal of chemicals and products considered necessary for society. Each day millions of pounds of toxic chemicals are used, stored, disposed of, and transported in and out of communities throughout the United States. U.S. corporations use about 65,000 different chemicals, and introduce over 5,000 new chemicals, each year. The majority of these chemicals have not undergone extensive testing to indicate whether they cause long-term health effects such as cancer and reproductive damage. Over 560 million tons of hazardous waste are generated by American industry annually—more than two tons for every resident. The Environmental Protection Agency (EPA) has identified over 30,000 uncontrolled toxic waste sites in the United States.

Most Americans assume that pollution and other environmental hazards are problems faced equally by everyone in our society. But a growing body of research shows that the most common victims of environmental hazards and pollution are minorities and the poor. A strong association between non-white and poor communities and the location of hazardous waste sites has been demonstrated (U.S. General Accounting Office 1983; United Church of Christ 1987; *National Law Journal* 1992). Studies on air quality have documented that minorities and the poor face higher levels of air pollution than other groups. Public health studies show that minorities are more severely affected by environmental hazards than whites.

While all people are threatened by environmental hazards and pollution, toxic chemicals, hazardous substances, and industrial pollutants are disproportionately located in non-white communities and people of color are more severely impacted by these hazards than their white counterparts. Disproportionate exposure to environmental hazards is part of the complex cycle of discrimination and deprivation faced by minorities in the United States....

Recent, pioneering studies provide compelling evidence that environmental quality is mediated by race and socioeconomic status, directly linked to dynamics of discrimination and racial inequality in America, and cannot be separated from issues of equity....

THEORETICAL PERSPECTIVES

The dominant theoretical perspective which is advanced in the disparate studies on the relationship between race, class and environmental quality is that environmental inequities are the result of institutionalized racism. Feagin and Eckberg's (1980) formulation on the dimensions of discrimination provide a framework for understanding how housing discrimination, redlining, residential segregation, market forces, discriminatory practices of federal agencies and local governments, and lack of political and economic power in minority communities produce environmental inequities in American society. They describe an interactive system which has seven dimensions: (a) motivation, (b) discriminatory action, (c) effects, (d) the relation between motivation and action, (e) the relation between action and effects, (f) the immediate organizational (institutional) context, and (g) the larger societal context.

There are two main formulations of this theoretical perspective; both explain how institutionalized racism results in minority communities being targeted as sites for environmentally hazardous industries and facilities. The first perspective emphasizes the functional link between racism, poverty and powerlessness. Minority communities are targeted for siting because they are poorer, less informed, less organized, and less politically influential (Bullard 1993).

To save time and money, companies seek to locate environmentally hazardous industries in communities which will put up the least resistance, which are less informed and less powerful politically, and are more dependent upon local job development efforts....

The second perspective focuses particularly on how the link between race and environmental inequity is essentially forged by segregated housing patterns which confine blacks, and other people of color, to poor communities over burdened by environmental risks....

In contrast to blacks, poor whites are more able to live in economically varied areas and, therefore, benefit from the clout of other (white) middle-class residents with whom they reside or live close to (Godsil, 1991). As a result of residential segregation, redlining, discriminatory land use, and zoning regulations, there is a disproportionate location of environmental hazards in minority communities. People of color face a disproportionate impact on their health and well-being from pollution and environmental hazards....

Both perspectives recognize that institutionalized racism is at the heart of environmental quality and that problems related to poor environmental quality are compounded by a series of problems also attributed to institutionalized racism....

A central question in the debate over the causes of environmental inequality is whether the differential levels of environmental quality are the result of class factors or racial dynamics—whether the bias of distribution of environmental hazards is a function of poverty rather than race. Are not minorities disproportionately targeted and impacted because they are disproportionately poor?

Clearly, poverty plays an important role. First, because of limited income, poor people cannot buy their way out of polluted neighborhoods. Second, because land values are lower in most poor neighborhoods, polluting industries seeking to reduce the costs of business are attracted to poor neighborhoods. Third, wealthier communities can use their political clout and resources—time, money, contacts and knowledge—to pressure city governments not to grant permits to polluting industries, in classic NIMBY fashion.[1] Fourth, because minorities are underrepresented in governing bodies, they tend to be less aware of policies which are being implemented and lack critical resources necessary to pressure government to protect their communities from hazards and threats (National Law Journal 1992). But a growing number of studies on the distribution of specific environmental hazards and environmental quality by race and income show that race is an independent factor, not reducible to class, in predicting the distribution of environmental hazards....

THE ENVIRONMENTAL
JUSTICE MOVEMENT

Obtaining environmental protection and improving environmental quality are inherently political issues, directly linked to the political process. Since the 1960s, the mainstream environmental movement has been one of the most visible and powerful social movements within the spectrum of political activism in the United States. It has built an impressive political base for environmental reform and has been instrumental in getting regulations and legislation passed to improve the environment. Since the 1970s environmental regulation has been one of the most important areas of U.S. domestic politics.

Communities of color have long been involved in the struggle for safe and healthy residential and occupational environments. Local level protest against inadequate sanitation and municipal service delivery, lead poisoning in urban dwellings, asbestos in schools and workplaces, rodents, and environmental pollutants ranging from agricultural pesticides to industrial chemicals are among those issues which have been of concern to communities of color and were consistently on the agenda of the Civil Rights Movement.

African Americans, Latinos, Asians, and Native Americans have long been concerned about environmental issues, but they have conceptualized their concerns as community, labor, economic, health, and civil rights issues rather than as "environmental" issues. At a roundtable organized by *Sierra* magazine, Carl Anthony, director of Urban Habitat, explained, "We've [people of color] all been involved in the struggle for environmental justice for a very long time, even if we didn't call it that."...

The event which focused national attention on environmental inequities and galvanized people of color and environmental activists to focus on the relationship between environmental hazards and minority communities occurred in 1982. That year, residents of Warren County, North Carolina protested the dumping of more than 32,000 cubic yards of soil contaminated with highly toxic polychlorinated biphenyl (PCBs) in their community. Several years earlier, in 1978, the contaminated soil had been illegally dumped along 210 miles of roadway which stretched into fourteen North Carolina counties. When the federal EPA and the state began clean up activities, the Governor decided to bury the contaminated soil in the city of Afton in Warren County. Warren County has the highest percentage of blacks in North Carolina (84%) and is one of the poorest counties in the state. Dr. Charles Cobb, of the United Church of Christ's Commission for Racial Justice, called the decision to bury the contaminated soil in Afton "attempted genocide" because the water table of Afton was only 5–10 feet below the surface and the residents of the community derived all of their drinking water from local wells.

When residents learned that PCBs from the landfill were likely to enter their drinking water supply they worked with the United Church of Christ and Southern Christian Leadership Conference to organize a massive protest against the siting. There was widespread protest, including nonviolent civil disobedience protest against the siting. Protesters came from all over the country and 500 people were arrested. The protest marked the first time anyone in the country had been jailed trying to halt a toxic waste landfill.

Although the campaign to halt the siting failed, the protest produced at least five significant outcomes. First, it caught the attention of activists in communities of color throughout the nation. Second, it brought about a convergence of civil rights and environmental rights. Third, it mobilized a nationally broad-based group to protest environmental inequities, Fourth, it marked the emergence of the environmental justice movement. . . .

Fifth, it led the Commission on Racial Justice to undertake its landmark study of the link between race and the location of hazardous waste sites in the United States which showed that race is the strongest factor related to the presence of hazardous wastes in residential communities throughout the United States (United Church of Christ 1987; Lee 1993).

The Commission's findings prompted Reverend Ben Chavis, then director of the UCC Commission, to use the term "environmental racism" to describe both the intentional and unintentional disproportionate imposition of environmental hazards in communities of color. Subsequently, the Commission formally defined the term to mean:

> . . . racial discrimination in environmental policymaking. It is racial discrimination in the enforcement of regulations and laws. It is the deliberate targeting of communities of color for toxic waste disposal and the siting of polluting industries. It is racial discrimination in the official sanctioning of the life-threatening presence of poisons and pollutants in communities of color. And, it is racial discrimination in the history of excluding people of color from the mainstream environmental groups, decisionmaking boards, commissions, and regulatory bodies (Bullard 1993:3).

Until the emergence of the environmental justice movement, the mainstream environmental movement's agenda was largely focused on protecting the natural world—endangered forests, nearly extinct species, polluted streams, global warming, and natural resource conservation—rather than on what Taylor (1993) has called "human interest ecology." . . .

In order to enhance their ability to effect policy at the local and national levels, leaders of the environmental justice movement challenged the mainstream environmental movement to address the causes and consequences of injustice in the environmental arena. . . .

Today, the environmental justice movement, which emerged to directly confront environmental racism, is composed of a national alliance of church, labor, civil rights, community groups, academics, and others. It is a community based movement led by people of color. According to Ben Chavis:

> Our guiding principle is that our work must be done from a grassroots perspective, and it must be multi-racial and multicultural . . . we're being inclusive, not exclusive. We're not saying to take the incinerators and the toxic waste dumps out of our communities and put them in white communities—we're saying they should not be in anybody's community . . . It's not a movement of retribution—it is a movement for justice. You can't get justice by doing an injustice on somebody else. When you have lived through suffering and hardship, you want to remove them, not only from your own people but from all people (Sierra Roundtable 1993:52).

NOTES

1. The term "NIMBY" refers to Not In My Back Yard.

REFERENCES

Bullard, Robert D. 1993. *Confronting Environmental Racism*. Boston, MA: South End Press.

Feagin, Joe R., and Douglas Lee Eckberg. 1980. "Discrimination, Motivation, Action, Effects, and Context." *Annual Review of Sociology* 6:1–20.

Godsil, Rachel D. 1991. "Remedying Environmental Racism." *Michigan Law Review* 90:394–425.

Lee, Charles. 1993. "Beyond Toxic Wastes and Race." pp. 41–52 in *Confronting Environmental Racism*, edited by Robert Bullard. Boston, MA: South End Press.

National Law Journal. 1992. "Unequal Protection: The Racial Divide in Environmental Law." *National Law Journal*.

Sierra Roundtable on Race, Justice, and the Environment. 1993. "A Place at the Table." *Sierra Magazine* (June): 50–96.

Taylor, Dorceta E. 1993. "Environmentalism and the Politics of Inclusion." pp. 53–62 in *Confronting Environmental Racism*, edited by Robert Bullard. Boston, MA: South End Press.

United Church of Christ Commission for Racial Justice. 1987. *Toxic Wastes and Race in the United States: A National Report on the Racial and Socio-Economic Characteristics of Communities with Hazardous Waste Sites*. New York: United Church of Christ.

U.S. General Accounting Office. 1983. *Siting of Hazardous Waste Landfills and Their Correlation with Racial and Economic Status of Surrounding Communities*. Washington, DC: GAO/RCED.

DISCUSSION QUESTIONS

1. Why does Pinderhughes think it is important to look at the quality of the environments where people live?

2. Define environmental racism.

3. How might the outcome of the protest in Warren County, North Carolina been different if the residents were White and middle class? What does that fact say about the forces necessary to challenge environmental practices?

4. How does understanding environmental racism expand the scope of the struggle for a safer environment?

39

Welfare Reform, Family Hardship, and Women of Color

BY LINDA BURNHAM

The stated intent of welfare reform was at least twofold: to reduce the welfare rolls and move women toward economic self-sufficiency. The first objective has been achieved: welfare rolls have declined dramatically since 1996. Welfare reform has stripped single mothers of any sense that they are entitled to government support during the years when they are raising their children.

Despite the "success" of welfare reform, research has repeatedly found that many women who move from welfare to work do not achieve economic independence. Instead, most find only low-paid, insecure jobs that do not lift their families above the poverty line. They end up worse off economically than they were on welfare: they work hard and remain poor. Others are pushed off welfare and find no employment. They have no reported source of income.

Women in transition from welfare to work—or to no work—face particular difficulties and crises related to housing insecurity and homelessness and food insecurity and hunger.

Low-income people in the United States faced a housing crisis long before the passage of the PRWORA.[1] In most states, the median fair-market cost of housing for a family of three is considerably higher than total income from a

From *Lost Ground: Welfare Reform, Poverty and Beyond*, ed. by Randy Albelda and Ann Withorn (Cambridge: South End Press, 2002), pp. 43–56. Reprinted by permission of the publisher.

1. Editor's note: Personal Responsibility and Work Opportunities Reconciliation Act of 1996 that introduced temporary aid for needy families.

Temporary Aid to Needy Families (TANF) grant (Dolbeare 1999). Further, as a consequence of two decades of declining federal support for public and subsidized housing, the great majority of both current and former TANF recipients are at the mercy of an unforgiving private housing market.

The withdrawal of the federal government's commitment to need-based income support adds a powerful destabilizing element to already tenuous conditions. The evidence that welfare reform is contributing to rising levels of housing insecurity and homelessness is piling up. . . .

Confronting the absurd and agonizing decision of whether to feed their children or house them, most mothers will use the rent money to buy food and then struggle to deal with the consequences. In one national study, 23 percent of former welfare recipients moved because they could not pay the rent (Sherman et al. 1998, 13). A New Jersey survey found that 15.8 percent of respondents who had had their benefits reduced or terminated in the previous 12 months had lost their housing (Work, Poverty and Welfare Evaluation Project 1999, 53). Furthermore, in Illinois, 12 percent of former recipients who were not working and 5 percent of former recipients who were working experienced an eviction (Work, Welfare and Families 2000, 25). . . .

Although the PRWORA was trumpeted as a step toward strengthening families, increased housing insecurity and homelessness have led to families being split apart. Most family shelters do not take men, so the fathers of two-parent families that become homeless must either go to a single men's shelter or make other housing arrangements. Many shelters also do not accommodate adolescent boys or older male teens. Family breakup may be required for a shelter stay.

The housing instability of poor women and their children has profound consequences, both for them and for society as a whole. Homelessness compromises the emotional and physical health of women and children, disrupts schooling, and creates a substantial barrier to employment. It widens the chasm between those who are prospering in a strong economy and those who fall ever farther behind. . . .

Like homelessness, the problems of food insecurity predate welfare reform. Low-income workers and welfare recipients alike have struggled for years to provide adequate food for themselves and their families. The robust economy of the late 1990s did not fundamentally alter this reality. Of families headed by single women, one in three experiences food insecurity and one in ten experiences hunger (Work, Welfare and Families 2000, 25).

Welfare reform has made women's struggles to obtain food for themselves and their families more difficult. Several studies document that former recipients cannot pay for sufficient food and that their families skip meals, go hungry, and/or use food pantries or other emergency food assistance. . . .

The Food Stamp Program is intended to ensure that no family goes hungry, but many families do not receive the food stamps to which they are entitled. Even before welfare reform, the rate of participation in the Food Stamp Program was declining more rapidly than the poverty rate. The number of people receiving food stamps dropped even more steeply later, from 25.5 million average monthly recipients in 1996 to 18.5 million in the first half of 1999 (U.S. General Accounting Office 1999, 46). The rate of participation is the lowest it has been in two decades, with a growing gap between the need for food assistance and families' use of food stamps.

Welfare reform has itself contributed to the underutilization of food stamps. . . . Confusion and misinformation on the part of eligibility workers, or their withholding of information, are also factors in the low participation of former recepients. . . . Among families who had left welfare, only 42 percent of those who were eligible for food stamps were receiving them (Zedlewski and Brauner 1999, 1–6).

Not surprisingly, demands for food from other sources are increasing. As the welfare rolls shrink, requests for food from charities rise. Catholic Charities reported a 38 percent rise in demand for emergency food assistance in 1998. In many cases, the demand for food goes unmet. As one report states, "The bottom line is that . . . for millions of households, workforce participation has been accompanied by hunger" (Venner, Sullivan, and Seavey 2000, 16).

WOMEN OF COLOR
AND IMMIGRANT WOMEN

Welfare reform is a nominally race-neutral policy suffused with racial bias, both in the politics surrounding its promulgation and in its impact. It may not have been the intent to racially target women of color for particular punishment, yet women of color and immigrant women have nonetheless been particularly hard hit in ways that were highly predictable.

Feminist theory has for some time recognized that the social and economic circumstances women of color must negotiate are shaped by the intersection of distinct axes of power—in this case primarily race, class, and gender. The relationships of subordination and privilege that define these axes generate multiple social dynamics that influence, shape, and transform each other, creating, for women of color, multiple vulnerabilities and intensified experiences of discrimination.

Welfare reform might legitimately be regarded as a class-based policy intended to radically transform the social contract with the poor. Poverty in the United States, however, is powerfully structured by racial and gender inequities. It is not possible, therefore, to institute poverty policy of any depth that does not also reconfigure other relations, either augmenting or diminishing race and gender inequalities. By weakening the social safety net for the poor, PRWORA necessarily has its greatest effect on those communities that are disproportionately represented among the poor. Communities of color and immigrant communities, already characterized by significantly higher levels of minimum-wage work, homelessness, hunger, and poor health, are further jeopardized by the discriminatory impact of welfare reform.

As a consequence of the historical legacy and current practices of, among other things, educational inequity and labor market disadvantage, patterns of income and wealth in the United States are strongly skewed along racial lines, for example, the disproportionate burden of poverty carried by people of color. . . .

Economic vulnerabilities due to race and ethnicity may be further compounded by disadvantages based on gender and immigration/citizenship status. Thus, for households headed by single women, the poverty rates are also stark. Over 21 percent of such white non-Hispanic households were below the poverty line in 1998, as compared to over 46 percent of black and 48 percent of Hispanic female-headed households (U.S. Census Bureau 1999). Immigrants, too, are disproportionately poor, with 18 percent below the poverty line as compared to 12 percent of the native born (U.S. Census Bureau 1999). Given the disproportionate share of poverty experienced by people of color, and the significant poverty of single-mother households, it is no surprise that the welfare rolls are racially unbalanced, with women of color substantially overrepresented. . . .

This racial imbalance has been cynically used for decades in the ideological campaign to undermine support for welfare—a crude but ultimately effective interweaving of race, class, gender, and anti-immigrant biases that prepared the consensus to "end welfare as we know it." Having been maligned as lazy welfare cheats and something-for-nothing immigrants, Latinas, African American women, and Asian women of particular nationality groups are now absorbing a punishing share of welfare reform's negative impacts. . . .

Beyond intensified impact due to disproportionate representation in the affected population, additional factors compound the disadvantages of women of color and immigrant women. One Virginia study found noteworthy differences in how caseworkers interact with black and white welfare recipients. A substantial 41 percent of white recipients were encouraged to go to school to earn their high school diplomas, while no black recipients were. A much higher proportion of whites than blacks found their caseworkers to be helpful in providing information about potential jobs (Gooden 1997). Other studies showed that blacks were removed from welfare for noncompliance with program rules at considerably higher rates than white recipients, while a higher proportion of the cases of white recipients were closed because they earned too much to qualify for welfare (Savner 2000).

Further, while welfare use is declining among all races, white recipients are leaving the welfare rolls at a much more rapid rate than blacks or Latinos. In New York City, for example, the number of whites on welfare declined by 57 percent between 1995 and 1998, while the rate of decline for blacks was 30 percent and that of Latinos 7 percent. White recipients have also been leaving the rolls at faster rates than minorities in states such as Illinois, Pennsylvania, Michigan, and Ohio. And nationally, the decline has been 25 percent for whites but only 17 percent for African Americans and 9 percent for Latinos (DeParle 1998, A1).

The causes of this phenomenon have been insufficiently studied, but some of the factors may include higher average educational levels among white recipients, greater concentrations of recipients of color in job-poor inner cities, racial discrimination in employment and housing, and discriminatory referral policies on the part of welfare-to-work caseworkers. . . .

Some of the most punitive provisions of PRWORA are directed at immigrants. The 1996 legislation banned certain categories of legal immigrants from a wide array of federal assistance programs, including TANF, food stamps, Supplementary Security Income, and Medicaid. In the year following passage, 940,000 legal immigrants lost their food stamp eligibility. Strong advocacy reversed some of the cuts and removed some restrictions, but legal immigrants arriving in the United States after the passage of the legislation are ineligible for benefits for five years. States have the right to bar pre-enactment legal immigrants from TANF and nonemergency Medicaid as well (National Immigration Law Center 1999).

These restrictions have had profound effects on immigrant communities. First of all, many immigrant women who are on welfare face significant barriers to meeting TANF work requirements. Perhaps the most formidable obstacles are limited English proficiency and low educational levels. . . .

Limited English, lack of education, and limited job skills severely restrict immigrant women's options in the job market, making it very difficult for them to comply with welfare-to-work requirements. Language problems also impede their ability to negotiate the welfare bureaucracy, which provides very limited or no translation services. These women lack information about programs to which they are entitled, and they worry about notices that come to them in English. When immigrant women recipients are able to find work, it is most often in minimum-wage or low-wage jobs without stability or benefits (Center for Urban Research and Learning 1999, 5; Equal Rights Advocates 1999, 31). . . .

Immigrant women recipients are also likely to experience severe overcrowding and to devote a huge portion of their income to housing. They share housing with relatives or with unrelated adults; live in garages or other makeshift, substandard dwellings; and worry constantly about paying the rent.

A more hidden, but still pernicious, impact of welfare reform has been the decline in applications for aid from immigrants who are eligible to receive it. . . . The intensive anti-immigrant propaganda that accompanied the passage of PRWORA and statewide anti-immigrant initiatives appears to have discouraged those who need and are entitled to aid from applying for it, surely undermining the health and welfare of immigrant women and their families. . . .

Undoing the damage of welfare reform . . . will require the promulgation and implementation of policies that restore and strengthen the social safety net for women and children while funding programs that support women along he path to economic self-sufficiency. In the absence of the political will for such a comprehensive reworking of U.S. social welfare policy, advocates for poor women and families face an extended, defensive battle to ameliorate the cruelest and most discriminatory effects of this radically regressive policy.

REFERENCES

Center for Urban Research and Learning. 1999. *Cracks in the System: Conversations with People Surviving Reform*. Chicago: Center for Urban Research and Learning, Loyola University, Howard Area Community Center, Organization of the NorthEast.

DeParle, Jason. 1998. Shrinking Welfare Rolls Leave Record High Share of Minorities. *New York Times,* 24 July.

Dolbeare, Cushing. 1999. *Out of Reach: The Gap Between Housing Costs and Income of Poor People in the United States*. Washington, DC: National Low-Income Housing Coalition.

Equal Rights Advocates. 1999. *From War on Poverty to War on Welfare: The Impact of Welfare Reform on the Lives of Immigrant Women*. San Francisco.

Gooden, Susan. 1997. Examining Racial Differences in Employment Status Among Welfare Recipients. In *Race and Welfare Report*. Oakland, CA: Grass Roots Innovative Policy Program.

National Immigration Law Center. 1999. *Immigrant Eligibility for Public Benefits*. Washington, DC.

Savner, Steve. 2000. Welfare Reform and Racial/Ethnic Minorities: The Questions to Ask. *Poverty & Race* 9(4): 3–5.

Sherman, Arloc, Cheryl Amey, Barbara Duffield, Nancy Ebb, and Deborah Weinstein. 1998. *Welfare to What: Early Findings on Family Hardship and Well-Being*. Washington, DC: Children's Defense Fund and National Coalition for the Homeless.

U.S. Census Bureau. 1999. *Poverty Thresholds in 1998 by Size of Family and Number of Related Children Under 18 Years*. Washington, DC.

U.S. General Accounting Office. 1999. *Food Stamp Program: Various Factors Have Led to Declining Participation*. Washington, DC.

Venner, Sandra H., Ashley F. Sullivan, and Dorie Seavey. 2000. *Paradox of Our Times: Hunger in a Strong Economy*. Medford, MA: Tufts University, Center on Hunger and Poverty.

Work, Poverty and Welfare Evaluation Project. 1999. Assessing Work First: What Happens After Welfare? Report for the Study Group on Work, Poverty and Welfare. Legal Services of New Jersey, New Jersey Poverty Research Institute, Edison, NJ.

Work, Welfare and Families. 2000. *Living with Welfare Reform: A Survey of Low Income Families in Illinois*. Chicago: Chicago Urban League and UIC Center for Urban Economic Development.

Zedlewski, Sheila R. and Sarah Brauner. 1999. *Are the Steep Declines in Food Stamp Participation Linked to Falling Welfare Caseloads?* Washington, DC: Urban Institute.

DISCUSSION QUESTIONS

1. According to Burnham, how do economic hardships disrupt family life?

2. Why does it appear that non-Hispanic White women are more successful in transitioning off of TANF than women of color?

3. Why is a supposedly race-neutral social policy like the Personal Responsibility and Work Opportunities Reconciliation Act having a negative impact on racial minorities and immigrants?

40

Child Welfare as a Racial Justice Issue

BY DOROTHY ROBERTS

My main intellectual and activist project over the last five years has been to explain child welfare policy as an issue of racial justice. Child welfare was once viewed as a social issue, but by the 1970s the main mission of public child welfare departments had become protecting children against maltreatment inflicted by pathological parents. Child welfare decision making has an atomistic focus, that zooms in on the situation of individual children and their families. If child welfare is discussed as a matter of rights at all, it is usually framed as a contest between children's rights and parents' rights, falsely assuming that the interests of parents and children in the child welfare system are always in opposition to each other and that the system treats all parents and children equally badly.

Strangely, criticisms of the child welfare system are not placed among the burning social justice issues of our day. I say strangely because anyone who is familiar with the child welfare system in the nation's large cities knows that it is basically an apartheid institution. Spend a day at any urban dependency court and you will see a starkly segregated operation. If you came with no preconceptions about the purpose of the child welfare system, you would have to conclude that it is an institution designed to monitor, regulate, and punish poor minority families, especially Black families.

From Dorothy Roberts, "Child Welfare as a Racial Justice Issue," 2002. Reprinted by permission of the author.

The number of Black children in state custody—those in foster care as well as those in juvenile detention, prisons, and other state institutions—is alarming.... Black children make up about two-fifths of the foster care population, although they represented less than one-fifth of the nation's children (Administration for Children and Families 2003). The color of child welfare is most apparent in big cities where there are sizeable Black and foster care populations. In Chicago, for example, almost all of the children in foster care are Black (Pardo 1999). The racial imbalance in New York City's foster care population is also mind-boggling: out of 42,000 children in the system at the end of 1997, only about 1000 were white (Guggenheim 2000). Black children in New York were 10 times as likely as White children to be in state protective custody. Although the total numbers are smaller, this racial disproportionality extends to cities and states where Black children are less visible.

State agencies treat child maltreatment in Black homes in an especially aggressive fashion. They are far more likely to place Black children than other children who come to their attention in foster care instead offering their families less traumatic assistance. A national study of child protective services by the U.S. Department of Health and Human Services reported that "[m]inority children, and in particular African American children, are more likely to be in foster care placement than receive in-home services, *even when they have the same problems and characteristics as white children*" (U.S. Dept. HHS 1997). Most white children who enter the system are permitted to stay with their families, avoiding the emotional damage and physical risks of foster care placement, while most Black children are taken away from theirs. Foster care is the main "service" state agencies provide to Black children. And once removed from their homes, Black children remain in foster care longer, are moved more often, receive fewer services, and are less likely to be either returned home or adopted than any other children (Courtney & Wong 1996; Jones 1997).

In some cases, protecting children requires immediately removing them from their homes. But the public often overlooks the costs to children of separating them from their families. In 2001, Judge Jack Weinstein of the Eastern District of New York issued a blistering condemnation of New York City's Administration for Children's Services for automatically removing children from mothers who were victims of domestic violence (*Nicholson v. Scoppetta* 2001). Judge Weinstein's decision is especially noteworthy for its rare judicial recognition of the harm inflicted on children by unnecessarily taking them from their parents. "It hardly needs to be added that the exact language of the Thirteenth Amendment covers protection of the children's rights," Judge Weinstein wrote. "They are continually forcibly removed from their abused mothers without a court adjudication and placed in a forced state custody in either state or privately run institutions for long periods of time. They are disciplined by those not their parents. This is a form of slavery."

A new politics of child welfare threatens to intensify state supervision of Black children. In the last several years, federal and state policy has shifted away from preserving families toward "freeing" children in foster care for adoption by terminating parental rights. The Adoption and Safe Families Act, passed in

1997, imposes an expedited time frame for state agencies to file petitions to terminate parental rights and give states a financial bonus for increasing the number of adoptions of children in foster care—$4,000 per child, $6,000 if the child is classified as having special needs. The campaign to increase adoptions has hinged on the denigration of foster children's parents, the speedy destruction of their family bonds, and the rejection of family preservation as an important goal of child welfare practice. Adoption is increasingly presented not as an option for a minority of foster children who cannot be reunited with their parents, but as the preferred outcome for all children in foster care. . . .

Welfare reform, by throwing many families deeper into poverty, heightens the risk that the most vulnerable children will be placed in foster care. A front page story in the *New York Times,* entitled "Side Effect of Welfare Law: the No-Parent Family," reported a study of census data in all 50 states that found that the rate of Black children in cities living without their parents has more than doubled as a result of welfare reform—an estimated 200,000 more Black children separated from their parents (Bernstein 2002).

In addition, tougher treatment of juvenile offenders, imposed most harshly on African American youth, is increasing the numbers incarcerated in juvenile detention facilities and adult prisons. These political trends are converging to address the deprivation of poor Black children by placing more of them in one form of state custody or another. Child welfare policy conforms to the current political climate, which embraces private solutions—and when those fail, punitive responses—to the seemingly intractable plight of America's isolated and impoverished inner cities. As welfare reform reduces the welfare rolls by promoting marriage and imposing sanctions for failing to find work, child welfare policy reduces the foster care rolls by terminating parental rights and promoting adoption of the "legal orphans" it creates.

The color of America's child welfare system undeniably shows that race matters to state interventions in families. So why have scholars and policymakers been slow to describe the disproportionate involvement of Black families in the system as a racial injustice? Let me suggest three related explanations.

First, there is profound confusion about the reasons for the system's racial disparity. The existing social science literature contains theories that attribute the racial disparity both to differences in the well being of children and to differences in the system's treatment of children. In other words, there is disagreement over whether the disparity stems from societal conditions *outside* the system, such as higher poverty rates among nonwhite children, or from racially biased practices *within* the system. Many experts believe that this distinction in causes—societal forces vs. child welfare practices—makes a crucial difference in how we should address the racial disparities (Courtney et al. 1998). If the cause of the system's racial imbalance is social and economic inequality, some say, we can't blame the system (Bartholet 1999).

This is related to a second reason for failing to see a racial justice issue—the concern for children's rights. If the child welfare system is simply reflecting inequities in children's living conditions, we would expect—we would even want—the state to intervene more often to protect Black children from the greater harm that they face. Indeed, wouldn't the government *violate* Black children's rights if it *failed* to intervene more often to protect them? . . .

A final stumbling block is the official understanding of racial discrimination. Under current civil rights jurisprudence, the racial disparity in the child welfare system may not constitute racial discrimination without a showing of racial motivation. The system is racist only if Black children are pulled out of their homes by bigoted caseworkers or as part of a deliberate government scheme to subjugate Black people. Any other explanation—such as higher rates of Black family poverty or unwed motherhood—negates the significance of race. . . .

But these views of the child welfare system, children's rights, and racial discrimination fail to take into account the political dimension of child welfare policy. To begin with, which harms to children are detected, identified as abuse or neglect, and punished is determined by inequities based on race, class, and gender. The U.S. child welfare system is and always has been designed to deal with the problems of poor families (Pelton 1989). The child welfare system hides the systemic reasons for poor families' hardships by attributing them to parental deficits and pathologies that require therapeutic remedies rather than social change. The harms caused to children by uncaring, substance-abusing, mentally unstable, absentee parents in middle-class and affluent families usually go unheeded. Although these children from privileged homes might spend years in psychotherapy, it is unlikely they will spend any time in foster care. Most child maltreatment charges are for neglect and involve poor parents whose behavior was a consequence of economic desperation as much as lack of caring for their children.

The racial disparity in the child welfare system also reflects a political choice about how to address child neglect. It is no accident that child welfare philosophy became increasingly coercive as Black children made up a greater and greater share of the caseloads. In the past several decades, the number of children receiving child welfare services has declined dramatically, while the foster care population has skyrocketed (U.S. Dept. HHS 1997). As the child welfare system began to serve fewer white children and more Black children, state and federal governments spent more money on out-of-home care and less on in-home services. This mirrors perfectly the metamorphosis of welfare once the welfare rights movement succeeded in making AFDC available to Black families in the 1960s. As welfare became increasingly associated with Black mothers, it became increasingly burdened with behavior modification rules and work requirements until the federal entitlement was abolished altogether in 1996. Both systems responded to their growing Black clientele by reducing their services to families while intensifying their punitive functions.

The child welfare system's reliance on a disruptive, coercive, and punitive approach also inflicts a political harm. American constitutional jurisprudence defines the harm caused by unwarranted state interference in families in terms of individual rights. Wrongfully removing children from the custody of their parents violates parents' due process right to liberty. The earliest cases interpreting the due process clause to protect citizens against government interference in their substantive liberty involved parental rights. But these explanations of harm do not account for the particular injury inflicted by racially disparate state intervention. The over-representation of Black children in the child welfare system, especially foster care, represents massive state supervision and dissolution of families. This interference with families helps to maintain the disadvantaged status of Black people in the United States. The child welfare system not only inflicts general harms disproportionately on Black families. It also inflicts a particular harm—a racial harm—on Black people as a group. . . .

State supervision of families is antithetical to the role families are supposed to play in a democracy, as critical components of civil society. Families are a principal form of "oppositional enclaves" that are essential to citizens' free participation in democratic institutions, to use Harvard political theorist Jane Mansbridge's term (Mansbridge 1996, 58). Family and community disintegration weakens Blacks' collective ability to overcome institutionalized discrimination and to work toward greater political and economic strength. Family integrity is crucial to group welfare and identity because of the role parents and other relatives play in transmitting survival skills, values, and self-esteem to the next generation. Placing large numbers of children in state custody interferes with the group's ability to form healthy and productive connections among its members. The system's racial disparity also reinforces the quintessential racist stereotype: that Black people are incapable of governing themselves and need state supervision.

The impact of state disruption and supervision of families is intensified when it is concentrated in inner-city neighborhoods. In 1998, one out of every ten children in Central Harlem had been taken from their parents and placed in foster care (Center for an Urban Future 1998, 6). In Chicago, almost all child protection cases are clustered in a few zip code areas, which are almost exclusively African American. The spatial concentration of child welfare supervision creates an environment in which state custody of children is a realistic expectation, if not the norm. . . .

The racial disparity in the foster care population should cause us to reconsider the state's current response to child maltreatment. The price of present policies that rely on child removal rather than family support falls unjustly on Black families and communities. In part because of narrow conceptions of racial discrimination and children's rights, judges, politicians, and the public have a hard time seeing this as a racial injustice. I propose that we figure out better ways of measuring and explaining this type of systemic, community-wide racial harm.

REFERENCES

Administration for Children and Families, U.S. Department of Health & Human Services. 2000. *Child Maltreatment 1998: Reports from the States to the National Child Abuse and Neglect Data System.* Washington, D.C.: U.S. Government Printing Office.

———— 2003. *The AFCARS Report: Preliminary FY 2001 Estimates as of March 2003.* http://www.acf.hhs.gov/programs/cb/publications/afcars/report8.htm.

Bernstein, Nina. 2002, July 29. "Side Effect of Welfare Law: The No-Parent Family." *New York Times*, p. 1.

Bartholet, Elizabeth. 1999. *Nobody's Children*. Boston: Beacon Press.

Center for an Urban Future, 1998. "Race, Bias, and Power in Child Welfare." *Child Welfare Watch*. Spring/Summer: 1.

Courtney, Mark E., et al. 1998. "Race and Child Welfare Services: Past Research and Future Directions." *Child Welfare* 75:99.

Courtney, Mark E. and Wong, Vin-Ling Irene. 1996. "Comparing the Timing of Exits from Substitute Care." *Child & Youth Services Review* 18:307.

Guggenheim, Martin. 2000. "Somebody's Children: Sustaining the Family's Place in Child Welfare Policy." *Harvard Law Review* 113: 1716.

Jones, Loring P. 1997. "Social Class, Ethnicity, and Child Welfare." *Journal of Multicultural Social Work* 6:123.

Males, Mike and Dan Macallair. 2000. *The Color of Justice: An Analysis of Juvenile Adult Transfers in California*. San Francisco: Justice Policy Institute.

Mansbridge, Jane. 1996. "Using Power/Fighting Power: The Polity." In *Democracy and Difference: Contesting the Boundaries of the Political,* edited by Seyla Benhabib, 46. Princeton: Princeton University Press.

Nicholson v. Scoppetta. 2001. 202 F.R.D. 377. United States District Court, Eastern District of New York.

Pardo, Natalie. 1999. "Losing Their Children." *Chicago Reporter* 28:1.

Pelton, LeRoy H. 1989. *For Reasons of Poverty: A Critical Analysis of the Public Child Welfare System in the United States*. New York: Praeger.

U.S. Department of Health and Human Services. 1997. *National Study of Protective, Preventive, and Reunification Services Delivered to Children and Their Families*. Washington, D.C.: U.S. Government Printing Office.

DISCUSSION QUESTIONS

1. What does Roberts mean by an apartheid institution? How does this apply to the child welfare systems in major cities?

2. How can children become legal orphans?

3. How does the impact of welfare policies on individual families influence the collective well-being of the Black community?

41

The Social Construction of the "Immoral" Black Mother

Social Policy, Community Policing, and Effects on Youth Violence

BY BETH E. RICHIE

Few areas of social life have been as contested in social policy debates as the concept of the family. Highly charged rhetoric about gender and generational relationships surrounds most recent proposals for reform. From nostalgic calls for conservative approaches by religious right-wing forces to seemingly progressive legislative intitiatives advocating gay/lesbian marriages, debates about family life are played out on various ideological templates. Even in progressive contexts, such as the recent reconsiderations of adolescent pregnancy, the problem has been constructed as the need to "strengthen fragile families." Similarly, in the field of public health, we see an emphasis on the family as the cornerstone of emotional and social well-being, examined via resiliency factors that emerge from particular forms of household arrangements. In these and other examples, current social policy reform is increasingly attached to the organization and meaning of the role of the family in contemporary society, and overall the constructs have a distinctively conservative tendency.

Motherhood, as a subcategory of the family debates, is constituted through a similar vast range of intellectual, political, and popular rhetoric, and with

From *Revisioning Women, Health, and Healing: Feminist, Cultural, and Technoscience Perspectives,*
ed. by Adele E. Clarke and Virginia L. Olesen (New York: Routledge, 1999), pp. 283–297.
Copyright © 1999. Reprinted by permission of Routledge/Taylor & Francis Books, Inc.

similar conservative undertones. While characterized by mixed conceptual frames (ranging from the best practice of motherhood to the healthiest type of relationships between mothers and children), still at the center of the idological debate are universalistic assumptions that revolve around "desirable" family forms, "appropriate" gender roles, and the maintenance of a separation between public and the private spheres. For example, policy makers continue to interrogate researchers about the effects on children when women work outside the home, and legislators argue about what single form family leave should take. . . .

Instead of being featured in the debates, most black women in low-income communities fall far outside the normative, hegemonic parameters of such discussions. With noted exceptions (Dickerson 1995), most considerations of the mothering that poor black women do is introduced into the political, social, and empirical debates from a very different social location. At best, their mothering is studied as a culturally distinct add-on to the dominant inquiries. In its worst and far more common form, low-income black women's mothering is used as a not-so-coded metaphor for much of what is wrong with contemporary society (Hill 1997). Black women are portrayed as creating pathological forms of families as "single heads of households," as draining public resources, or as breeding too many children who pose physical, social and economic risks to others. Their mothering is viewed as something quite different from the mothering efforts of other groups—as a category of activities enacted in such dissimilar ways from the dominant model that they are constructed as confusing, atypical, and dysfunctional. Ultimately, I will argue here, this outside position renders black women's mothering immoral, if not criminal, in the perspective of those who formulate and enforce social policies.

My argument here is that, worse than simply ignoring the role that mothering assumes in poor black women's lives, the current analytical and ideological framework does great harm to these women, their children, and their communities. Rather than seeking to understand and then address the social needs of black women and their families within the contexts in which we actually live, current conceptualizations ignore the specificity of the micro processes of mothering and misinterpret key behaviors and actions of mothers. The social policies that ensue reinforce such conceptualizations pathologizing and stigmatizing effects. The overall result is increased marginalization, structural disenfranchisement, hypersurveillance and overregulation of poor black women's mothering in new and profound ways. The particular case that I will use to argue this point concerns the consequences that social policy on youth violence has on black women's mothering.

I frame this discussion with findings from a study of twenty-four adult women who are female caretakers of adolescent children in a low-income community in a major urban area where, like many other cities, youth violence is a devastating social and public health problem. The broader research project of which this is a part examines how youth violence is distinctively and decidedly gendered in nature, and how the interventions designed to address these problems ignore this important dimension. Hence the problem of youth violence, typically constructed as a problem that affects young men of color, is neither linked to the issue of gender violence nor understood to have any effects on girls and women when, in actuality, it certainly does. . . .

METHODOLOGY

My study was designed to explore the impact of youth violence on mothers and mothering, using the life-history interview technique to elicit data on the ways that black women thought and felt about their experiences. The life-history method was selected because it is particularly useful in gathering information about stigmatized, uncomfortable, or difficult circumstances in the subjects' lives. Compared to other, more structured qualitative methods, conducting life-history interviews offers a more open and intense opportunity to learn about the subjects' backgrounds, opinions, and feelings, as well as the meanings they give to both the mundane and exceptional experiences in their lives. Mothering is obviously in this realm. . . .

The people to be interviewed were drawn from populations of black women whose children are involved in or at serious risk of experiencing or witnessing violence in the private or public spheres of their lives. Twenty-four women agreed to participate, ranging in age from nineteen to sixty-nine years old. Included in the sample were (1) mothers or guardians of pregnant or parenting adolescents; (2) mothers who resided in public housing, subsidized housing or public shelters; and (3) primary caretakers of children detained in institutional settings for juveniles. . . .

Four basic areas were covered by the interviews, beginning with an open-ended question: "Tell me the things about yourself that are important to you." Next the women were asked about factors they felt influenced their role as mothers: "What is it like raising children in your household and neighborhood?" Third, I sought to capture their experiences and perceptions of gender and youth violence and how it affected their lives: "In what ways does violence or the threat of violence affect you and your family or neighborhood?" Last, I asked how their life might be different in the future, what they wanted, dreamed for, hoped for, and expected for themselves and their children. . . .

THE FIVE THEMES

Theme 1: The Diminished Ability to Parent
Due to Limited Economic and Social Supports
within the Context of Urban Decay

The context within which the women lived was marked by their economic marginalization. They described the following characteristics of their world. First, observation and experience had led them to conclude that "doing the right thing for your children" would not necessarily work for them as members of a marginalized ethnic group. They described feeling that, as low-income women whose attempts to mother had repeatedly failed, they somehow fell outside of society's parameters of goodness or fairness. This sense translated into feelings of powerlessness, frustration, and discontent with their own mothering abilities. They were typically self-blaming even when, paradoxically, they articulated an insightful analysis of the social conditions that led to their marginalized position as women and as mothers.

A second dimension of this theme was the degree to which their household fabric has been limited by changes in their efficacy as adults in the social world. Their household composition changes frequently, usually in response to economic shifts, and this often limits important intergenerational contact. They and other adults lose jobs or are only marginally employed. Their families double up in inadequate housing. They simply do not have the resources to perform their parenting roles as well as they desire. Successful role models for both their children and themselves were limited, and extended family networks were quite tenuous.

While many of the women interviewed grew up in poor families, the effects of persistent multigenerational poverty are taking a toll on them. For while poverty may not be new, the level and the nature of hostile public sentiment, the prolonged feelings of despair, and the extent of the violence in their communities are new. They did not grow up watching and knowing of their friends being killed the way their children do, and this huge experiential gap has left most of them unprepared to help their children make sense of these tragic events or offer much support.

The women interviewed for this study also described how public socializing systems are failing them and their children. Schools are considered dangerous, rigid places where the mothers described feeling as alienated as the young people do. One woman said, "They look much like prisons, and I feel like they are holding my child captive there for some crime of going to school. I have no rights as a visitor, and definitely no input into what happens there."

A broader exploration of structural conditions reveals that community institutions and most public spaces are decaying. Businesses, movie theaters, libraries, and parks are closing, and services at hospitals and mental health facilities are being cut back ... The women reported that a decline in availability of public transportation (buses have changed their routes and cabs won't stop to transport them) has left them isolated within their communities. ...

The women understand these isolating and confining strategies as symbolic of the larger community's fear of them and their children. Many schools have metal detectors through which children must pass. Surprisingly, so do laundromats, video arcades, and the music stores young people frequent. Gated retail establishments favor merchandizing large bottles of beer and candy packaged like liquor over fresh produce. In these and other obvious ways, raising children is limited by perverse environmental conditions, lack of social support, symbolic fear, and persistent economic decay. ...

Theme 2: The Constant Fear of Losing
Children to Public Agencies

This second theme can best be characterized by the words of one of the women, who said, "The state is actually raising our children, and as far as I can see they are not doing a very good job. Our job as mothers has therefore become to keep running from child protection, from truant and probation officers, from social workers and the like who are trying to take our kids from us. Family values, not! It's like the slave days . . . they want to take our kids." This was one of many moving testimonies to how women are struggling to escape the intervention of authorities and maintain their custodial rights.

The phenomenon of women being surveilled and monitored in their domestic activities as mothers has an important relationship to the problem of youth violence, which is obviously also a problem of policing. The impact of feeling monitored as a mother while your children are being policed is profound. One informant described it as a "land mine, where you are constantly chasing your kid through dangerous streets hoping you will catch him before the police do. In the meantime, though, you have to watch out for yourself too."

This impression of mothers being scrutinized while they themselves are at risk takes several forms. One form is related to the increasing public anxiety related to the safety of children whose mothers are being battered. On the one hand, this attention is important and long overdue. Yet in a more problematic sense, we see how concern for women has been placed in conflict with the needs of their children, thus positioning advocates for battered women at odds with child protective service workers in some communities.

A second manifestation of the policing of women's mothering is the rigid monitoring of women whose children have been identified as at risk of abuse because of a series of early juvenile offenses. Against the backdrop of the national trend to hold parents accountable, women whose children are in more trouble face increased jeopardy themselves. Paradoxically, the women in this study described feeling that when the "authorities are watching," their children feel even *more* inclined to act out, especially when custody issues are pending. The children then can manipulate their mothers, knowing that they are likely to get away with undercutting her parental authority. One woman said, "It's like the kids *know* what they are doing. And I find myself begging my kids to behave rather than rearing them in any strong way. I don't have any dignity left when the kids know that my ability to mother them has been called into question by outsiders.". . .

The relationships of mothers with other (non–child-specific) public institutions are also important. Most of the women interviewed considered law enforcement agencies dangerous, public assistance programs adversarial, and human services typically unhelpful. These women don't feel there is much of a safety net that they can trust or depend on to support their families. Most described profound despair and were disheartened. Yet they persist in trying to raise their children with very limited resources and in dangerous isolation.

Theme 3: The Fear of Abuse and Injury

One of the consequences of women's . . . continued attempts to enhance the safety of their children is the considerable risk of violence the women themselves face. This finding had specific gendered dimensions. The women were at risk because they were women and mothers. The responses demonstrated keen awareness that their neighborhood or "the block" is dangerous for all community members. However, they accepted and espoused the rhetoric that considered boys and men at particular risk, and therefore they themselves took particular risks for their male children as an extension of their mothering role.

For example, the women described trying to intervene with other young people when their children were in trouble. The combined mistrust of outside agencies, the sense of community loyalty (which emerged as the fourth theme), and the subjective desire to enact some degree of agency in their family and community life compelled some of the women I interviewed to try to resolve conflicts on their children's behalf. This left them extremely vulnerable. In almost half of the reported cases the women were injured by young men when they tried to protect their children. These assaults usually involved a weapon. . . .

Theme 4: Generalized, Culturally-Constructed Loyalty to Black Young Men

As one informant said, "The puddle is muddied by the position of black men in society, especially the 'endangered species' [meaning young black men]. But as a community we are as sick as our secrets." This powerful statement suggests that given the well-known effects that violence, poverty, racism, and lack of opportunity have had on black boys, it can be very difficult and problematic to raise the issue of the condition of black girls and the compromised positions of their mothers. More broadly, the frequently expressed sentiments of the women in this study suggest that the nature of gender relationships in the black community are complicated by cultural loyalties. The rhetoric sounds like this: "Men are vulnerable to societal abuse and women have had more opportunity than they." "Boys are the endangered species, and girls need to be more responsible." "It's black mothers who are raising these sons but no one pays attention to us."

These sentiments represent the opinion of a considerable segment of black communities in this country, and the extent to which this culturally constructed loyalty interacts with and is influenced by mothering warrants further investigation. In this study, it suggested a skewed set of community priorities bolstered by a simplistic public policy agenda that not only ignores the vulnerability of women and girls but also particularly punishes mothers for attempting to protect their daughters.

Theme 5: Involvement in Prevention
Initiatives and Community Activism

The fifth theme concerns the problematic nature and outcomes of women's community activism to prevent youth violence. There is a long history of documentation and analysis of black women's activism that emerged, in part, from the unique position we've assumed vis-à-vis the labor force, constitutional rights, social justice initiatives, and reform movements. This literature has generally concluded that black women's community work has been an important source of empowerment and expression of agency.

The interviews revealed a different picture. The women's accounts of their actual experiences were full of powerlessness, a sense of failure, increased risk of injury and fear for their safety, and renewed pessimism regarding their ability to accomplish the role of mothering in ways they desired. In a troubling sense, what has historically been a source of liberation for black women has become, in the face of these contemporary problems, actually a way to further marginalize women and stigmatize their inability to protect and nurture their children. Now this is in the public sphere as well as at home. The combination of structural conditions and hostile relationships between outside agencies and community groups contaminates these initiatives and causes them to fail. Most regrettably, women are set up as scapegoats here as well.

This conclusion emerged from several accounts of women who had been convinced to report their children's criminal activities in exchange for some help or leniency. They quickly learned that with current enhanced prosecution practices, their children are facing very significant prison terms. Others described how their initial enthusiasm for working with the violence-prevention program associated with a law enforcement agency were tempered when they felt compelled to "set kids up." Many reported feeling alienated from their families and neighbors and afraid of retaliation because of their assumed cooperation with police. Three who testified about their role as community liaisons and mentors found themselves quoted in a legislative report supporting repressive welfare reforms. A simple case of tokenism? Perhaps. Certainly these stories indicate the clash between the women's subjective need to feel competent and recognized in their roles as mothers and the objective limitations of their power in the social worlds within which they live. . . . These findings describe more than the failure of programs to successfully engage low-income black women in violence prevention initiatives. Such programs set women up as local targets even more than they were in the first place.

CONCLUSION

In this article, I have attempted to explore how the micro process of black women's mothering is constrained by stigmatization, persistent social problems, and misguided social policy. . . . Uninformed social policy, which ignores such structural conditions, has profound and unchallenged effects on black women's efforts at the micro processes of mothering. Intervention programs are misguided, pathologically oriented, and dangerous for black women. In the case of youth violence, they have further stigmatized women and punished black mothers.

REFERENCES

Dickerson, B. (ed.). 1995. *African American Single Mothers: Understanding Their Lives and Families.* Thousand Oaks: Sage Publications.

Hill, R. 1997. Social Welfare Policies and African American Families. Pp. 349–63 in Harriette McAdoo (ed.), *Black Families.* Thousand Oaks: Sage Publications.

DISCUSSION QUESTIONS

1. According to the women Ritchie interviewed, how does living in a decaying urban community complicate mothering?

2. Why do many of the Black women Ritchie interviewed feel powerless as mothers?

3. How does the poverty these mothers experienced differ from the poverty that their children face?

42

Salsa and Ketchup

Transnational Migrants Straddle Two Worlds

BY PEGGY LEVITT

The suburb, with its expensive homes with neatly trimmed lawns and sport-utility-vehicles, seems like any other well-to-do American community. But the mailboxes reveal a difference: almost all are labeled "Patel" or "Bhagat." Over the past two decades, these families moved from the small towns and villages of Gujarat State on the west coast of India, first to rental apartments in northeastern Massachusetts and then to their own homes in subdivisions outside Boston. Casual observers watching these suburban dwellers work, attend school, and build religious congregations might conclude that yet another wave of immigrants is successfully pursuing the American dream. A closer look, however, reveals that they are pursuing Gujarati dreams as well. They send money back to India to open businesses or improve family homes and farms. They work closely with religious leaders to establish Hindu communities in the United States, and also to strengthen religious life in their homeland. Indian politicians at the state and national level court these emigrants' contributions to India's political and economic development.

The Gujarati experience illustrates a growing trend among immigrants to the United States and Europe. In the 21st century, many people will belong to two societies at the same time. Researchers call those who maintain strong, regular ties to their homelands and who organize aspects of their lives across national borders "transnational migrants." They assimilate into the country that receives them, while sustaining strong ties to their homeland. Assimilation and transnational relations are not mutually exclusive: they happen simultaneously and influence each other. More and more, people earn their living, raise their family, participate in religious communities, and express their political views across national borders.

Social scientists have long been interested in how newcomers become American. Most used to argue that to move up the ladder, immigrants would have to abandon their unique customs, language and values. Even when it became acceptable to retain some ethnic customs, most researchers still assumed that connections to homelands would eventually wither. To be Italian American or Irish American would ultimately have much more to do with the immigrant experience in America than with what was happening back in Italy or Ireland. Social scientists increasingly recognize that the host-country experiences of some migrants remain strongly influenced by continuing ties to their country of origin and its fate.

These transnational lives raise fundamental issues about 21st century society. What are the rights and responsibilities of people who belong to two nations? Both home- and host-country governments must decide whether and how they will represent and protect migrants and what they can demand from them in return. They may have to revise their understandings of "class" or "race" because these terms mean such different things in each country. For example, expectations about how women should balance work and family vary considerably in Latin America and in the United States. Both home- and host-country social programs may have to be reformulated, taking into account new challenges and new opportunities that arise when migrants keep one foot in each of two worlds.

TWO CASES: DOMINICANS
AND GUJARATIS IN BOSTON

My research among the Dominican Republic and Gujarati immigrants who have moved to Massachusetts over the past three decades illustrates the changes that result in their origin and host communities. Migration to Boston from the Dominican village of Miraflores began in the late 1960s. By the early 1990s, nearly two-thirds of the 550 households in Miraflores had relatives in the Boston area, most around the neighborhood of Jamaica Plain, a few minutes from downtown. Migration has transformed Miraflores into a transnational village. Community members, wherever they are, maintain such strong ties to each other that the life of this community occurs almost simultaneously in two places. When someone is ill, cheating on their spouse, or finally granted a visa, the news spreads as fast on the streets of Jamaica Plain, Boston as it does in Miraflores, Dominican Republic.

Residents of Miraflores began to migrate because it became too hard to make a living at farming. As more and more people left the fields of the Dominican Republic for the factories of Boston, Miraflores suffered economically. But as more and more families began to receive money from relatives in the United States (often called "remittances"), their standard of living improved. Most households can now afford the food, clothing, and medicine for which previous generations struggled. Their homes are filled with the TVs, VCRs, and other appliances their migrant relatives bring them. Many have been able to renovate their houses, install indoor plumbing, even afford air conditioning. With money donated in Boston and labor donated in Miraflores, the community built an aqueduct and baseball stadium, and renovated the local school and health clinic. In short, most families live better since migration began, but they depend on money earned in the United States to do so.

Many of the Mirafloreños in Boston live near and work with one another, often at factories and office-cleaning companies where Spanish is the predominant language. They live in a small neighborhood, nestled within the broader Dominican and Latino communities. They participate in the PTA and in the neighborhood organizations of Boston, but feel a greater commitment toward community development in Miraflores. They are starting to pay attention to elections in the United States, but it is still Dominican politics that inspires their greatest passion. When they take stock of their life's accomplishments, it is the Dominican yardstick that matters most.

The transnational character of Mirafloreños' lives is reinforced by connections between the Dominican Republic and the United States. The Catholic Church in Boston and the Church on the island cooperate because each feels responsible for migrant care. All three principal Dominican political parties campaign in the United States because migrants make large contributions and also influence how relatives back home vote. No one can run for president in the Dominican Republic, most Mirafloreños agree, if he or she does not campaign in New York. Conversely, mayoral and gubernatorial candidates in the northeastern United States now make obligatory pilgrimages to Santo Domingo. Since remittances are one of the most important sources of foreign currency, the Dominican government instituted policies to encourage migrants' long-term participation without residence. For example, under the administration of President Leonel Fernández (1996-2000), the government set aside a certain number of apartments for Dominican emigrants in every new construction project it supported. When they come back to visit, those of Dominican origin, regardless of their passport, go through the customs line for Dominican nationals at the airport and are not required to pay a tourist entry fee.

RELIGIOUS TIES

The people from Miraflores illustrate one way migrants balance transnational ties and assimilation, with most of their effort focused on their homeland. The Udah Bhagats, a sub-caste from Gujarat State, make a different set of choices. They are more fully integrated into certain parts of American life, and their homeland ties tend to be religious and cultural rather than political. Like Gujaratis in general, the Udah Bhagats have a long history of transnational migration. Some left their homes over a century ago to work as traders throughout East Africa. Many of those who were forced out of Africa in the 1960s by local nationalist movements moved on to the United Kingdom and the United States instead of moving back to India. Nearly 600 families now live in the greater Boston region.

The Udah Bhagats are more socially and economically diverse than the Mirafloreños. Some migrants came from small villages where it is still possible to make a good living by farming. Other families, who had moved to Gujarati towns a generation ago, owned or were employed by small businesses there. Still others, from the city of Baroda, worked in engineering and finance before migrating. About half of the Udah Bhagats now in Massachusetts work in factories or warehouses, while the other half work as engineers, computer programmers or at the small grocery stores they have purchased. Udah Bhagats in Boston also send remittances home, but for special occasions or when a particular need arises, and the recipients do not depend on them. Some still own a share in the family farm or have invested in Gujarati businesses, like one man who is a partner in a computer school. Electronics, clothing, and appliances from the United States line the shelves of homes in India, but the residents have not adopted western lifestyles as much as the Miraflorenos. The Gujarati state government has launched several initiatives to stimulate investment by "Non-Resident Gujaratis," but these are not central to state economic development policy.

In the United States, both professional and blue-collar Gujaratis work alongside native-born Americans; it is their family and religious life that is still tied to India. Some Bhagat families have purchased houses next door to each other. In an American version of the Gujarati extended family household, women still spend long hours preparing food and sending it across the street to friends and relatives. Families gather in one home to do *puja,* or prayers, in the evenings. Other families lives in mixed neighborhoods, but they too spend much of their free time with other Gujaratis. Almost everyone still speaks Gujarati at home. While they are deeply grateful for the economic opportunities that America offers, they firmly reject certain American values and want to hold fast to Indian culture.

As a result, Udah Bhagats spend evenings and weekends at weddings and holiday celebrations, prayer meetings, study sessions, doing charitable work, or trying to recruit new members. Bhagat families conduct these activities within religious organizations that now operate across borders. Rituals, as well as charitable obligations, have been redefined so they can be fulfilled in the United States but directly supervised by leaders back in India. For example, the Devotional Associates of Yogeshwar or the Swadhyaya movement requires followers back in Gujarat to dedicate time each month to collective farming and fishing activities; their earnings are then donated to the poor. An example of such charitable work in Boston is families meeting on weekends to assemble circuit boards on sub-contract for a computer company. For the Udah Bhagats, religious life not only reaffirms their homeland ties but also erects clear barriers against aspects of American life they want to avoid. Not all Indians are pleased that Hindu migrants are so religious in America. While some view the faithful as important guardians of the religious flame, others, claim that emigrants abroad are the principal underwriters of the recent wave of Hindu nationalism plaguing India, including the Hindu-Muslim riots that took place in Ahmedabad in 2002.

THE RISE OF TRANSNATIONAL MIGRATION

Not all migrants are transnational migrants, and not all who take part in transnational practices do so all the time. Studies by Alejandro Portes and his colleagues reveal that fewer than 10 percent of the Dominican, Salvadoran, and Colombian migrants they surveyed regularly participated in transnational economic and political activities. But most migrants do have occasional transnational contacts. At some stages in their lives, they are more focused on their country of origin, and at other times more committed to their host nation. Similarly, they climb two different social ladders. Their social status may improve in one country and decline in the other.

Transnational migration is not new. In the early 1900s, some European immigrants also returned to live in their home countries or stayed in America while being active in economic and political affairs at home. But improvements in telecommunications and travel make it cheaper and easier to remain in touch than ever before. Some migrants stay connected to their homelands daily through e-mail or phone calls. They keep their fingers on the pulse of everyday life and weigh in on family affairs in a much more direct way than their earlier counterparts. Instead of threatening the disobedient grandchild with the age-old refrain, "wait until your father comes home," the grandmother says, "wait until we call your mother in Boston."

The U.S. economy welcomes highly-educated, professional workers from abroad, but in contrast to the early 20th century, is less hospitable to low-skilled industrial workers or those not proficient in English. Because of poverty in their country of origin and insecurity in the United States, living across borders has become a financial necessity for many less skilled migrant workers. At the same time, many highly skilled, professional migrants choose to live transnational lives; they have the money and know-how to take advantage of economic and political opportunities in both settings. These days, America tolerates and even celebrates ethnic diversity—indeed, for some people, remaining "ethnic" is part of being a true American—which also makes long-term participation in the homeland and putting down roots in the United States easier.

Nations of origin are also increasingly supportive of long-distance citizenship, especially countries that depend on the remittances and political clout of migrants. Immigrants are no longer forced to choose between their old and new countries as they had to in the past. Economic self-sufficiency remains elusive for small, non-industrialized countries and renders them dependent on foreign currency, much of it generated by migrants. Some national governments actually factor emigrant remittances into their macro-economic policies and use them to prove credit-worthiness. Others, such as the Philippines, actively promote their citizens as good workers to countries around the world. Transnational migrants become a key export and their country of origin's main connection to the world economy. By footing the bill for school and road construction back home, transnational migrants meet goals that weak home governments cannot. The increasingly interdependent global economy requires developing nations to tie themselves more closely to trade partners. Emigrant communities are also potential ambassadors who can foster closer political and economic relations.

THE AMERICAN DREAM GOES TRANSNATIONAL

Although few immigrants are regularly active in two nations, their efforts, combined with those of immigrants who participate occasionally, add up. They can transform the economy, culture and everyday life of whole regions in their countries of origin. They transform notions about gender relations, democracy, and what governments should and should not do. For instance, many young women in Miraflores, Dominican Republic no longer want to marry men who have not migrated because they want husbands who will share the housework and take care of the children as the men who have been to the United States do. Other community members argue that Dominican politicians should be held accountable just like Bill Clinton was when he was censured for his questionable real estate dealings and extramarital affairs.

Transnational migration is therefore not just about the people who move. Those who stay behind are also changed. The American-born children of migrants are also shaped by ideas, people, goods, and practices from outside—in their case, from the country of origin—that they may identify with during particular periods in their lives. Although the second generation will not be involved with their ancestral homes in the same ways and with the same intensity as their parents, even those who express little interest in their roots know how to activate these connections if and when they decide to do so. Some children of Gujaratis go back to India to find marriage partners and many second-generation Pakistanis begin to study Islam when they have children. Children of Mirafloreños born in the United States participate actively in fund-raising efforts for Miraflores. Even Dominican political parties have established chapters of second-generation supporters in the United States.

Transnational migrants like the Mirafloreños and the Udah Bhagats in Boston challenge both the host and the origin nations' understanding of citizenship, democracy, and economic development. When individuals belong to two countries, even informally, are they protected by two sets of rights and subject to two sets of responsibilities? Which states are ultimately responsible for which aspects of their lives? The Paraguayan government recently tried to intercede on behalf of a dual national sentenced to death in the United States, arguing that capital punishment is illegal in Paraguay. The Mexican government recently issued a special consular ID card to all Mexican emigrants, including those living without formal authorization in the United States. More than 100 cities, 900 police departments, 100 financial institutions, and 13 states accept the cards as proof of identity for obtaining a drivers' license or opening a bank account. These examples illustrate the ways in which countries of origin assume partial responsibility for emigrants and act on their behalf.

Transnational migration also raises questions about how the United States and other host nations should address immigrant poverty. For example, should transnationals qualify for housing assistance in the United States at the same time that they are building houses back home? What about those who cannot fully support themselves here because they continue to support families in their homelands? Transnational migration also challenges policies of the nations of origin. For example, should social welfare and community development programs discriminate between those who are supported by remittances from the United States and those who have no such outside support? Ideally, social programs in the two nations should address issues of common concern in coordination with one another.

There are also larger concerns about the tension between transnational ties and local loyalties. Some outside observers worry when they see both home country and U.S. flags at a political rally. They fear that immigrants' involvement in homeland politics means that they are less loyal to the United States. Assimilation and transnational connections, however, do not have to conflict. The challenge is to find ways to use the resources and skills that migrants acquire in one context to address issues in the other. For example, Portes and his colleagues find that transnational entrepreneurs are more likely to be U.S. citizens, suggesting that becoming full members of their new land helped them run successful businesses in their countries of origin. Similarly, some Latino activists use the same organizations to promote participation in American politics that they use to mobilize people around homeland issues. Some of the associations created to promote Dominican businesses in New York also played a major role in securing the approval of dual citizenship on the island.

These are difficult issues and some of our old solutions no longer work. Community development efforts directed only at Boston will be inadequate if they do not take into account that Miraflores encompasses Boston and the island, and that significant energy and resources are still directed toward Miraflores. Education and health outcomes will suffer if policymakers do not consider the many users who circulate in and out of two medical and school systems. As belonging to two places becomes increasingly common, we need approaches to social issues that not only recognize, but also take advantage of, these transnational connections.

REFERENCES

Guarnizo, Luis, Alejandro Portes, and William Haller. "Assimilation and Transnationalism: Determinants of Transnational Political Action among Contemporary Migrants." *American Journal of Sociology* 108 (2003): 1211–48.

Portes, Alejandro, William Haller, and Luis Guarnizo. "Transnational Entrepreneurs: The Emergence and Determinants of an Alternative Form of Immigrant Economic Adaptation." *American Sociological Review* 67 (2002): 278–298.

DISCUSSION QUESTIONS

1. How are 21ˢᵗ century immigrants different from previous generations?

2. Having read this article, do you think immigrants have to give up a former national identity to become assimilated into a new society?

3. Compare and contrast the appearances of Dominican and Gujarat immigrants.

43

Residential Segregation and Neighborhood Conditions in U.S. Metropolitan Areas

BY DOUGLAS S. MASSEY

S ocial scientists have long studied patterns of racial and ethnic segregation because of the close connection between a group's spatial position in society and its socioeconomic well-being. Opportunities and resources are unevenly distributed in space; some neighborhoods have safer streets, higher home values, better services, more effective schools, and more supportive peer environments than others. As people and families improve their socioeconomic circumstances, they generally move to gain access to these benefits. In doing so, they seek to convert past socioeconomic achievements into improved residential circumstances, yielding tangible immediate benefits and enhancing future prospects for social mobility by providing greater access to residentially determined resources.

Throughout U.S. history, racial and ethnic groups arriving in the United States for the first time have settled in enclaves located close to an urban core, in areas of mixed land use, old housing, poor services, and low or decreasing socioeconomic status. As group members build up time in the city, however, and as their socioeconomic status rises, they have tended to move out of these enclaves into areas that offer more amenities and improved conditions—areas in which majority members are more prevalent—leading to their progressive spatial assimilation into society.

From *America Becoming: Racial Trends and Their Consequences*, vol. 1, ed. by Neil Smelser, William J. Wilson and Faith Mitchell (Washington, DC: National Academies Press, 2000), pp. 391–424. Copyright © 2001 by the National Academy of Sciences, courtesy of the National Academies Press.

The twin processes of immigrant settlement, on the one hand, and spatial assimilation, on the other, combine to yield a diversity of segregation patterns across groups and times, depending on the particular histories of in-migration and socioeconomic mobility involved. Groups experiencing recent rapid in-migration and slow socioeconomic mobility tend to display relatively high levels of segregation, whereas those with rapid rates of economic mobility and slow rates of in-migration tend to be more integrated.

When avenues of spatial assimilation are systematically blocked by prejudice and discrimination, however, residential segregation increases and persists over time. New minorities arrive in the city and settle within enclaves, but their subsequent spatial mobility is stymied, and ethnic concentrations increase until the enclaves are filled, whereupon group members are forced into adjacent areas, thus expanding the boundaries of the enclave (Duncan and Duncan, 1957). . . .

Discriminatory barriers in urban housing markets mean individual Black citizens are less able to capitalize on their hard-won attainments and achieve desirable residential locations. Compared with Whites of similar social status, Blacks tend to live in systematically disadvantaged neighborhoods, even within suburbs (Schneider and Logan, 1982; Massey and Denton, 1992).

In a very real way, barriers to spatial mobility are barriers to social mobility; and a racially segregated society cannot logically claim to be "color bind." The way a group is spatially incorporated into society is as important to its socioeconomic well-being as the manner in which it is incorporated into the labor force. It is important, therefore, that levels and trends in residential segregation be documented so that this variable can be incorporated fully into research and theorizing about the causes of urban poverty. . . .

EXPLAINING THE PERSISTENCE OF RACIAL SEGREGATION

A variety of explanations have been posited to account for the unusual depth and persistence of Black segregation in American cities. One hypothesis is that racial segregation reflects class differences between Blacks and Whites—i.e., because Blacks, on average, have lower incomes and fewer socioeconomic resources than Whites, they cannot afford to move into White neighborhoods in significant numbers. . . .

This explanation has not been sustained empirically, however. When indices of racial segregation are computed within categories of income, occupation, or education, researchers have found that levels of Black-White segregation do not vary by social class. . . . According to Denton and Massey (1988), Black families annually earning at least $50,000 were just as segregated as those earning less than $2,500. Indeed, Black families annually earning more than $50,000 were more segregated than Hispanic or Asian families earning less than $2,500. In other words, the most affluent Blacks appear to be more segregated than the poorest Hispanics or Asians; and in contrast to the case of Blacks, Hispanic and

Asian segregation levels fall steadily as income rises, reaching low or moderate levels at incomes of $50,000 or more (Denton and Massey, 1988).

Another explanation for racial segregation is that Blacks prefer to live in predominantly Black neighborhoods, and that segregated housing simply reflects these preferences. This line of reasoning does not square well with survey evidence on Black attitudes, however. Most Blacks continue to express strong support for the ideal of integration. When asked on opinion polls whether they favor "desegregation, strict segregation, or something in-between," Blacks answer "desegregation" in large numbers (Schuman et al., 1985). . . .

Black respondents are not only committed to integration as an ideal, survey results suggest they also strongly prefer it in practice. When asked about specific neighborhood racial compositions, Blacks consistently select racially mixed areas as most desirable. Although the most popular choice is a neighborhood that is half-Black and half-White, as late as 1992, nearly 90 percent of Blacks in Detroit would be willing to live in virtually any racially mixed area (Farley et al., 1994). . . .

If it were up to them, Blacks would live in racially mixed neighborhoods. But it is not up to them only, of course; their preferences interact with those of Whites and, thus, produce the residential configurations actually observed. Even though Blacks may prefer neighborhoods with an even racial balance, integration will not occur if most Whites find this level of racial mixing unacceptable . . . .

SEGREGATION AND THE
CONCENTRATION OF POVERTY

The past two decades have been hard on the socioeconomic well-being of many Americans. The structural transformation of the U.S. economy from goods production to service provision generated a strong demand for workers with high and low levels of schooling, but offered few opportunities for those with modest education and training. In the postindustrial economy that emerged after 1973, labor unions withered, the middle class bifurcated, income inequality grew, and poverty spread; and this new stratification between people was accompanied by a growing spatial separation between them. The stagnation of income proved to be remarkably widespread, and inequality rose not only for minorities—Blacks, Hispanics, and Asians—but also for non-Hispanic Whites. As a result of their continued racial segregation, however, the spatial concentration of poverty was especially severe for Blacks. High levels of income inequality paired with high levels of racial or ethnic segregation result in geographically concentrated poverty, because the poverty is localized in a small number of densely settled, racially homogenous, tightly clustered areas, often in an older, urban core abandoned by industry. Had segregation not been in place, the heightened poverty would be distributed widely throughout the metropolitan area (Massey and Denton, 1993). By 1990, 83 percent of poor inner-city Blacks lived in neighborhoods that were at least 20 percent poor (Kasarda, 1993). . . .

Because more than 70 percent of urban Blacks are highly segregated but 90 percent of all other groups are not, the population of poor experiencing high concentrations of poverty is overwhelmingly Black. Given the interaction between racial segregation and the changing socioeconomic structure of American society, the issue of race cannot be set aside to focus on the politics of race versus class. . . .

THE CONSEQUENCES OF CONCENTRATED POVERTY

The argument that the prevalence of concentrated poverty among Blacks decisively undermines the life chances of the Black poor was first made forcefully by Wilson (1987). He argued that class isolation, through a variety of mechanisms, reduced employment, lowered incomes, depressed marriage, and increased unwed childbearing *over and above any effects of individual or family deprivation.* . . .

In 1988, the Rockefeller and Russell Sage Foundations funded the Social Science Research Council (SSRC) to establish a program of research into the causes and consequences of persistent urban poverty. One SSRC subcommittee—the Working Group on Communities and Neighborhoods, Family Processes, and Individual Development—met regularly over the next eight years to conceptualize and then implement a program of research to determine how concentrated poverty affected social and cognitive development. . . .

These empirical analyses clearly show that socioeconomic inequality is perpetuated by mechanisms operating at the neighborhood level, although the specific pathways are perhaps more complex than Wilson or others imagined. Not only do neighborhood effects vary in their nature and intensity at different stages of the life cycle, they are often conditioned by gender, mediated by family processes, and possibly interactive in how they combine with individual factors to determine social outcomes. Despite these complexities, however, research permits three broad generalizations.

- First, neighborhoods seem to influence individual development most powerfully in early childhood and late adolescence.

- Second, the spatial concentration of affluence appears to be more important in determining cognitive development and academic achievement than the concentration of poverty.

- Third, the concentration of male joblessness affects social behavior more than cognitive development, particularly among Blacks.

These effects persist even after controlling for unobserved heterogeneity. Thus, Wilson's (1987) theory is basically correct—there is something to the hypothesis of neighborhood effects.

One of the most important disadvantages transmitted through prolonged exposure to the ghetto is educational failure. Datcher (1982) estimates that moving a poor Black male from his typical neighborhood (66 percent Black with an average annual income of $8,500) to a typical White neighborhood (86 percent White with a mean income of $11,500) would raise his educational attainment by nearly a year. . . . Crane (1991) likewise shows that the dropout probability for Black teenage males increases dramatically as the percentage of low-status workers in the neighborhood increases. Residence in a poor neighborhood also decreases the odds of success in the labor market. . . .

The quantitative evidence thus suggests that any process that concentrates poverty within racially isolated neighborhoods will simultaneously increase the odds of socioeconomic failure within the segregated group. People who grow up and live in environments of concentrated poverty and social isolation are more likely to become teenage parents, drop out of school, achieve low educations, earn lower adult incomes, and become involved with crime—either as perpetrator or victim.

REFERENCES

Crane, J. 1991. The epidemic theory of ghettos and neighborhood effects on dropping out and teenage childbearing. *American Journal of Sociology* 96:1226–1259.

Datcher, L. 1982. Effects of community and family background on achievement. *The Review of Economics and Statistics* 64: 32–41.

Denton, N., and D. Massey. 1988. Residential segregation of Blacks, Hispanics, and Asians by socioeconomic status and generation. *Social Science Quarterly* 69:797–817.

Duncan, O., and B. Duncan. 1957. *The Negro Population of Chicago: A Study of Residential Succession.* Chicago: University of Chicago Press.

Farley, R., C. Steeh, M. Krysan, T. Jackson, and K. Reeves. 1994. Stereotypes and segregation: Neighborhoods in the Detroit area. *American Journal of Sociology* 100:750–780.

Kasarda, J. 1993. Inner city concentrated poverty and neighborhood distress: 1970–1990. *Housing Policy Debate* 4(3):253–302.

Massey, D., and N. Denton. 1992. Racial identity and the spacial assimilation of Mexicans in the United States." *Social Science Research* 213: 235–260.

Massey, D., and N. Denton. 1993. *American Apartheid: Segregation and the Making of the Underclass.* Cambridge: Harvard University Press.

Schneider, M., and J. Logan. 1982. Suburban racial segregation and Black access to local public resources. *Social Science Quarterly* 63:762–770.

Schuman, H., C. Steeh, and L. Bobo. 1985. *Racial Attitudes in America: Trends and Interpretations.* Cambridge: Harvard University Press.

Wilson, W. 1987. *The Truly Disadvantaged: The Inner City, the Underclass, and Public Policy.* Chicago: University of Chicago Press.

DISCUSSION QUESTIONS

1. Why does Massey think it is important to study racial and ethnic segregation?

2. Why does Massey suggest that residential location is not just a matter of income and individual preferences?

3. How can residential segregation, especially when coupled with poverty, become a barrier to social mobility?

44

Race in American Public Schools

Rapidly Resegregating School Districts

BY ERICA FRANKENBERG AND CHUNGMEI LEE

INTRODUCTION

In 1954, the U.S. Supreme Court handed down the historic *Brown v. Board of Education* decision outlawing state-mandated separate schools for black and white students. Since that decision, hundreds of American school districts, if not more, have attempted to implement desegregation plans. . . . We are now almost 50 years from the initial Supreme Court ruling banning segregation and more than a decade into a period in which the U.S. Supreme Court has authorized termination of desegregation orders. These plans are being dissolved by court orders even in some communities that want to maintain them; in addition, some federal courts are forbidding even voluntary desegregation plans. Given this context, it is crucial to continue to mark the progress of these policies and examine how their presence or absence affects the schooling experience for all students.

Nationally, segregation for blacks has declined substantially since the pre-*Brown* era and reached its lowest point in the late 1980s. For Latinos, the story has been one of steadily rising segregation since the 1960s and no significant desegregation efforts outside of a handful of large districts. These changes in segregation patterns are happening in the context of an increasingly diverse public school enrollment. In particular, the 2000 Census shows an extraordinary growth of Latino population in the past decade. This change in overall

From The Civil Rights Project at Harvard University, August 2002, pp. 2–22. Reprinted by permission. http://www.civilrightsproject.harvard.edu/

population is reflected in the school population as well. High birth rates, low levels of private school enrollment and increased immigration of Latinos have resulted in a rise of Latino public school enrollment, which is now more than 7 million. Nationwide, the Latino share of public school enrollment has almost tripled since 1968, compared to an increase of just 30% in black enrollment and a decrease of 17% in white enrollment during the same time period. A smaller percent of students attend private schools than a half-century ago and white private school enrollment is lowest in the South and West where whites are in school with higher proportions of minority students. Yet, little attention has been paid to the results of these two trends—rising segregation and increasing diversity—on the racial composition of our public schools.

RESEARCH QUESTIONS

This study examines segregation trends in large school districts across the country and addresses the following key questions:

- Are metropolitan countywide districts, which had shown considerable integration through the mid-1980s,[1] still integrated?

- To what extent are children in central city school districts segregated from children of other races?

- Are there effects of the dramatic increase in minority enrollment in large suburban systems? . . .

Patterns of segregation by race are strongly linked to segregation by poverty, and poverty concentrations are strongly linked to unequal opportunities and outcomes. Since public schools are the institution intended to create a common preparation for citizens in an increasingly multiracial society, this inequality can have serious consequences. Given that the largest school districts in this country (enrollment greater than 25,000) service one-third of all school-aged children, it is important to understand at a district level the ways in which school segregation, race, and poverty are intersecting and how they impact these students' lives. In our analysis we focus on two important components, race and segregation.

DATA AND METHODS

We analyze enrollment data collected by the U.S. Department of Education in the NCES Common Core of Data from the school year 2000–01, examining the 239 school districts with total enrollment greater than 25,000.[2]

---

[1]Gary Orfield and Frank Monfort, *Racial Change and Desegregation in Large School Districts: Trends through the 1986–87 School Year*, (Alexandria: National School Boards Association, 1988).

[2]Due to the fact that enrollment data disaggregated by race was not available for the Tennessee districts on CCD, we used the data as reported by the Tennessee Department of Education.

Using exposure indices, we calculate the racial isolation of both black and Latino students from white students; that is, we calculate the percent of white students in school of typical black and Latino students.[3] We also investigate the racial isolation of white students to determine whether their schooling experience is becoming more integrated as the minority share of the public school enrollment continues to increase. To do so, we calculate the percentage of black students and the percentage of Latino students in school of the average white student. We use this measure because it reports the actual racial composition of the school, and desegregated schools have been shown to have educational and diversity benefits for their students. . . .

Additionally, this study looks specifically at districts that have, at various times, been under court-mandated desegregation plans. We examine districts in each of several categories pertaining to designs of desegregation plans: busing within city, magnet plans, city-suburban desegregation, no plan, court rejected city-suburban, and partial or complete unitary status declared by mid-1980s. We compare exposure of black students to white students, since most desegregation plans were primarily concerned with the segregation of blacks from whites. We compute the 2000 exposure indices for these districts to identify any trends among districts, based on the type of desegregation the district did (or did not) have, as well as to compare the 1988 and 2000 exposure indices.

FINDINGS

The racial trend in the school districts studied is substantial and clear: *virtually all* school districts analyzed are showing lower levels of inter-racial exposure since 1986, suggesting a trend towards resegregation, and in some districts, these declines are sharp. As courts across the country end long-running desegregation plans and, in some states, have forbidden the use of any racially-conscious student assignment plans, the last 10–15 years have seen a steady unraveling of almost 25 years worth of increased integration. From the early 1970s to the late 1980s, districts in the South had the highest levels of black-white desegregation in the nation; from 1986–2000, however, some of the most rapidly resegregating districts for black students' exposure to whites are in the South. Some of these districts maintained a very high level of integration for a quarter century or more until the desegregation policies were reversed.

Other findings include:

- Many of the districts experiencing the largest changes in black–white exposure are also having similar changes in Latino exposure to whites.

---

[3]Exposure index is a measure of the proportion of a particular racial group in the school of the average student of another group. For example, a black-white exposure index of 23% indicates that there are 23% white students in the school of the average black student. If a district is perfectly integrated, the exposure index is a summary measure: it describes the average exposure of one group to another among all schools in a given district.

- Districts that show the least resegregation in black-white exposure are mostly in the South, likely due to lingering effects of desegregation plans in districts where the plans have been dissolved and the continuing impacts of plans still in place.

- The lowest levels of black-white exposure are in districts with either no desegregation plan or where the courts rejected a city-suburban plan. The highest exposure rates are in districts with city-suburban plans, even though all of these districts have since been declared unitary and show a trend toward resegregating.

- Despite an increasingly racially diverse public school enrollment, white students in over one-third of the districts analyzed became more segregated from black and/or Latino students.

As attention to civil rights is waning, it is even more important to document the segregation in our public schools in order to inform educational policy discussions on racial segregation and its related effects on public school children, particularly when these students attending racially isolated and unequal schools will be punished for not achieving at high levels.

We find that since 1986, in almost every district examined, black and Latino students have become more racially segregated from whites in their schools. The literature suggests that minority schools are highly correlated with high-poverty schools and these schools are also associated with low parental involvement, lack of resources, less experienced and credentialed teachers, and higher teacher turnover—all of which combine to exacerbate educational inequality for minority students.[4] Desegregation puts minority students in schools with better opportunities and higher achieving peer groups. . . .

DISTRICTS WITH VARIOUS TYPES
OF DESEGREGATION PLANS

Since the Supreme Court issued *Brown II* in 1955 giving district courts discretion to craft desegregation plans unique to each school system, school districts have used a variety of plans to desegregate schools. Some districts tried to encourage voluntary desegregation by creating magnet schools in inner-city areas, while many others, including the vast majority of southern districts, had mandatory desegregation that included busing in urban districts. Some mandatory desegregation plans applied only to the city, and other plans included city-suburban

---

[4]Gary Orfield, *Schools More Separate Consequences of a Decade of Resegregation* (Cambridge, MA: The Civil Rights Project Harvard University, July 2001) Janet Ward Schofield. "Review of Research on School Desegregation's Impact on Elementary and Secondary School Students," in *Handbook of Research on Multicultural Education*, ed. James Banks and Cherry McGee Banks (New York: Simon & Schuster Macmillan, 1995), pp. 597–617; Gary Orfield, Susan Eaton, and the Harvard Project on School Desegregation, eds., *Dismantling Desegregation: The Quiet Reversal of Brown v. Board of Education* (New York: New Press, 1996).

remedies where the city and the suburbs were in a single district. The lowest exposure rates of blacks to whites are in districts with either no plan or where the courts rejected a city-suburban plan. With falling white enrollment in most central city districts, plans that are limited to only desegregating within the city will never have the opportunity to produce any significant desegregation because of the small percentage of white students in these districts.

Although desegregation plans were often faulted for creating white flight, the results in Table 1 [on page 352] show that this was not true of all types of desegregation plans. In fact, the districts in Table 1 with *any* form of desegregation at one point have higher exposure rates than those with no plan or where city-suburban plans were rejected. Black students in districts with no plan or where city-suburban desegregation was rejected are highly segregated from white students. Thus, the effects of having no desegregation plan resulted in lower exposure for blacks to whites than any white flight that might have resulted from desegregation plans. Certain forms of desegregation efforts such as magnet schools[5] and busing (within a city) have had a somewhat muted positive impact on desegregation levels. The highest exposure of blacks to whites—both in 1988 and 2000—are in districts with city-suburban plans. Black students in these districts attend schools that are at least one-quarter white (and over 40% white in two of the districts).

The Supreme Court desegregation decisions of the 1990s lessened the burden that school systems were required to meet to prove that they had fully desegregated. As a result, the districts in Table 1 have all been released from desegregation requirements, and not surprisingly, all show decreasing levels of black-white exposure. Even though all of the districts in this table have since been declared unitary,[6] the integration levels for districts that had city-suburban plans still remain at least three times greater than those districts that had no desegregation plan, or where the court rejected city-suburban plans. . . . As we think about possible solutions to reverse the resegregation of the last fifteen years, particularly with predominantly minority central city school districts, the obvious and long-lasting impact of city-suburban plans might be worth replicating. . . .

CONCLUSION

While the public school enrollment reflects the country's growing diversity, our analysis of the nation's large school districts indicates a disturbing pattern of growing isolation. We find decreasing black and Latino exposure to white students is occurring in almost every large district as well as declining white exposure to blacks and Latinos in almost one-third of large districts. Black and

---

[5]Magnet schools are a form of public school choice which has gained increasing popularity as a desegregation remedy but has recently been struck down by the courts for using race-conscious admissions in order to meet their goals.

[6]Unitary might best be understood as the opposite of a 'dual' system, in which a school district, in essence, operates two separate systems, one black and one white. A unitary district is assumed to be one that has repaired the damage caused by generations of segregation and overt discrimination.

Table 1 Exposure of Blacks to Whites in Districts with Various Desegregation Plans[7]

DISTRICT	1988	2000
	Busing within city	
Columbus, OH	47.9	26
Cleveland, OH	21.7	9.7
Minneapolis, MN	51.8	20.9
Denver, CO	34.9	19.4
Boston, MA	20.4	11.2
	All/Part Plan Dismissed	
Los Angeles, CA	11	8
Dallas, TX	11.3	5.1
Norfolk, VA	31.6	24
Oklahoma City, OK	33.7	20.6
Austin, TX	32.3	19.3
Washington, D.C.	1.5	2.1
	Magnet Plans	
Kansas City, KS	19.6	10.4
Milwaukee, WI	29.9	13.1
Cincinnati, OH	29.1	16.7
Philadelphia, PA	11.6	8.7
Chicago, IL	4.8	3
	Busing, Magnet, Voluntary Suburban	
St. Louis, MO	14.8	13.2
	No Plan	
New York, NY	9.8	6.6
Atlanta, GA	3.9	3
Baltimore, MD	9.4	5.9
DeKalb County, GA	23.4	7.4
	City-Suburban	
Indianapolis, IN	46.7	27.2
Broward County, FL	36.1	23.7
Hillsborough County, FL	58.5	39.5
Clark County, FL	64.4	40.2
Nashville, TN	52.3	41.1
Duval County, FL	44.1	36
	Court Rejected City-Suburban	
Detroit, MI	6.0	2.1
Houston, TX	10.1	6.3
Richmond, VA	9.5	5.5

[7]All classification of desegregation plans are by what type of plan they had in 1988. Many of these districts are no longer operating under any desegregation plan.

Latino students display high levels of segregation from white students in many districts. This is due in part to small white percentages in these districts. However, even when white students are only a small percentage of total enrollment they tend to be concentrated in a few schools, which results in lower exposure of black and Latino students to white students even further.

The isolation of blacks and Latinos has serious ramifications: this isolation is highly correlated with poverty, which is often strongly related to striking inequalities in test scores, graduation rates, courses offered and college-going rates. Virtually no attention is being paid to this troubling pattern in the current discussion of educational reform even though it is very strongly related to many outcomes the reformers wish to change.

Recent Civil Rights Project studies of a number of cities have found important educational and civic benefits for students who attend diverse schools. However, the Supreme Court desegregation decisions of the 1990s relaxed the judicial standards school districts had to meet to be released from court oversight, and many school districts are no longer under desegregation plans. Further, school systems that wish to pursue voluntary desegregation measures by reducing racial isolation and/or to promote diversity in their schools must prove that this is both a "compelling governmental interest" and that the plan is narrowly tailored; lower court decisions have split as to whether these are compelling interests.

Many Americans believe that there is nothing that can be done about these problems and that desegregation efforts have failed. This article suggests that a great deal was done, particularly in the South, and that, after a series of court decisions sharply limiting desegregation rights, it is being undone, even in large districts where the desegregation was substantial and long lasting. Interracial exposure can simply not occur in districts that do not have different racial groups present. Perhaps it is time for communities, educational leaders and our courts to consider whether or not there is a better alternative to the system of increasingly separate and unequal schools we are creating in our large districts.

DISCUSSION QUESTIONS

1. What role did court-ordered desegregation play in interracial exposure for students in public schools?

2. Which desegregation plans were the most effective and why?

3. What do Frankenberg and Lee mean by resegregation? Why is it important to study these trends?

45

Race–Gender Experiences and Schooling

Second-Generation Dominican, West Indian, and Haitian Youth in New York City

BY NANCY LOPEZ

INTRODUCTION

At the dawn of the twenty-first century, a gender gap in educational attainment has emerged in the USA. Women from all racial and ethnic groups are attaining higher levels of schooling than men.... Although this phenomenon is relatively new among groups that have been racialized as 'white,' women from groups who have been defined as racial minorities have historically reached higher levels of education than their male counterparts. During the 1990s, African-American women were still twice as likely to obtain a college degree as men. In the Boston public high school graduating class of 1998, it was estimated that there were 100 Black and Hispanic males for every 180 Black and Hispanic females going to a four-year college. In New York City public high schools, where the majority of the student population is Black, Latino, and Asian (86%), more Black and Latina women graduated than men during the 1990s....

From *Race Ethnicity and Education* 5(1): 67–89, 2002. Copyright © 2002. Reprinted by permission of the Taylor & Francis Group.

In an effort to grapple with the question of why women attain higher levels of education than men, I examine the life histories of second-generation Caribbean youth by placing race and gender processes at the center of my analysis. A guiding premise of the study is that race and gender are socially constructed processes that are overlapping, intertwined, and inseparable. Accordingly, the leading questions for the study are, how do racialization processes differ for men and women? How do they intersect in the school setting? . . .

My attempt to pioneer an explanation for the race–gender gap in education among groups that have been defined as racial minorities in the USA begins with two central concepts: *race–gender experiences* and *race–gender outlooks*. Race–gender experiences are the episodes in which men and women undergo racial(izing) and gender(ing) processes in a variety of social spaces, including but not limited to public spaces, schools, work and family life. Over time, these repeated experiences have a cumulative effect on youth outlooks toward education. Race–gender outlooks are the life perspectives articulated by second-generation youth about education and social mobility. Race–gender outlooks emerge as responses to the cumulative race–gender experiences the second generation are subjected to during their youth and young adulthood. . . .

NEO-COLONIALISM AND CARIBBEAN MIGRATION TO THE USA

Although Dominican, Haitian, and anglophone West Indian immigrants speak different languages (i.e. Spanish, Haitian Creole, patua/English), they share a common history of European colonization, the decimation of indigenous populations, and the subsequent importation of Africans as slaves. Since the mid-nineteenth century, Caribbean nations also share a common economic and political relationship to the USA. . . . Through trade agreements such the Caribbean Basin Initiative (1983), free trade zones, and the International Monetary Fund policies, the USA exercises considerable economic and political hegemony in these countries. Paradoxically, although the Caribbean has become a popular tourist destination for American vacationers, in many respects these small island nations have become the backyard of the USA as well as a source of plentiful cheap labor and an expanding market for U.S. goods. . . .

RACE–GENDER STIGMA AS LIVED EXPERIENCE

The ways in which a given immigrant group is assigned racial meaning have important consequences for the life chances of the second generation in terms of housing, schooling, and labor market opportunities. Because they are predominantly of African phenotype, the overwhelming majority of Caribbean immigrants are defined as members of racial groups that fall to the bottom of the U.S. White/Black racial pyramid. . . .

I asked participants to discuss how they felt their ethnic group was seen in the USA. Orfelia, a 21 year-old Dominican woman who grew up in Corona, Queens, reflected the media's role in creating and circulating stereotypical negative images of racial 'others':

> If you put on the news, anyone who does anything bad, if he's not Black, he's Hispanic and that makes us look bad. It makes us look shameful. You watch the news and you see that when any white guy does something, you won't see their face. They might just say it and that's all. But if it's a Dominican, a Hispanic, a Black, they put him on for about two minutes, so that you can know him.

Orfelia poignantly describes how, regardless of intention, the national and local narrative portrayed in the media has 'symbolically tainted' communities of African phenotype as the source of crime (Wacquant, 1997). . . . In the collective consciousness of most people in the USA, the profile of a criminal and drug dealer is a dark-skinned man (Davis, 1997). . . .

On a daily basis, men of African phenotype are subjected to numerous micro-aggressions in public spaces that stem from the hegemonic view of dark-skinned men as hoodlums and criminal suspects. When I asked Mark, a 24 year-old West Indian man who lived in the Brownsville public housing project in Brooklyn, if he felt safe in his neighborhood, he remarked, 'the police are the only people I fear when I'm walking down the street.' On his way home from school one night, Mark faced a scenario that resonated with the experiences of the men:

> That night I was the only Black person who came out of the train station. As I exited, the police approached me and started questioning me about a shootout that had just occurred. I explained I was in college and showed my ID and they finally let me go.

Mark stated that he feels extremely uncomfortable when women, regardless of race, instantaneously clutch their purses and walk away from him. Mark further mentioned that one of the ways in which he coped with the hoodlum narrative was by wearing a walkman and listening to music so that he may distract himself from the many 'hate stares' that he is subjected to on a daily basis.

In part because Caribbean neighborhoods are portrayed as the drug capitals of the world, intensive police surveillance and extensive racial profiling have become the normal state of affairs in these communities. For instance, John, a 25 year-old Dominican man who grew up in Washington Heights, casually commented that 'From their patrol cars, the police, they snap pictures of you while you are walking down the streets.' Men also reported that some neighborhood storeowners have plainly asked them to leave their stores. Social rejection by service providers, such as cab drivers who refuse to pick them up, make them pay in advance, or drop them off blocks away from their final destination, are examples of the routine experiences young men of African phenotype negotiated throughout their youth and young adulthood. Although women also spoke about experiencing these types of incidents, they noted that these micro-aggressions were more likely to happen to them when they were in the company of men.

The racialization of women of African phenotype is quite different from that of their male counterparts in that it was linked to stigmatizing narratives about their sexuality. Nicole, an 18 year-old West Indian honors high school student, who lived in St Albans, Queens, commented:

> What I hate is how people view African-Americans up here. It is really hard for me when I'm on the street with my little sister because every-body is looking at me. I don't want them to think I'm some girl who just went out and sleeps around. So, I tell my little sister in a loud voice, 'We're going to see Mommy now!' I make it obvious that she is not my daughter. . . .

Nicole's social critique of how women who are defined as racial others are racialized as sexually promiscuous 'welfare queens' is a testament to how per-ceptions of race, gender and the welfare state are intertwined. However, what is important here is how Nicole responds to these experiences. When I asked her about her own plans for the future, Nicole spoke about becoming a med-ical doctor because she did not want to become another 'teenage-mother sta-tistic.' Thus, Nicole carves her identity, as well as her views about schooling and social mobility, against the backdrop of negative hegemonic stereotypes about 'urban girls' (Leadbeater & Way, 1996).

RACE–GENDER HIGH SCHOOL LESSONS

On a brisk winter morning, I began my fieldwork at Urban High School. One of the first things I noticed was that scaffolding enveloped the 85 year-old school building. Although the school was designed for a maximum of 2500 students, with a population of over 4000, Urban High School suffered from severe over-crowding. Twenty-eight makeshift trailer classrooms 'accommodated' all incom-ing ninth graders, who were expected to meet high academic standards. . . . Upon entering the main building, all students must walk through the massive, airport-style, full-body metal detectors that line the entrance. Security guards are ubiquitous—positioned in the hallways, lunchroom, and every corner of the building, using bullhorns to direct the traffic of students changing classes. Not surprisingly, teachers and students alike refer to the trailer classrooms as Riker's island—a jail located in New York City. . . .

While sitting in the security office, I often witnessed young men who had been involved in fights being whisked away in handcuffs as security guards muttered angrily under their breath about their plans to testify against these students in court. The U.S. punishment industry has indeed made the crimi-nalization of low-income Black and Latino youth who attend overcrowded, urban, public schools a 'normal' occurrence (Davis, 1997). Thus, one of the worrisome by-products of school overcrowding that I see is the forging of a pipeline between low-income, racially stigmatized public schools and the burgeoning prison industrial complex.

Mr. Green's economics class for seniors provides a snapshot of how race and gender meanings intersect in the classroom setting. One morning, Juan, a Dominican young man, walked in late to class and Mr. Green, a self-identified biracial man in his early twenties who could be categorized as phenotypically white, demanded that Juan remove his hat immediately. School rules state that no students may wear a hat inside the school building. Juan retorted that if he had to do away with his hat then the women in the class also had to remove their hats. Mr. Green responded, 'Ladies can wear their hats because it's fashion.' After threatening to send him to the principal's office. Mr. Green reluctantly asked the women to remove their hats, and Juan likewise obliged. However, by the end of class, the 'ladies' had their hats back on without a word from Mr. Green. Later that month, I found Juan in the college office, and I asked him about why he was not coming to class. Juan explained that he stopped coming to class because of problems with Mr. Green.

Interestingly, the same so-called 'oppositional' behavior from young women was not sanctioned as harshly. On another occasion, Ani, a class clown who, like Juan, was regularly late for class, joked about Mr Green's resemblance to the television personality and comedian, Pee Wee Herman. Partly because Mr Green did indeed resemble the actor, of course the entire class overflowed with laughter, including Mr Green. In disbelief, a young man sitting beside me in the back of the classroom turned to another young man and whispered. 'Imagine if we had done that, he would have kicked us out of the class!'

Notwithstanding the reality that young men are generally more rambunctious than their female counterparts, both men and women teachers were generally more lenient towards young women who transgressed school rules, were late to class, and missed homework, than they were towards young men. Given these informal practices, it is not surprising then that Latino and Black men comprise a disproportionate number of students who drop out, are discharged, expelled, and placed in special education programs for emotionally and academically challenged youth.

While men's narratives were peppered with stories of institutional hostility, school expulsions, problems with school personnel, tempered only by an occasional positive relationship with a gym teacher, women spoke about having friendly relationships with their teachers. Rosy, a 21 year-old West Indian woman who grew up in Flatbush, Brooklyn, reminisced:

> I was playing around my second year in high school. And my math teacher kept behind me. I used to dress real boy—like a boy! With a hat on backwards and everything. She said, 'You know, Rosy, you can be a lot more if you just apply yourself and dress like a young lady.' So I did . . . By her telling me all this she kind of changed my whole attitude and I just changed.

Rosy explained that her grades improved drastically after following her teacher's advice. Therefore, at many overcrowded schools, being a good student is sometimes related to social behavior and conforming to gender roles, rather than to academic performance (Sedlak et al., 1986).

While spending time in the security office at Urban High School, I asked Mr. Peña, a Dominican male security guard in his thirties, about his interactions with female students. . . . Before Mr. Peña even had an opportunity to respond, another Latino male security guard joked that they are not allowed to make physical contact with female students, only female security guards were allowed to do that. In contrast, male security guards were allowed to chase, manhandle, and apprehend male students. Since there were only two female security guards in the entire school compared to over two dozen male security personnel, the informal institutional practice was to police the men, but not the women. In due course, the problematic student was profiled as male. . . .

Compared to men, women were more active participants in the classroom. In Mr. Hunter's Global Studies course for sophomores, Yocasta, a Dominican young woman who usually sat in the front row, boasted that she did her homework last night because she wanted to earn an A: "I'll do a report if that's what it takes to get a 99." Women were not interested in doing just enough work to pass their classes. They strove for academic excellence and they were proud to verbalize their efforts to study and earn good grades. One morning, in an American History class for juniors, Ms. Gutierrez asked students to share their aspirations with the rest of the class. While young men often spoke about owning businesses, women consistently spoke about college. Brimming with a mixture of pride and excitement, Lissette, a Dominican young woman in the class, asserted, "I want to go to college to be a lawyer and become a strong independent woman" On several visits to the college office, I noted that there were always more women present than men. Women also volunteered to organize school activities, such as the senior proms, cultural festivals, and dances. In this light, women's experience can be described as simultaneous institutional engagement and ghettoization. . . .

TOWARD A RACE–GENDER FRAMEWORK OF SCHOOLING

I began this article with the question of why women, particularly those from communities that have been defined as racial minorities, attain higher levels of education than their male counterparts. I approached this question by exploring the outlooks of men and women from the largest new immigrant group in New York City—second-generation Caribbean young adults from the Dominican Republic, the anglophone Caribbean and Haiti. . . .

Men narrated instances in which they were racially stigmatized as hoodlums, were assumed to be problem students in school. . . . In response to these race–gender experiences, men expressed vacillating attitudes toward schooling.

While men acknowledged the importance of continuing their schooling in the restructured economy, their concrete lived experiences with race and gender processes throughout their lives led them to articulate doubts about their prospects for social mobility. Women were also racially discredited as sexually promiscuous 'mamasitas' and welfare queens. However, when compared to their male counterparts, women reported fewer problems with teachers at school. . . . In part because of the adult responsibilities women learned to manage at a young age, they had a more firmly established gender identity than their male counterparts. Moreover, because of shifting gender roles and changing family structures, women did not see themselves depending on a man. This reality led women to articulate their sense of womanhood in terms of attaining independence through pursuing their education. Therefore, men's ambivalence and women's optimism towards education were products of their differing and cumulative *race–gender experiences,* their responses to these experiences, and their perceptions of their prospects for social mobility.

NOTES

1. Urban High School is a pseudonym used to protect the identity of the school. All of the names of students included in this study were also changed.

REFERENCES

Davis, A. 1997. Race and criminalization: Black Americans and the punishment industry, pp. 264–279, in: W. Luriano (Ed.) *The House that Race Built* (New York, Vintage).

Leadbeater, B. & Way, N. (Eds) (1996) *Urban girls: Resisting stereotypes, creating identities* (New York, New York University Press).

Sedlak, M., Wheeler, C., Pullin, D. & Cusick, P. (1986) *Selling students short: Classroom bargains and academic reform in the American high school* (New York, Teachers College Press).

Wacquant, L. (1997) Three pernicious premises in the study of the American ghetto, *International Journal of Urban and Regional Research*, 20, pp. 341–353.

DISCUSSION QUESTIONS

1. What does Lopez mean by a gender gap in educational attainment?
2. According to Lopez, how do media images of Black men influence the public lives of the Caribbean men she interviewed?
3. What does Lopez mean by oppositional behavior? What does gender mean in terms of how teachers and school officials address such behavior among students?
4. What key factors account for why young Caribbean women are more likely to embrace school than are young Caribbean men?

46

The Significance of Race and Gender in School Success among Latinas and Latinos in College

BY HEIDI LASLEY BARAJAS AND JENNIFER L. PIERCE

Assimilation in American society has long been a central concern of sociologists (Glazer and Moynihan 1963; Gordon 1964; Park 1950; Rumbaut and Portes 1990). In Robert Park's original and influential formulation, the process of assimilation or the acceptance of the dominant culture's norms and values comes about through an immigrant group's contact with a new culture. This concept is not only central to research on recent immigrants but to studies in the sociology of education where it is considered key to understanding the success or failure of students from racial ethnic minority and white working-class backgrounds. Students who succeed, . . . do so because they have assimilated to the dominant norms and values such as individualism, while those who fail do not. Thus, the path to student success is paved through the process of assimilation to an individualistic and meritocractic understanding of the social world.

Several assumptions inform this understanding of student success. First, success is predicated on assimilation. If students do not conform to the mainstream culture, they will fail. Such an assumption precludes other possible definitions of success, such as students who may be successful academically but are still strongly tied to a culture and an identity that is not white, Anglo-Saxon, Protestant, and individualistic. . . .

This article provides an empirical and theoretical challenge to the logic of the conventional assimilationist argument by looking at the success of Latino

From *Gender & Society* 15(6): 859–878. Copyright © 2001 Sociologists for Women in Society. Reprinted by permission of Sage Publications, Inc.

students in college in a midwestern region of the United States. Currently, the high school drop-out rate of young Latinos nationwide is 46 percent (McMillen 1995). While the literature in the sociology of education suggests that students of color must adopt white middle-class behaviors to succeed, our research demonstrates that Latino students construct paths through the terrain of discrimination and prejudice they encounter in schools in much more complex and varied ways. . . . We specifically selected a group of *successful* Latino students, a group that has rarely been studied, because we were interested in addressing theoretical questions that this particular student population could help us answer.

We will demonstrate that their paths to success did not follow the typical assimilationist trajectory predicted by the literature. Furthermore, there are *gendered* patterns through which these students construct paths to success in college. Young Latinas in this study navigate successfully through and around negative stereotypes of Hispanics by maintaining positive definitions of themselves and by exphasizing their group membership as Latinas. Furthermore, their positive self-definition is reinforced through supportive relationships with other Latinas earlier in high school and now in college. On the other hand, young Latino men who also see themselves as part of a larger cultural group tend to have less positive racial and ethnic identities than the women. Typically, they are supported by mentors, such as white athletic coaches, and tend to see themselves as having "worked hard," thus they draw from the meritocractic ethos of sports and regard their success in more individualistic terms. While successful Latinas do not assimilate in the ways predicted by the literature, the young men in this study accept the individualistic and meritocractic ethos of the dominant culture, but not without a psychological price. . . .

Data were collected by the first author during a two-year period from 1996 to 1998 through a mentor program called "The Bridge" at a large U.S. research university that we call Midwestern University. Latino college students volunteered to participate in the program and mentored Latinos in local area high schools. All 45 college student mentors and 27 high school student mentees who participated in the program were interviewed. Among the college students, 31 were young women and 14 were young men. Their ages ranged from 18 to 25. Among the high school students, 11 were women and 16 were men. Students who participated in the study came from various Hispanic backgrounds, primarily Mexican, Puerto Rican, and Honduran. The majority were from second- or third-generation immigrant and poor or working-class families. . . .

LATINAS: SUCCEEDING THROUGH
RELATIONSHIPS WITH OTHERS

When asked why they wanted to mentor to Latino kids, the young women in this study were prompted to speak candidly about their experiences in the larger social world and how these experiences informed their school experiences and their desire to become mentors. More than two-thirds of the Latinas said they enrolled in the mentor program because they had a strong desire to help someone like themselves. For example, Emilia, a 23-year-old university senior from Latin America describes herself as having lived two lives: one as a poor daughter of a single mom in Latin America and another as the privileged stepdaughter of a white father in the United States. Emilia grew up in Latin America and came to the United States after her mother married an American working for the government. . . . She recounted the following story:

> My stepfather works for the government. When I was around high school age, he was transferred to an office [in another country]. He went first, and my mother, me, and my brother followed a short time later. We were at the airport in New York waiting to get on the plane. . . . Well, the man at the counter called for all family members of these government officials to begin boarding. My mom and my brother and I went to the door. But the man at the counter stopped us and told us this was boarding for special people and that we needed to wait. My mom tried to explain that we were family members, but he just wouldn't listen [she begins to cry]. I just remember him being so rude. He just assumed that because we are brown, because we weren't white, that we could not be family members of a government official. . . .

Emilia's early experiences with discrimination prompted her desire to work with other Latinos so that they could learn that "brown people are successful" too. Moreover, like the other mentors, Emilia found the program to be a safe space for her. She enjoyed being part of a group where positive meanings were attached to brownness, she liked working with other Latinos, and she liked teaching others how to navigate the treacherous waters of a college that was unwelcoming to its students of color.

For Jennifer, a 25-year-old student of Mexican heritage and a senior in college, the mentor experience produced a heightened awareness about her own community. . . .

> I have always lived in West Town. I have always lived around mostly Latinos and I never thought about it. I know what some people think about Mexicanos, but I never let it brother me. Then I started working in the elementary school with the teachers. I didn't know how much need there is out there. I mean, I never saw how little my community has—like resources, opportunities. And other people like them, the kids need to see people like them who are educated, who are going college. These kids are smart, they just don't have what other kids have. Going to college has really opened my eyes as to what other people have. . . .

Although Jennifer is aware of her difference from white students, she describes herself as a Latina in very positive ways throughout her interview—"I know what some people think . . . but I never let it bother me." Her conscious understanding of what being different meant appears to have changed when Jennifer worked in her own community. She was taught by her family and chose to think positively about her Mexicano background, and she did not use her difference as a way to explain her own difficulties in getting through school. However, after attending college, she became aware of the privileges people who were not from West Town enjoyed and became attuned to the lack of resources that were available in her own community. . . .

The majority of mentors had similar reports about the importance of positive relationships with other Latinas in their lives. Marta, a 20-year-old Chicana and a college junior, thought the most important contribution of mentoring is the fact that it is relational, particularly because Latino backgrounds are so varied. . . .

Marta believes that having a Latino mentor for Latino kids helps them to see themselves in positive ways, but this is only important to a point because each student is different. Her mentee is from Mexico, and she herself is from the southwestern region of the United States. . . . Despite her recognition of differences, Marta believes the most important part of mentoring high school Latinos is to help them understand why they are seen as different in school and to establish a real relationship with them. . . .

Through her work with K-12 Latinos and her own experiences, she recognized that going to school at all levels is a family choice for Latinos, rather than an individual one. For Marta, social class and gender play an important part in how Latinos "think about themselves, and too, how other [white] people think about you." What frustrates her, however, is how little school personnel know about the dynamics of many Latino families, particularly poor families. . . .

Several other Latina mentors expressed their disappointment with school authorities who do not understand the fact that Latino families make decisions about education for different reasons than white families do. Many said that school authorities consider going to school a taken-for-granted decision, failing to realize that for many poor and migrant Latino families, one child going to school may be a financial sacrifice for the entire family. High school and college attendance require money for clothes, school materials, lunch money, and transportation. Paying tuition or living expenses at college is rarely a possibility.

Like Jennifer and Marta, Gina emphasizes the importance of positive relationships to survive being considered different in school. A high-achieving college senior, Gina talked freely about being raised by her single mother on the West Coast and living with her extended family: her grandmother, her aunt, and her aunt's daughter. . . .

She knows that being recognized as other opens the possibility of "thinking of myself negatively." However, she doesn't allow others to racialize her in negative ways. When asked how she handles the way she is seen as different by school authorities and mainstream peer culture, she says,

> I also feel like a misfit in [this Midwestern state]? I mentioned that it is obvious I am different because, well, I had someone ask, "What kind of food do you eat?" *Excuse me?* The same kind of food *you* eat." I understand what they are getting at, but it is kind of insulting sometimes. People don't mean to be harmful, though sometimes they do and sometimes they just ask me questions because they are curious. I say it is not appropriate.

I don't know. I think it has made me think about not having a day that I see my mom and grandma struggle. My mom and grandma are really strong women and so, being a woman, yeah, that affects me. We are doing fine. . . .

Gina maintains that she chooses how to behave rather than allowing others to define her behavior to fit their assumptions. She does not allow herself "to feel oppressed by it." Futhermore, she emphasizes her "positive strong identity," something she hopes to convey to the mentee she works with.

These vignettes demonstrate how young Latinas maintain positive self-definitions and self-valuations in the face of racial discrimination, prejudice, and pejorative stereotypes. As Hill Collins (1990, 140–44) pointed out, when Black women have a safe space, they are able to create such definitions for one another. For these Latinas, safe spaces are created in relationships with friends, family, and community including association with other successful Latino students in spaces such as Latino organizations. These relationships with cultural translators become spaces in which Latinas learn positive meanings and valuations that counter the negative significations operating in schools. In addition, Latinas create new relationships as mentors and in the mentor program because they share what they have learned about being successful Latinas, and they add to their own positive self-understandings by acting as role models.

LATINOS: SUCCEEDING
THROUGH ATHLETICS

Like Latinas, young Latino men talked about being made to feel different at school and refrained from talking directly about race or labeling school experiences as acts of prejudice, discrimination, or racism. They also discussed their desire to mentor and to help others like themselves. Unlike their Latina counterparts, however, these young men tended to talk about themselves in very singular ways, as individuals who worked very hard. Their focus was on ways they, as individuals, were able to change their attitudes about school and achieve school success because of support from a coach who was typically a white male. . . .

Given that individualism and meritocracy are central American cultural ideals (Bellah et al. 1985), it is not surprising that Latinos held fast to these ideas. All students are socialized to accept the notion that the character and desire of the individual determines their destiny and that everyone will be rewarded for the hard work they perform. For young Latinos, these ideas were further encouraged through their participation in school athletics, and because most of these students were successful athletes, the notion that they were successful because they worked hard was strongly reinforced.

Ricky, an 18-year-old college freshman, was typical of the majority of young Latinos in this study. In high school, he experienced isolation from others like himself. He comments:

I was not the type to have really good friends that I hang out with, that I call, things like that. I just had friends. People that I talk to. They were not really my type. I just don't like to get all personal, on a personal level with people because sometimes, I don't know, I just feel . . . I feel that sometimes you just find more differences and things you don't agree with that person. . . .

The marginalization Ricky experienced in high school was common to almost all of the Latinos in this study. Few had friends who were Latino, and fewer still had close friends among other students of color or among white students in high school or in college. . . .

When asked why he was so motivated to succeed in school, he said that his junior year of high school, he started working out of family necessity. That year he turned "away from school," but got back "on track" through sports. His senior year, he was recruited by a suburban high school to wrestle. Although he continued to work part-time, the coaches, acting as mentors, helped Ricky focus on both wrestling and school. At the same time, Ricky was greatly influenced by his new peers at school.

I saw the success other people were having . . . how they kept going in 10th grade, 11th grade. And then I saw myself, and I was like wow, I dropped out of the race. . . . Most of my influence comes from the economic status that we are at, and like the way our lives are, and I just don't want to be like that when I grow up. I want to get out of school, get a job, buy a house, buy a car, you know, pay for all my things. Just live a normal life, and I know that a lot of the minority students are in the same situation. . . .

Ricky attributes his success to his own initiative in taking advantage of the opportunities offered through sports. Furthermore, sports reinforces the idea that school success is based on merit and that these advantages are open to everyone equally. Consequently, he believes that any problems must lie with the individual or the individual's family background. . . .

However, when talking about his mentor experience in the elementary school in the neighborhood where he grew up, Ricky contradicts himself:

I think that a big part that [school] plays a role in shaping their [mentees'] character—because I was sitting there at school and I was looking around the walls, looking at pictures, and just the way the school was built. The resources that they had, classrooms, the desk, computers, it's like amazing. It's not fair. It's not equal. And it's all in the other school [where he had transferred]. It's amazing the difference, those kids have amazing resources compared to these kids. Over here, you basically have a teacher—and like the teacher has to purchase teaching aids herself. I felt bad just because there is such a difference there. And they are the ones that need most of the help. . . .

The opportunity to return to a K-12 institution as a college student changed Ricky's perspective. As he reflects more on his opportunities, he discusses the ways they were made available to him. For instance, he thinks that one of the reasons he was able to go to a different and better school is because his mother drove him there every day. He was also able to participate in school activities because the white coach and athletic director made sure that he obtained financial waivers. "Everyone was making school and everything more convenient for me. They wanted me to succeed, also they made it easier. They helped me out and I took advantage [of the opportunity]."...

Brian, a 20-year-old college junior, and Reuben, a 22-year-old senior, both equate learning to succeed in school with their participation in sports. The difference between their experiences and Ricky's is that for Brian and Reuben, sports in their high school years was only one of many opportunities they had. As a swimmer, Reuben learned to compete and developed confidence about his abilities. Although he continued to swim in college, he did not hesitate to give it up when his swimming schedule interfered with his course work and extracurricular activities. He says,

> Swimming in high school, and even in college, was important. But what really made a difference is that my mom always taught me to try different things. Giving up swimming was a decision, but it wasn't like giving it up left me with nothing. I just moved on to the next thing—which is traveling and writing. My Mom, and my Dad too in a different way, encouraged me to try whatever I wanted. I guess what I am trying to say is that success is one thing, but having the experience is the important thing....

There are obvious social class differences in the lives of Ricky and Reuben, and their discussions about school and education reflect these differences. Reuben comes from a middle-class background with professional parents who both have an extensive education. Furthermore, Reuben's father is white and his mother is Latina.... Ricky, on the other hand, comes from a working-class, single-parent home where a high school diploma is considered a great accomplishment.

For Ricky, participation in sports was the opportunity to succeed. Had he not been exceptional at his sport, the opportunity would not have been there. On the other hand, participating in sports for Reuben was one among many choices for success. Had Reuben failed to excel in sports, he would not have been viewed as a failure by his parents, and this one failed opportunity would not have denied him success.

CONCLUSION

... Despite the negative stereotypes they faced, successful Latinas found ways to carve out safe spaces through their relationships with other Latinas and to maintain a positive sense of racial ethnic identity. Consequently, their success in school did not entail giving up their ethnic identity. On the other hand, as men, Latinos experienced certain opportunities and advantages through sports that most of the young women did not. Specifically, sports provided them with a valuable mentor such as a coach who encouraged them to do well in sports and academics. In addition, competition through sports supported and reinforced the notion that they alone were responsible for their success. At the same time, however, these young men often paid a psychological price for their conformity to these norms. The majority had strongly ambivalent feelings about their racial ethnic identities, and although they often associated with other Latinos on campus, they had less social support and shared understanding for being "different."

The gendered differences we have highlighted speak to the significance of race and gender as categories of analysis that operate together to produce divergent experiences for young Latinas and Latinos. While both Latino women and men faced racial prejudice, discrimination, and exclusion throughout their school years, young women were able to insulate themselves through supportive relationships with other Latinas in high school and in college, while young men were able to transcend some of these obstacles through participation in sports. Early in their schooling, Latinas sought out and found cultural translators who aided them in becoming bicultural, while Latinos found models from the dominant group who encouraged mainstream success but did not help them learn how to navigate between dominant and minority group cultures. These gendered strategies for success suggest that relationships and connection to others are more important to these young women and girls as Gilligan (1982) and others have argued. On the other hand, athletic ability is more highly valued and encouraged for boys in American culture than for girls regardless of race or ethnicity. Hence, participation in athletics becomes a vehicle for success for these racial ethnic minority boys, but not for girls.

Significantly, however, in contrast to studies that suggest that women's focus on relationships inhibits competitive achievement, our findings demonstrate how Latinas used relationship as a path to success. . . . Latinas experience a chilly climate in classrooms both as women and as members of a racial ethnic minority. However, rather than succumb to the pressures of this gendered and raced dynamic, they seek out protective relationships, support, and encouragement where they can achieve a positive sense of racial ethnic identity that they carry with them from high school to college. As members of a racial ethnic minority, young Latinos also encounter a chilly classroom, but as men they are encouraged to participate in sports, which becomes a springboard to success. However, once these young men enter college, the gendered advantages promised by sports diminish and race begins to take on more significance in their lives. Because they lacked cultural translators, they had not developed strong positive Latino identities in high school and found themselves at once confused and ambivalent about their racial identity, about other Latinos, and about the general fate of members from their own racial ethnic minority group. In this way, our analysis highlights how the privileges of masculinity promised through sport did not shield them from the psychological injuries and disadvantages shaped by race.

REFERENCES

Bellah, Robert, Richard Madsen, William Sullivan. Ann Swidler, and Steven Tipton. 1985. *Habits of the heart: Individualism and commitment in American life*. Berkeley and Los Angeles: University of California Press.

Chase, Susan. 1995. *Ambiguou empowerment: The work narratives of women school superintendents*. Amherst: University of Massachusetts Press.

Gilligan, Carol. 1982. *In a different voice: Psychological theory and women's development*. Cambridge, MA: Harvard University Press.

Glazer, Nathan, and Daniel P. Moynihan. 1963. *Beyond the melting pot: The Negroes, Puerto Ricans, Jews, Italians, and the Irish of New York City*. Cambridge: MIT Press.

Gordon, Milton. 1964. *Assimilation in American life: The role of race, religion and national origins*. New York: Oxford University Press.

Hill Collins, Patricia 1990. *Black feminist thought*. New York: Routledge.

McMillen, Mary. 1995. *National Center for Educational Statistics: Drop-out report*. Washington, DC: Government Printing Office.

Park, Robert. 1950. *Race and culture*. Glencoe, IL: Free Press.

Rumbaut, Rubén, and Alexandro Portes. 1990. *Immigrant America*. Berkeley: University of California Press.

DISCUSSION QUESTIONS

1. How does assimilation theory explain educational success?
2. Why do the Latinas interviewed by Barajas and Pierce offer to mentor other students of color?
3. How do sports help young Latinos to achieve educational success?
4. Why are young Latinos more likely to embrace the ideology of assimilation than young Latinas?

47

Keeping Them in Their Place

The Social Control of Blacks Since the 1960s

BY EDUARDO BONILLA-SILVA

All domination is, in the last instance, maintained through social control strategies. For example, during slavery whites used the whip, overseers, night patrols, and other highly repressive practices along with some paternalistic ones to keep blacks "in their place." After slavery was abolished whites felt threatened by free blacks, hence very strict written and unwritten rules of racial contact (the Jim Crow laws) were developed to specify "the place" of blacks in the new environment of "freedom." And, as insurance, lynching and other terroristic forms of social control were used to guarantee white supremacy. In contrast, as Jim Crow practices subsided, the control of blacks is today chiefly attained through state agencies (e.g., the police, the criminal court system, and the FBI). Manning Marable describes the new system of control:

> The informal, vigilante-inspired techniques to suppress Blacks were no longer practical. Therefore, beginning with the Great Depression, and especially after 1945, white racists began to rely almost exclusively on the state apparatus to carry out the battle for white supremacy. Blacks charged with crimes would receive longer sentences than whites convicted of similar crimes. The police forces of municipal and metropolitan areas received a carte blanche in their daily acts of brutality against Blacks. The

Federal and state government carefully monitored Blacks who advocated any kind of social change. Most important, capital punishment was used as a weapon against Blacks charged and convicted of major crimes. The criminal justice system, in short, became a modern instrument to perpetuate white hegemony. Extralegal lynchings were replaced by "legal lynchings" and capital punishment.[1]

In the following sections, I review data on social control to see how well they fit Marable's interpretation of post-civil rights dynamics.

THE STATE AS ENFORCER OF RACIAL ORDER

Data on arrest show that the contrast between black and white arrest rates since 1950 has been striking. The black rate increased throughout this period reaching almost 100 per 1,000 by 1978 compared to 35 per 1,000 for whites. In terms of how many blacks are incarcerated, we found a pattern similar to their arrest rates. Although blacks have always been overrepresented in the inmate population, this overrepresentation has skyrocketed since the late 1940s. In 1950, blacks were 29 percent of the prison population. Ten years later, their proportion reached 38 percent. By 1980, blacks made up 47 percent of the incarcerated population, six times that of whites. Today the incarceration rate of blacks has "stabilized" to constitute around 50 percent of the prison population.[2]

This dramatic increase in black incarceration has been attributed to legislative changes in the penal codes and the "get tough" attitude in law enforcement fueled by white fear of black crime. Furthermore, the fact that blacks are disproportionately convicted and receive longer sentences than whites for similar crimes contributes to their overrepresentation in the penal population. . . . This disparity in sentencing, in conjunction with the complex ways in which race works out in the criminal justice system, may explain why, although blacks made up 31 percent of those arrested in 1995, their incarceration rate was close to 50 percent. In comparison, whites constituted 67 percent of those arrested but had an incarceration rate of 50 percent.[3]

OFFICIAL STATE BRUTALITY AGAINST BLACKS

Police departments grew exponentially after the 1960s, particularly in large metropolitan areas with large concentrations of blacks. . . . Despite attempts in the 1970s and 1980s to reduce the friction between black communities and police departments by hiring more black police officers and, in some cases, even hiring black chiefs of police, "there has been little change in the attitudes of blacks toward the police, especially when the attitudes of black respondents are compared to those of white respondents.[4] A 1996 report by the Joint Center for Political Economic Studies confirmed this trend: 43 percent of blacks polled believed that police brutality and harassment of blacks were a serious problem where they lived. These numbers double when the black population polled resides in urban areas.

The level of police force used with blacks has always been excessive. However, since the police became the primary agent of social control of blacks, the level of violence against them has skyrocketed. For example, in 1975, 46 percent of all the people killed by the police in official action were black. That situation has not changed much since. Robert C. Smith reported recently that of the people killed by the police, over half are black; the police usually claim that when they killed blacks it was "accidental" because they thought that the victim was armed although in fact the victims were unarmed in 75 percent of the cases; there was an increase in the 1980s in the use of deadly force by the police and the only ameliorating factor was the presence of a sensitive mayor in a city; and in the aftermath of the Rodney King verdict, 87 percent of civilian victims of police brutality reported in the newspapers of fifteen major U.S. cities were black, and 93 percent of the officers involved were white. . . .

CAPITAL PUNISHMENT AS MODERN FORM OF LYNCHING

The raw statistics on capital punishment seem to indicate racial bias prima facie: "Of 3984 people lawfully executed since 1930 [until 1980], 2113 were black, over half of the total, almost five times the proportion of blacks in the population as a whole.[5] However, social scientific research on racial sentencing has produced mixed results. . . .

There is a substantial body of research showing that blacks charged with murdering whites are more likely to be sentenced to death than with any other victim-offender dyad. Similarly, blacks charged with raping white women receive the death sentence at a much higher rate than whites charged with raping white women. The two tendencies were confirmed by Spohn in a 1994 article using data for Detroit in 1977 and 1978: "Blacks who sexually assaulted whites faced a greater risk of incarceration than either blacks or whites who sexually assaulted blacks or whites who sexually assaulted whites; similarly, blacks who murdered whites received longer sentences than did offenders in the other two categories.[6]

The most respected study on race and death penalty, carried out by David C. Baldus to support the claim of Warren McClesky, a black man convicted of murdering a white police officer in 1978, found that there was a huge disparity in the imposition of the death penalty in Georgia. The study found that in cases involving white victims and black defendants, the death penalty was imposed 22 percent of the time whereas with the reverse dyad, the death penalty was imposed in only 1 percent of the cases. Even after controlling for a number of variables, blacks were 4.3 times as likely as whites to receive a death sentence. In a 1990 review of 28 studies on death penalty sentencing, 23 of the studied showed that the fact that victims are white "influences the likelihood that the defendant will be charged with a capital crime or that death penalty will be imposed.[7]

It should not surprise anyone that in a racialized society, court decisions on cases involving the death penalty exhibit a race effect. Research on juries suggests that they tend to be older, more affluent, more educated, more conviction-prone, and more white than the average in the community. . . .

Preliminary data from the Capital Jury Study[8]—ongoing interviews with more than 1,000 jurors who have served in death penalty trials in 14 states—reveal that deep-seated prejudice finds its way into the jury room. The following three statements by some of the jurors interviewed in this study chillingly illustrate this point:

> He [the defendant] was a big man who looked like a criminal. . . . He was big an' black an' kind of ugly. So, I guess when I saw him I thought this fits the part.

> You know, if they'd been white people. I would've had a different attitude. I'm sorry that I feel that way.

> Just a typical nigger. Sorry, that's the way I feel about it.[9]

HIGH PROPENSITY TO ARREST BLACKS

Blacks complain that police officers mistreat them, disrespect them, assume that they are criminals, violate their rights on a consistent basis, and are more violent when dealing with them than when dealing with whites. Blacks and other minorities are stopped and frisked by police in "alarmingly disproportionately numbers."[10] Why is it that minorities receive "special treatment" from the police? Studies on police attitudes and their socialization suggest that police officers live in a "cops' world" and develop a cop mentality. That cops' world is a highly racialized one: minorities are viewed as dangerous, prone to crime, violent, and disrespectful. Various studies have noted that the racist attitudes that police officers exhibit have an impact in their behavior toward minorities. . . . In terms of demographic bias, research suggests that because black communities are overpatrolled, officers patrolling these areas develop a stereotypical view of residents as more likely to commit criminal acts and are more likely to "see" criminal behavior than in white communities.

Thus it is not surprising that blacks are disproportionately arrested compared to whites. It is possible to gauge the level of overarrest endured by blacks by comparing the proportion of times that they are described by victims as the attackers with their arrest rates. Using this procedure, Farai Chideya contends,

> For virtually every type of crime, African-American criminals are arrested at rates above their commission of the acts. For example, victimization reports indicated that 33 percent of women who were raped said that their attacker was black; however, black rape suspects made up fully

43 percent of those arrested. The disproportionate arrest rate adds to the public perception that rape is a "black" crime.[11]

Using these numbers, the rate of overarrest for blacks in cases of rape is 30 percent. As shocking as this seems to be, the rate for cases wherein the victim is white is even higher. . . .

POST–CIVIL RIGHTS SOCIAL CONTROL AND THE NEW RACISM

Police brutality, overarrest, racial profiling, and many of the other social control mechanisms used to keep blacks "in their (new) place" in the contemporary United States are not overwhelmingly covert. . . . These practices are invisible to vast numbers of U.S. citizens. They are rendered invisible in four ways. First, because the enforcement of the racial order from the 1960s onward has been institutionalized, individual whites can express a detachment from the racialized way in which social control agencies operate in the United States. Second, because these agencies are legally charged with maintaining order in society, their actions are deemed neutral and necessary. Thus, it is no surprise that whereas blacks mistrust the police in surveys, whites consistently support them. Third, journalists and academicians investigating crime are central agents in the reproduction of distorted views on crime. Few report the larger facts of crime in the United States (e.g., most crime is committed by whites: so-called white-collar crime costs us ten times as much as street crime; youth crime, which accounts for most crime, is directly connected to the "structure of opportunity" youngsters face. . . . Instead, thanks to their efforts, "The public's perception is that crime is violent, Black, and male, [trends that] have converged to create the *criminal blackman*.[12] Finally, incidents that seem to indicate racial bias in the criminal justice system are depicted by white-dominated media as isolated incidents. For example, cases that presumably expose the racial character of social control agencies (e.g., the police beating of Rodney King, . . . the acquittal or lenient sentences received by officers accused of police brutality, etc.) are viewed as "isolated" incidents and are separated from the larger social context in which they transpire. Therefore, these mechanisms fit my claim about the new racism because they are largely undetected and ignored.

NOTES

1. Manning Marable. *How Capitalism Underdeveloped Black America* (Boston: South End Press, 1983), pp. 120–121.

2. Gerald Jaynes and Robin M. Williams, *A Common Destiny* (Washington, DC: National Academies Press, 1989). 457–459. Trend data suggest that the arrest rate for blacks and whites has stabilized. Sixty-seven to 70 percent of those arrested are whites and 29 to 31 percent are black.

3. K. Russell, *The Color of Crime,* (New York. New York University Press), 114.

4. Mark S. Rosentraub and Karen Harlow, "Police Policies and the Black Community: Attitude Toward the Police," pp. 107–121 in *Contemporary Public Policy Perspectives and Black Americans,* edited by Mitchell F. Rice and Woodrow Jones, Jr. (Westport, CT and London: Greenwood Press, 1984), p. 119.

5. Samuel R. Gross and Robert Mauro, *Death and Discrimination: Racial Disparities in Capital Sentencing* (Boston: Northeastern University Press, 1989).

6 Cassia Spohn, "Crime and the Social Control of Blacks: Offender/Victim Race and the Sentencing of Violent Offenders," pp. 249–268 in *Inequality, Crime, and Social Control,* edited by George S. Bridges and Martha A. Myers (Boulder, San Francisco, and Oxford: Westview Press. 1994), 264.

7. Derrick Bell, *Race, Racism, and American Law* (Boston, Toronto, and London: Little, Brown and Company, 1992), pp 332–333.

8. William J. Bowers, Maria Sandys, and Benjamin D. Steiner, "Foreclosed Impartiality in Capital Sentencing: Jurors' Predispositions, Guilt Trial Experience, and Premature Decision Making." *Cornell Law Review* 83 (1998): 1476–1556.

9. Amnesty International, *Killing with Prejudice: Race and Death Penalty in the USA, 1999.* Available online at http://www.amnesty-usa.org/rightsforall/dp/race (May 6, 2001).

10. Bell, *Race, Racism, and American Law,* 340.

11. Farai Chideya, *Don't Believe the Hype,* (New York: Penguin Books 1995), 194.

12. K. Russell. *The Color of Crime,* 114.

DISCUSSION QUESTIONS

1. What does Bonilla-Silva mean when he says that racial minorities receive special treatment from the police and the courts?

2. Why do many scholars see the death penalty as modern-day lynching?

3. Why do Black people and White people have different views of the police?

48

Watching the Canary

BY LANI GUINIER
AND GERALD TORRES

"To my friends, I look like a black boy. To white people who don't
know me I look like a wanna-be punk. To the cops I look like a criminal." Niko, now fourteen years old, is reflecting on the larger implications of his daily journey, trudging alone down Pearl Street, backpack heavy
with books, on his way home from school. As his upper lip darkens with the
first signs of a moustache, he is still a sweet, sometimes kind, unfailingly polite
upper-middle-class black boy. To his mom and dad he looks innocent, even
boyish. Yet his race, his gender, and his baggy pants shout out a different, more
alarming message to those who do not know him. At thirteen, Niko was aware
that many white people crossed the street as he approached. Now at fourteen,
he is more worried about how he looks to the police. After all, he is walking
while black.

One week after Niko made these comments to his mom, the subject of
racial profiling was raised by a group of Cambridge eighth graders who were
invited to speak in a seminar at Harvard Law School. Accompanied by their
parents, teachers, and the school principal, the students read essays they had
written in reaction to a statement of a black Harvard Law School student
whose own arrest the year before in New York City had prompted him to
write about racial profiling.[1] One student drew upon theories of John Locke

to argue that "the same mindset as slavery provokes police officers to control black people today." Another explained a picture he had drawn showing a black police officer hassling a black woman because the officer assumed she was a prostitute. Black cops harass black people too, he said aloud. "It just seems like all the police are angry and have a lot of aggression coming out." A third boy concluded that when the cops see a black person they see "the image of a thug." Proud that he knew the *American Heritage Dictionary*'s definition of a thug—a "cut-throat or ruffian"—he concluded that the cops are not the key to understanding racial profiling. Nor did he blame the white people who routinely crossed the street as he approached. If what these white people see is a thug, "they would normally want to pull their purse away." He blamed the media for this "psychological enslavement," as well as those blacks who allowed themselves to be used to "taint our image."

One boy spoke for fifteen minutes in a detached voice, showing little emotion; but he often strayed from his prepared text to describe in great detail the story of relatives who had been stopped by the police or to editorialize about what he had written. Only after all the students left did the professor discover why the boy had talked so long—and why so many adults had shown up for this impromptu class.

Several of the boys, including the one who had spoken at length, had already had personal encounters with the police. Just the week before, two of the boys had been arrested and had spent six hours locked in separate cells. . . .

WATCHING THE CANARY

Rashid and Jonathan (not their real names) are the sons of a lawyer and a transit employee, respectively. "Why don't you arrest *them*?" one of the boys asked the officer, referring to the white kids walking in the same area. "We only have two sets of cuffs," the officer replied. These cops knew whom to take in: the white kids were innocent; the black boys were guilty.

In the words of one of their classmates, black boys like Rashid and Jonathan are viewed as thugs, despite their class status. Aided by the dictionary and the media, our eighth-grade informant says this is racial profiling. Racial profiling, he believes, is a form of "psychological enslavement." . . .

But these black boys are not merely victims of racial profiling. They are canaries. And our political-race project asks people to pay attention to the canary. The canary is a source of information for all who care about the atmosphere in the mines—and a source of motivation for changing the mines to make them safer. The canary serves both a diagnostic and an innovative function. It offers us more than a critique of the way social goods are distributed. What the canary lets us see are the hierarchical arrangements of power and privilege that have naturalized this unequal distribution.

. . . We have urged those committed to progressive social change to watch the canary—and to assure the most vulnerable among us a space to experiment with democratic practice and discover their own power. Even though the canary is in a cage, it continues to have agency and voice. If the miners were watching the canary, they would not wait for it to fall off its perch, legs up. They would notice that it is talking to them. "I can't breathe, but you know what? You are being poisoned too. If you save me, you will save yourself. Why is that mine owner sending all of us down here to be poisoned anyway?" The miners might then realize that they cannot escape this life-threatening social arrangement without a strategy that disrupts the way things are.

What would we learn if we watched these particular two black boys? First, we would discover that from the moment they were born, each had a 30 percent chance of spending some portion of his life in prison or jail or under the supervision of the criminal justice system. . . . Among black men between the ages of 18 and 30 who drop out of high school, more become incarcerated than either go on to attend college or hold a job.[2] . . .

In the United States, if young men are not tracked to college and they are black or brown, we wait for their boredom, desperation, or sense of uselessness to catch up with them. We wait, in other words, for them to give us an excuse to send them to prison. The criminal justice system has thus become our major instrument of urban social policy.

David Garland explains that imprisonment has ceased to be the incarceration of individual offenders and has instead become "the systematic imprisonment of whole groups of the population"—in this case, young black and Latino males in large urban municipalities. Or as the political scientist Mary Katzenstein observes, "Policies of incarceration in this country are fundamentally about poverty, about race, about addiction, about mental illness, about norms of masculinity and female accommodation among men and women who have been economically, socially, and politically demeaned and denied."[3] . . .

But how does this "race to incarcerate" happen disproportionately to young black and Latino boys? Why is it that increasingly the nation's prisons and jails have become temporary or permanent cages for our canaries? One reason is that white working-class youth enjoy greater opportunities in the labor market than do black and Latino boys, owing in part to lingering prejudice. . . .

A second reason for the disproportionate impact of incarceration on the black and brown communities is the increased discretion given to prosecutors and police officers and the decreased discretion given to judges, whose decisions are exposed to public scrutiny in open court, unlike the deals made by prosecutors and police. Media sensationalism and political manipulation around several high profile cases (notably Willy Horton and Polly Klaas) led to mandatory minimum sentences in many states. Meanwhile, laws such as "three strikes and you're out" channeled unreviewable discretion to prosecutors, who decide which strikes to call and which to ignore. . . .

A third and, according to some commentators, the most important explanation for the disproportionate incarceration of black and Latino young men is the war on drugs. In this federal campaign—one of the most volatile issues in contemporary politics—drug users and dealers are routinely painted as black or Latino, deviant and criminal. This war metaphorically names drugs as the enemy, but it is carried out in practice as a massive incarceration policy focused on black, Latino, and poor white young men. It has also swept increasing numbers of black and Latina women into prison. . . .

Presidents Ronald Reagan and George Bush had a distinct agenda, according to Marc Mauer: "to reduce the powers of the federal government," to "scale back the rights of those accused of crime," and to "diminish privacy rights."[4] Their goal was to shrink one branch of government (support for education and job training), while enlarging another (administration of criminal justice). Mauer concludes that the political and fiscal agendas of both the Reagan and first Bush administrations were quite successful. They reduced the social safety net and government's role in helping the least well off. Their success stemmed, in part, from their willingness to "polarize the debate" on a variety of issues, including drugs and prison.

Racial targeting by police (racial profiling) works in conjunction with the drug war to criminalize black and Latino men. Looking for drug couriers, state highway patrols use a profile, developed ostensibly at the behest of federal drug officials, that suggests black and Lations are more likely to be carrying drugs. The disproportionate stops of cars driven by blacks or Latinos as well as the street sweeps of pedestrians certainly helps account for some of the racial disparity in sentencing and conviction rates. And because much of the drug activity in the black and Latino communities takes place in public, it is easier to target. . . .

A fourth explanation for the high rates of incarceration of black and brown young men is the economic boon that prison-building has brought to depressed rural areas. Prison construction has become—next to the military—our society's major public works program. And as prison construction has increased, money spent on higher education has declined, in direct proportion. Moreover, federal funds that used to go to economic or job training programs now go exclusively to building prisons. . . .

A fifth explanation is the need for a public enemy after the Cold War. Illegal drugs conveniently fit that role. President Nixon started this effort, calling drugs "public enemy number one." George Bush continued to escalate the rhetoric, declaring that drugs are "the greatest domestic threat facing out nation" and are turning our cities "into battlegrounds." By contrast, the use and abuse of alcohol and prescription drugs, which are legal, rarely result in incarceration. . . .

When drunk drivers do serve jail time, they are typically treated with a one- or two-day sentence for a first offense. For a second offense they may face a mandatory sentence of two to ten days. Compare that with a person arrested and convicted for *possession* of illegal drugs. Typical state penalties for a first-time offender are up to five years in prison and one to ten years for a second offense. . . .

We do not, by any means, claim to have exhaustively researched the criminal justice implications of racial profiling, the war on drugs, or our nation's mass incarceration policies. What we do claim is that canary watchers should pay attention to these issues if they want to understand what is happening in the United States. The cost of these policies is being subsidized by all taxpayers; one immediate result is that government support for other social programs has become an increasingly scarce resource.

NOTES

1. Bryonn Bain, "Walking While Black," *The Village Voice,* April 26, 2000, at 1, 42. Bain and his brother and cousin were arrested, held overnight and then released, with all charges eventually dropped, after the police in New York City, looking for young men who were throwing bottles on the Upper West Side, happened upon Bain et al. as they exited a Bodega. Bain, at the time, had his laptop and law books in his backpack, because he was enroute to the bus station where he intended to catch a bus back to Cambridge. Bain's essay in *The Voice* generated 90,000 responses.

2. Bruce Western and Becky Pettit, "Incarceration and Racial Inequality in Men's Employment," 54 *Industrial and Labor Relations Review* 3 (2000).

3. "Remarks on Women and Leadership: Innovations for Social Change," sponsored by Radcliffe Association, Cambridge, Massachusetts, June 8, 2001. In her talk, Katzenstein cities David Garland. "Introduction: The Meaning of Mass Imprisonment," 3(1) *Punishment and Society* 5–9 (2001).

4. Marc Mauer. (1999) *Race to Incarcerate,* New York: New Press.

DISCUSSION QUESTIONS

1. Does being middle class protect Black and Latino youth from racial profiling?

2. Why do Guinier and Torres use the metaphor of the canary to talk about the young Black boys, Rashid and Jonathan?

3. How do Guinier and Torres see the expansion of criminal justice related to the status of education and job training?

49

The Uneven Scales
of Capital Justice

BY CHRISTINA SWARNS

In 1972, the U.S. Supreme Court declared the death penalty unconstitutional. The Court found that because the capital-punishment laws gave sentencers virtually unbridled discretion in deciding whether or not to impose a death sentence, "The death sentence [was] disproportionately carried out on the poor, the Negro, and the members of unpopular groups."

In 1976, the Court reviewed the revised death-penalty statutes—which are in place today—and concluded that they sufficiently restricted sentencer discretion such that race and class would no longer play a pivotal role in the life-or-death calculus. In the 28 years since the reinstatement of the death penalty, however, it has become apparent that the Court was wrong. Race and class remain critical factors in the decision of who lives and who dies.

Both race and poverty corrupt the administration of the death penalty. Race severely disadvantages the black jurors, black defendants, and black victims within the capital-punishment system. Black defendants are more likely to be executed than white defendants. Those who commit crimes against black victims are punished less severely than those who commit crimes against white victims. And black potential jurors are often denied the opportunity to serve on death-penalty juries. As far as the death penalty is concerned, therefore, blackness is a proxy for worthlessness.

From *The American Prospect*, June 18, 2004. Reprinted by permission.

Poverty is a similar—and often additional—handicap. Because the lawyers provided to indigent defendants charged with capital crimes are so uniformly undertrained and undercompensated, the 90 percent of capitally charged defendants who lack the resources to retain a private attorney are virtually guaranteed a death sentence. Together, therefore, race and class function as an elephant on death's side of the sentencing scale.

When and how does race infect the death-penalty system? The fundamental lesson of the Supreme Court's 1972 decision to strike down the death penalty is that discretion, if left unchecked, will be exercised in such a manner that arbitrary and irrelevant factors like race will enter into the sentencing decision. That conclusion remains true today. The points at which discretion is exercised are the gateways through which racial bias continues to enter into the sentencing calculation.

Who has the most unfettered discretion? Chief prosecutors, who are overwhelmingly white, make some of the most critical decisions vis-à-vis the death penalty. Because their decisions go unchecked, prosecutors have arguably the greatest unilateral influence over the administration of the death penalty.

Do prosecutors exercise their discretion along racial lines? Unquestionably yes. Prosecutors bring more defendants of color into the death-penalty system than they do white defendants. For example, a 2000 study by the U.S. Department of Justice reveals that between 1995 and 2000, 72 percent of the cases that the attorney general approved for death-penalty prosecution involved defendants of color. During that time, statistics show that there were relatively equal numbers of black and white homicide perpetrators.

Prosecutors also give more white defendants than black defendants the chance to avoid a death sentence. Specifically, prosecutors enter into plea bargains—deals that allow capitally charged defendants to receive a lesser sentence in exchange for an admission of guilt—with white defendants far more often than they do with defendants of color. Indeed, the Justice Department study found that white defendants were almost twice as likely as black defendants to enter into such plea agreements.

Further, prosecutors assess cases differently depending upon the race of the victim. Thus, the Department of Justice found that between 1995 and 2000, U.S. attorneys were almost twice as likely to seek the death penalty for black defendants accused of killing nonblack victims than for black defendants accused of killing black victims.

And, finally, prosecutors regularly exclude black potential jurors from service in capital cases. For example, a 2003 study of jury selection in Philadelphia capital cases, conducted by the Pennsylvania Supreme Court Commission on Race and Gender Bias in the Justice System, revealed that prosecutors used peremptory challenges—the power to exclude potential jurors for any reason aside from race or gender—to remove 51 percent of black potential jurors while excluding only 26 percent of nonblack potential jurors. Such bias has a long history: From 1963 to 1976, one Texas prosecutor's office instructed its lawyers to exclude all people of color from service on juries by distributing a memo containing the following language: "Do not take Jews, Negroes, Dagos,

Mexicans or a member of any minority race on a jury, no matter how rich or how well educated." This extraordinary exercise of discretion harms black capital defendants because statistics reveal that juries containing few or no blacks are more likely to sentence black defendants to death.

Such blatant prosecutorial discretion has significantly contributed to the creation of a system that is visibly permeated with racial bias. Black defendants are sentenced to death and executed at disproportionate rates. For example, in Philadelphia, African American defendants are approximately four times more likely to be sentenced to death than similarly situated white defendants. And nationwide, crimes against white victims are punished more severely than crimes against black victims. Thus, although 46.7 percent of all homicide victims are black, only 13.9 percent of the victims of executed defendants are black. In some jurisdictions, all of the defendants on death row have white victims; in other jurisdictions, having a white victim exponentially increases a criminal defendant's likelihood of being sentenced to death. It is beyond dispute, therefore, that race remains a central factor in the administration of the death penalty.

Socioeconomic status also plays an inappropriate yet extremely influential role in the determination of who receives the death penalty. The vast majority of the people who are sentenced to death and executed in the United States come from a background of poverty. Indeed, as noted by the Supreme Court in 1972, "One searches our chronicles in vain for the execution of any member of the affluent strata of this society. The Leopolds and Loebs are given prison terms, not sentenced to death."

The primary reason for this economic disparity is that the poor are systematically denied access to well-trained and adequately funded lawyers. "Capital defense is now a highly specialized field requiring practitioners to successfully negotiate minefield upon minefield of exacting and arcane death-penalty law," according to the Pennsylvania commission. "Any misstep along the way can literally mean death for the client." It is therefore critical that lawyers appointed to represent poor defendants facing death possess the requisite compensation, training, and skill to mount a meaningful challenge to the government's case.

Unfortunately, few if any of the defendants on death row are provided with lawyers possessing the requisite skills and resources. Instead, poorly trained and underfunded court-appointed lawyers who provide abysmal legal assistance typically represent those death-sentenced prisoners. Tales of the pathetic lawyering provided by appointed counsel to their capitally charged clients are legion. Perhaps the most famous example is that of Calvin Burdine, whose court-appointed lawyer slept through significant portions of his trial. Another example is the case of Vinson Washington, whose court-appointed lawyer suggested to the defense psychiatrist that Vinson "epitomized the banality of evil." Death-sentenced defendants are so frequently provided with poor representation that, in 2001, Supreme Court Justice Ruth Bader Ginsberg commented that she had never seen a death-penalty defendant come before the Supreme Court in search of an eve-of-execution stay "in which the defendant was well-represented at trial."

One reason that appointed counsel perform so poorly is that they are grossly undercompensated. In some cases, capital-defense attorneys have been paid as little as $5 an hour. Not surprisingly, these paltry rates of compensation have yielded an equally paltry quality of representation. As was succinctly noted by the 5th U.S. Circuit Court of Appeals in its review of the quality of representation provided by a court-appointed lawyer to a capitally charged defendant in Texas: "The state paid defense counsel $11.84 per hour. Unfortunately, the justice system got only what it paid for."

Lawyers appointed to handle capital trials also often lack the expertise necessary to appropriately defend capitally charged defendants. Many states fail to provide appointed counsel with the training necessary to handle these complex cases, and many fail to impose minimum qualifications for lawyers handling capital cases. As a result, capital defendants have been represented by lawyers with absolutely no experience in criminal, much less capital, law. Although the American Bar Association has promulgated standards for the representation of indigent defendants charged with capital offenses, and although those guidelines have been endorsed by the Supreme Court, no death-penalty jurisdiction has implemented a system that meets these requirements. Thus, lawyers without meaningful training or expertise in the area of capital punishment continue to represent defendants facing death.

Because race and class continue to play a powerful role in the administration of the death penalty, it is clear that the current system is as broken today as it was in 1972. As the Supreme Court explained at the time, "A law that stated that anyone making more than $50,000 would be exempt from the death penalty would plainly fall, as would a law that in terms said that blacks, those who never went beyond the fifth grade in school, those who made less than $3,000 a year, or those who were unpopular or unstable should be the only people executed. A law which in the overall view reaches that result in practice has no more sanctity than a law which in terms provides the same."

Because the current death-penalty law, while neutral on its face, is applied in such a manner that people of color and the poor are disproportionately condemned to die, the law is legally and morally invalid.

DISCUSSION QUESTIONS

1. Why did the Supreme Court declare the death penalty unconstitutional in 1972?

2. What does discretion mean and why is it important in the criminal justice system? How does it influence who ends up on a jury?

3. Why does Swarns look at both race and social class in exploring the experiences of individuals in the criminal justice system?

50

Detaining Minority Citizens, Then and Now

BY ROGER DANIELS

Sixty years ago, on February 19, 1942, in the aftermath of Pearl Harbor, Franklin Delano Roosevelt signed Executive Order 9066. Although it mentioned no ethnic group by name, it was the instrument by which some 120,000 Japanese Americans, more than two-thirds of them American citizens, were incarcerated, without indictment or trial, in ten desolate concentration camps in the interior of the United States.

In the months since the destruction of the World Trade Center and damage to the Pentagon, we are told that more than 1,000 aliens, largely of Middle Eastern nationalities, have been locked up. Many commentators have compared the two cases—some seeing a disturbingly similar pattern in the reaction against a feared nonwhite population, others praising what they see as the relative moderation of today's government.

Historical analogies are always tricky, particularly when one of the things being compared is a current event. Contemporary history is, after all, a contradiction in terms. Nevertheless, the ways in which our memory of what was done to Japanese Americans has evolved over six decades can shed some light on our contemporary situation.

In early 1942, there was almost no negative reaction to Roosevelt's order. The American Civil Liberties Union refused to protest, and Congress passed

From *Chronicle of Higher Education*, Feb. 15, 2002, p. 10. Reprinted by permission of the author.

without a single negative vote a law imposing criminal penalties on anyone refusing to comply with the order. . . .

Later during World War II, however, government attitudes to Japanese Americans began to soften: For example, by 1943 those who were American citizens were allowed to volunteer for military service, and, a year later, the draft was reimposed on them—even on those still behind barbed wire. And while the U.S. Supreme Court handed down, with three dissents, the now-discredited *Korematsu v. United States* decision in December 1944—in effect sanctioning the government's right to put American citizens in concentration camps without indictment or trial—at the same time, indeed on the same day, it affirmed in the companion Ex parte Mitsuye Endo decision that citizens of undoubted loyalty should be free to come and go as they pleased.

After the war, President Harry S. Truman staged a welcome-home ceremony for members of the fabled Japanese American 442nd Regimental Combat Team and told the Nisei soldiers that "you have fought prejudice and won." Two years later, Truman successfully urged Congress to pass the Japanese American Evacuation Claims Act of 1948, which appropriated $38-million for all property of Japanese Americans that had been damaged or lost during their exile. That was nowhere near the true value, but no further government redress was made for nearly 40 years.

In the meantime, historians all but forgot what Eugene V. Rostow, in 1945, had called "our worst wartime mistake." In what was probably the most-liberal, general American-history text published in the 1950s, Richard Hofstadter and two collaborators, who devoted almost a full page to American intolerance in 1917–18, could say only: "Since almost no one doubted the necessity for the war, there was much less intolerance than there had been in World War I, although large numbers of Japanese Americans were put into internment camps under circumstances that many Americans were later to judge unfair or worse."

In the same decade, a security-minded Congress passed the Emergency Detention Act of 1950, authorizing, in the event of any future, "Internal Security Emergency," the apprehension and confinement of persons who "probably will engage in, or probably will conspire with others to engage in, acts of espionage or of sabotage. . . ." Passed over Truman's veto and never invoked, the act nevertheless stayed on the books for 17 years. Then, in the civil-rights-conscious and protest-prone 1960s, a campaign for repeal was mounted, culminating in a law sponsored by two Japanese American legislators from Hawaii in 1971. Five years later, as a kind of bicentennial confession of political sin, President Gerald R. Ford used the 34th anniversary of Executive Order 9066 to issue a proclamation marking a "sad day in American history."

However, at the start of the 1970s a few Japanese American activists were already beginning to urge the Japanese American Citizens League to support a campaign for some kind of more tangible public atonement. The struggle was long and difficult. First, dubious Japanese Americans, who feared a backlash, had to be persuaded; then a presidential Commission on the Wartime Relocation and Internment of Civilians had to be created to study the matter.

In 1983, that body called for a formal apology and a tax-free payment of $20,000 to each surviving internee.

And that, finally, set off a national examination of the wartime atrocity. After five years of debate, and formal opposition from the Reagan administration, the Civil Liberties Act of 1988 passed Congress and received Ronald Reagan's signature—although no money was appropriated, and no payments were made or individual apologies issued, until late 1990. The last of more than 80,000 payments was made early in the Clinton administration. The total cost was some $1.6-billion.

One of the recurrent themes—heard at first intermittently, more strongly since Ford's 1976 proclamation—has been: This kind of thing must never happen again. Anti-Asian activities have not been completely absent in recent decades, but few would disagree with the notion that there is less prejudice against Asians and other nonwhites than there was six decades ago. Yet, although nothing even resembling the mass roundup of Japanese Americans in 1942 has taken place, there has, nevertheless, been a persistent willingness to blame groups of "others" for the actions of individuals.

That is why, despite its phased-in atonements for the events of 1942, the U.S. government could continue to behave as if the civil rights of some persons—mostly aliens—were less to be regarded than those of others. During the crisis triggered by the takeover of the American Embassy in Tehran, for example, the Carter White House asked the Immigration and Naturalization Service to give it the names and addresses of all Iranian students in the United States—something the inept and underfinanced bureaucracy could not do. Then the administration asked colleges and universities to provide such lists, and most obediently complied. (Harvard was a notable exception).

During the Reagan administration, the inhumane treatment of illegal Haitian immigrants, incarcerated in Miami, was eventually enjoined by a federal judiciary unconstrained by a wartime emergency; the administration of the first President George Bush moved some of those immigrants to the American military base at Guantanamo Bay, thus largely avoiding the jurisdiction of federal courts, a policy continued by the Clinton and present Bush administrations. In a similar action singling out a group of people, just prior to the start of Desert Storm federal agents interrogated Muslim leaders in the United States—both citizens and aliens. (The questioning largely ceased after complaints; the government lamely explained that the interviews were to provide protection for those questioned.)

Today, we once again read about the detention of resident aliens for questioning; about plans to bypass normal legal procedures and create military tribunals to try "any individual who is not a United States citizen"; about federal requests to colleges and universities for the names of all foreign students. But when compared with what was done to Japanese. Americans during World War II, government actions before and after September 11 do not seem to amount to very much. Indeed, many media commentators have objected that even to mention them in connection with the massive violations of civil liberties by the Roosevelt administration is inappropriate.

That is an evasion: the kind of evasion that has allowed us to offer apologies for the actions we have taken against those whom we perceive to be outsiders, and then do the same thing to a different group. Time and again, scholars (if not the government) have eventually acknowledged that we, as a nation, have violated the spirit of our Constitution. Time and again, we have gone on to violate it again.

If that history isn't enough to give us pause about where our current actions might lead, consider this: Roosevelt's order was issued in wartime; the subsequent government actions have occurred in societies in which both racial discrimination and xenophobia have been greatly reduced, and they have often been accompanied (as Roosevelt's was) by reassuring statements from the highest levels of government about religious, racial, and ethnic egalitarianism. Even in the best of times, we have been able to justify abridging some peoples' rights.

Moreover, despite those laudable statements about nondiscrimination from the top of the chain of command, practice at the lower levels of government has often been blatantly biased. Nowhere has that been more evident than in the immigration service. And in no instance that I know has any lower government official been punished or even reprimanded for improper actions. As I was writing last week, we were still hearing of the roundup of suspects by law-enforcement officials and the assertions by INS officials of their plans to concentrate on "Middle Eastern" students and others who overstay their visas.

Moreover, private groups like airlines have forced citizens and aliens who look like the "enemy" to leave flights for which they had tickets—sometimes even winning praise for doing so. "I was relieved at the story of the plane passengers a few weeks ago who refused to board if some Mideastern-looking guys were allowed to board," Peggy Noonan, a contributing editor, wrote in *The Wall Street Journal*. "I think we're going to require a lot of patience from a lot of innocent people. . . . And you know, I don't think that's asking too much." Even more disturbing than such blathering is the fact that the cabinet officer responsible for aviation, Secretary of Transportation Norman Mineta—himself a child victim of wartime incarceration, who often recounts going off in his Cub Scout uniform to a 1942 government roundup—has done nothing about such blatant discrimination, at least not in public.

Optimists assure us that a mass incarceration of American citizens in concentration camps will not recur. But reflection on our past suggests we ought not to be so sanguine. To be sure, it was not just the disaster at Pearl Harbor, but the subsequent sequence of Japanese triumphs that triggered Executive Order 9066. But shouldn't we then ask, If successful terrorist attacks hadn't abated after September 11, would the current government reaction have been so moderate?

Many Japanese Americans, the only group of our own citizens ever incarcerated because of who they were as opposed to what they did, argue that the optimists are wrong and that what has happened before can surely, given the right triggering circumstances, happen again. As a student of Japanese American history, I can only agree with them.

DISCUSSION QUESTIONS

1. What actions did the government take with regard to Japanese immigrants and Japanese citizens after the bombing of Pearl Harbor in 1941?
2. What restrictions did Arab Americans face in the aftermath of the events on September 11, 2001?
3. What current issues do Arab Americans face as citizens of the United States?

InfoTrac College Edition: Bonus Reading

http://infotrac.thomsonlearning.com

You can use the InfoTrac College Edition feature to find an additional reading pertinent to this section. You can also search on the author's name or keywords to locate the following reading:

Reed, Diane F. and Edward L. Reed. 1997. "Children of Incarcerated Parents." *Social Justice* 24 (Fall): 152–170.

Reed and Reed focus on the plight of the children of jailed men and women. These young people face many obstacles. Reed and Reed document how harsher laws have resulted in over one million people being held in jails and prisons. These inmates are not just isolated individuals; they are connected to families and many of them are parents. Their segregated status only sustains the cycle of disadvantage for their children, who are often considered either delinquent or candidates for foster care. And these children are disproportionately children of color.

STUDENT EXERCISES

1. Think about your first experience of finding employment. Did you start in the informal economy, like babysitting or doing yardwork, or the formal economy, like the Burger Barn or the mall? At what age did you start to work? How much money did you earn and what did you do with it? What role did your community or your family network play in your securing of employment? Having thought about this, share your experiences with the whole class, or in groups of eight or ten, so that there will be some diversity of experiences to be heard. How do you think your own work experience is related to your race, gender, social class, and residential location? What lessons do you learn from this exercise about how race, gender, social class, residential location, and other social factors are related to the young person's job search?

2. In their article "Race in American Public Schools," Frankenberg and Lee discuss the different plans for desegregation employed in major metropolitan areas. Did your public school system institute any such practices to alter the racial composition of the schools? If so, what type of plan was it? Was there court-ordered or voluntary busing? Did the school district develop magnet schools to promote diversity? Look at contemporary newspaper accounts or documents from the school board or school district to see what policies might have influenced your experience. Then write an essay on how your experience with school segregation or integration has (or has not) exposed you to students from races and social class backgrounds different from your own.

PART VII

Mobilizing for Change: Looking Forward and Learning from the Past

R acial hierarchies, like all social hierarchies, are unstable and fluid. They change over time, but *how* do they change? They can change because the political economy needs different groups of workers with particular skills, which can result in dominant groups changing their views of others. For example, African Americans were barred from industrial work in one era but then were recruited for those same jobs when immigration was curtailed. Access to industrial work changed not only their opportunities, but brought them into the cities in great numbers and gave them the resources to create new representations of themselves, as in the flowering of the Harlem Renaissance (Marks 1989).

More often, however, change occurs because of actions taken by oppressed people. Dominant groups use their economic and political power to develop ideologies that make the current racial order appear natural. Disadvantaged groups in turn mobilize to not only challenge those ideologies but even to change laws and create a more inclusive and just society. Such actions show us that although systems of inequality may be established by those with the most power, they are not just quietly accepted. Rather they are contested by oppressed groups and their allies. For example, notions of racial and cultural inferiority justified slavery, the mistreatment of American Indians, and the denial of rights to Chinese immigrants. Members of these groups had different interpretations of their situations. For example, many slaves rejected the system, escaping to form Maroon colonies in the South, and some even engaged in armed rebellions. American Indians knew that European settlers had guns while they had only bows and arrows and knew that the laws of the land gave advantages to White people, those with economic and political power. Still, they continued to contest and challenge these views. Denied the opportunity to become citizens, the Chinese used the state and federal court systems to fight the erosion of their rights as human beings (Takaki 1993). Among all oppressed racial and ethnic groups, there is a long history of active resistance, even as the elite of the nation used the power of government to control those who would not conform.

Racism is a tool for exclusion, but it also can be used to build community. People who are denied participation in the mainstream society build their own communities; segregation itself can end up fostering community building. Such communities are places from which people question the controlling images and negative stereotypes that the dominant group employs to justify inequality. Within their own communities, young people of color can learn affirming messages about themselves, as well as structural explanations for their lack of resources and opportunities. Such messages contradict explanations from the ruling system of power that often define those with fewer resources as less able. Alternative explanations motivate oppressed people and those who support oppressed people's goals, so that *all* people have opportunities, not just the advantaged group. Although some members of oppressed groups internalize the negative messages from the dominant group, most develop a counter (or an oppositional) perspective.

Earlier, in Part II, Michael Omi and Howard Winant ("On Racial Formation") presented the concept of *racial projects*, defined as organized efforts to distribute social and economic resources along racial lines. However, racial projects can also be organized to redistribute resources with equality in mind as disadvantaged people challenge interpretation of themselves as undeserving. The history of resistance reveals racial projects waged by disadvantaged groups often to build and sustain community, so that they can survive and raise the next generation. With these actions, they challenge dominant, controlling images with new ones that better represent themselves (Omi and Winant 1994; Takaki 1993). Sharing new images with the majority groups and others in the society challenges the ideological justification for their unequal treatment.

Throughout this volume, we have seen that African Americans, Native Americans, Chinese Americans, Japanese Americans, and the many Latino groups have been displeased with their status in U.S. society. Earlier European immigrant groups and their descendants were also unhappy with their status as exploited workers denied full participation in the society. Many survived in ethnic enclaves, but some found options for mobility during World War II, even while people of color remained economically and socially marginalized. In the twentieth century, people of color used different strategies to initiate social change. People waged legal struggles to abolish laws legalizing racial segregation and to create new ones that protected them from discrimination. For example, the National Association for the Advancement of Colored People (NAACP) brought together a legal team that worked to equalize salaries for teachers in segregated school systems, to guarantee Black workers' rights, and to challenge the system of Jim Crow segregation. Other racial groups also worked through the courts. As laws changed, so did expectations, but stakeholders in the system do not change immediately. As Martin Luther King, Jr. wrote in his famous "Letter from a Birmingham Jail," no one gives up privilege willingly (Williams 1987).

New arrangements are not simply a result of changes in the law. They take determined people actively pressing for their rights. Part VII looks at how people mobilize to bring those rights into being, thereby pushing the nation to face up to its mission to be a society in which all people are given, not just promised, rights. It is through **social movements,** groups that organize and act to bring about social change, that our nation has been reshaped. Is the nation now ready to move forward toward more equality?

Writing as the twenty-first century approached, Aldon Morris ("The Genius of the Civil Rights Movement") has reviewed the history of resistance that launched a major grassroots movement, as ordinary African American men and women as well as their leaders engaged in nonviolent social protests, often, however, having to confront violent resistance. They boycotted buses and stores, held sit-ins at lunch counters and movie theaters, and marched for voting rights. Their very actions challenged mainstream images of inferiority and presumed satisfaction with second-class citizenship. There was power in that resistance. It forced a nation to look at the consequences of an imposed segregation and to take action. In addition to building support for civil rights legislation, the Civil Rights Movement was an inspiration for other groups, both within the United States and abroad.

The Black freedom struggle is well known, but African Americans have not been the only racial minority to use civil disobedience for social change. Media images of American Indians as either savages or noble environmentalists mask Indians' long struggle for social justice. In an essay in Part V, Russell Thornton outlined their quest for sovereignty and treaty rights. Resistance among indigenous people did not end after they were restricted to reservations. The American

Indians have continued in various ways to keep Native American communities and cultures alive. Black Americans' protests in southern cities were covered by the national and international media, while American Indians' protests have been less visible to the wider public—except for the 1969 occupation of Alcatraz Island in California by Native American students. Diane-Michele Prindeville's research examining Native American and Hispanic women ("Identity and the Politics of American Indian and Hispanic Women Leaders") shows the continuing activism of Native American and Hispanic women. She shows that race and gender are central to the women's identities, and that both race and gender motivate them to act on behalf of their communities.

When social change occurs, it is often the result of social movements that emerge when people make the decisions to change their lives and work with others to reshape institutions. Vincent Harding ("Signs . . . Signs . . . Turn Visible Again") talks about the need for signposts to help us sort out the challenges we currently face. Harding believes that learning about heroes of the nineteenth and twentieth centuries can inspire a new generation to work for social justice. Changing the structures of domination that support racism requires much effort. Many people reevaluated their lives before participating in the Civil Rights Movement. Their final decisions were based on an understanding of the system of racial oppression. While we hear often about a few leaders, we see from Harding's article that the Civil Rights Movement included people of different racial and ethnic groups working as allies. The actions of those in the past can motivate us, but today's efforts to establish racial justice must be based on understanding the changing racial landscape and how new racial dynamics might call for different strategies for building coalitions.

In Part V, Suzanne Oboler ("It Must Be a Fake!") voiced concerns about the Hispanic label and social racism that stigmatized Latinos. Here, Eduardo Bonilla-Silva ("We Are All Americans!: The Latin Americanization of Racial Stratification in the USA") also addresses what the enhanced diversity of the nation means for racial stratification and dynamics between groups. Like Oboler, Bonilla-Silva notes that internal racism is mirroring the global scene as we move from a biracial to a triracial system. In this new system, some non-White groups are becoming "honorary White folks," much as the immigrants from Eastern and Southern Europe at the end of the nineteenth century were White in terms of citizenship rights, but were treated as racial inferiors. Asians and Latinos might act as a buffer between Black and White groups. They can also promote a notion of color-blindness, while White supremacy remains intact. These twenty-first-century dynamics can complicate building broad coalitions for ending racial injustices.

How do we work to end racial injustice and build a society that is a community of equals? We can begin by acknowledging our racial legacy and the impact of race on both institutions and individuals. Even if the media pushes a color-blind framework, we have learned from this volume that race matters and our racial identities are significant. However, powerful institutions and people still benefit from racial stratification and they use their resources to distort that reality. When we see the power arrangements and systemic racism in the society, we can begin to dismantle it. As in the past, changing the landscape requires a great deal of effort. Reading this book has given you information that challenges many of the myths and misinformation we often have about other groups. Now we have to think about working for change.

Jacqueline Johnson, Sharon Rush and Joe Feagin ("Reducing Inequalities") discuss the need for institutional change and antiracism work on the part of many to build a new society. Action is necessary on many fronts. The public educational system can play a role in redressing patterns of segregation that keep us from knowing each other. Media constructions of people of color are distortions that are kept alive when people lack direct contact as equals. The commitment to building an anti-racist society requires cognitive and emotional work for us all. In the essay by Chesler, Peet, and Sevig in Part III ("Blinded by Whiteness") we learned about racial identity and how learning about privilege can help people make new decisions. If we change both institutions and individual behavior, we can work to build an anti-racist society. People of color can reject ideologies of inferiority. White people can learn new insights into their own situation and that of others when they form friendships across racial–ethnic lines. Such relationships can help them recognize their own privileges and the cost of those privileges for others. Johnson, Rush and Feagin's vision is a powerful one that can be shared with others who are interested in actively shifting the landscape towards greater racial equality.

REFERENCES

Marks, Carole. 1989. *Farewell—We're Good and Gone: The Great Black Migration*. Bloomington: Indiana University Press.

Omi, Michael and Howard Winant. 1994. *Racial Formation in the United States*. New York: Routledge.

Takaki, Ronald. 1993. *A Different Mirror: A History of Multicultural America*. Boston: Little, Brown and Company.

Williams, Juan. 1987. *Eyes on the Prize: America's Civil Rights Years, 1954–1965*. New York: Penguin Books.

51

The Genius of the Civil Rights Movement

Can It Happen Again?

BY ALDON MORRIS

I t is important for African Americans, as well as all Americans, to take a look backward and forward as we approach the turn of a new century, indeed a new millennium. When a panoramic view of the entire history of African Americans is taken into account, it becomes crystal clear that African American social protest has been crucial to Black liberation. In fact, African American protest has been critical to the freedom struggles of people of color around the globe and to progressive people throughout the world.

The purpose of this essay is: 1) to revisit the profound changes that the modern Black freedom struggle has achieved in terms of American race relations; 2) to assess how this movement has affected the rise of other liberation movements both nationally and internationally; 3) to focus on how this movement has transformed how scholars think about social movements; 4) to discuss the lessons that can be learned from this groundbreaking movement pertaining to future African American struggles for freedom in the next century.

It is hard to imagine how pervasive Black inequality would be today in America if it had not been constantly challenged by Black protests throughout each century since the beginning of slavery. The historical record is clear that slave resistance and slave rebellions and protest in the context of the Abolitionist movement were crucial to the overthrow of the powerful slave regime.

From Aldon Morris, "The Genius of the Civil Rights Movement: Can It Happen Again?"
Reprinted by permission of the author.

The establishment of the Jim Crow regime was one of the great tragedies of the late nineteenth and early twentieth centuries. The overthrow of slavery represented one of those rare historical moments where a nation had the opportunity to embrace a democratic future or to do business as usual by reinstalling undemocratic practices. In terms of African Americans, the White North and South chose to embark along undemocratic lines.

For Black people, the emergence of the Jim Crow regime was one of the greatest betrayals that could be visited upon a people who had hungered for freedom so long: what made it even worse for them is that the betrayal emerged from the bosom of a nation declaring to all the world that it was the beacon of democracy.

The triumph of Jim Crow ensured that African Americans would live in a modern form of slavery that would endure well into the second half of the twentieth century. The nature and consequences of the Jim Crow system are well known. It was successful in politically disenfranchising the Black population and in creating economic relationships that ensured Black economic subordination. Work on wealth by sociologists Melvin Oliver and Thomas Shapiro (1995), as well as Dalton Conley (1999), are making clear that wealth inequality is the most drastic form of inequality between Blacks and Whites. It was the slave and Jim Crow regimes that prevented Blacks from acquiring wealth that could have been passed down to succeeding generations. Finally, the Jim Crow regime consisted of a comprehensive set of laws that stamped a badge of inferiority on Black people and denied them basic citizenship rights.

The Jim Crow regime was backed by the iron fist of southern state power, the United States Supreme Court, and white terrorist organizations. Jim Crow was also held in place by white racist attitudes. As Larry Bobo has pointed out, "The available survey data suggests that anti–Black attitudes associated with Jim Crow were once widely accepted . . . [such attitudes were] expressly premised on the notion that Blacks were innately, intellectually, culturally, and temperamentally inferior to Whites" (Bobo, 1997:35). Thus, as the twentieth century opened, African Americans were confronted with a powerful social order designed to keep them subordinate. As long as the Jim Crow order remained intact. The Black masses could breathe neither freely nor safely. Thus, nothing less than the overthrow of a social order was the daunting task that faced African Americans during the early decades of the twentieth century.

The voluminous research on the modern civil rights movement has reached a consensus: That movement was the central force that toppled the Jim Crow regime. To be sure, there were other factors that assisted in the overthrow including the advent of the television age, the competition for Northern Black votes between the two major parties, and the independence movement in Africa which sought to overthrow European domination. Yet it was the Civil Rights Movement itself that targeted the Jim Crow regime and generated the great mass mobilizations that would bring it down.

What was the genius of the Civil Rights Movement that made it so effective in fighting a powerful and vicious opposition? The genius of the Civil Rights Movement was that its leaders and participants recognized that change could occur if they were able to generate massive crises within the Jim Crow order–crises of such magnitude that the authorities of oppression must yield to the demands of the movement to restore social order. Max Weber defined power as the ability to realize one's will despite resistance. Mass disruption generated power. That was the strategy of nonviolent direct action. By utilizing tactics of disruption, implemented by thousands of disciplined demonstrators who had been mobilized through their churches, schools, and voluntary associations, the Civil Rights Movement was able to generate the necessary power to overcome the Jim Crow regime. The famous crises created in places like Birmingham and Selma, Alabama, coupled with the important less visible crises that mushroomed throughout the nation, caused social breakdown in Southern business and commerce, created unpredictability in all spheres of social life, and strained the resources and credibility of Southern state governments while forcing white terrorist groups to act on a visible stage where the whole world could watch. At the national level, the demonstrations and repressive measures used against them generated foreign policy nightmares because they were covered by foreign media in Europe, the Soviet Union, and Africa. Therefore what gave the mass-based sit-ins, boycotts, marches, and jailing their power was their ability to generate disorder.

As a result, within ten years—1955 to 1965—the Civil Rights movement had toppled the Jim Crow order. The 1964 Civil Rights Bill and the 1965 Voting Rights Act brought the regime of formal Jim Crow to a close.

The Civil Rights Movement unleashed an important social product. It taught that a mass-based grassroots social movement that is sufficiently organized, sustained, and disruptive is capable of generating fundamental social change. In other words, it showed that human agency could flow from a relatively powerless and despised group that was thought to be backward, incapable of producing great leaders.

Other oppressed groups in America and around the world took notice. They reasoned that if American Blacks could generate such agency they should be able to do likewise. Thus the Civil Rights movement exposed the agency available to oppressed groups. By agency I refer to the empowering beliefs and action of individuals and groups that enable them to make a difference in their own lives and in the social structures in which they are embedded.

Because such agency was made visible by the Civil Rights movement, disadvantaged groups in America sought to discover and interject their agency into their own movements for social change. Indeed, movements as diverse as the student movement, the women's movement, the farm worker's movement, the Native American movement, the gay and lesbian movement, the Environmental movement, and the disability rights movement all drew important lessons and inspiration from the Civil Rights Movement. From that movement other groups discovered how to organize, how to build social movement organizations, how to mobilize large numbers of people, how to devise appropriate tactics and strategies, how to infuse their movement activities with cultural creativity, how to confront and defeat authorities, and how to unleash the kind of agency that generates social change.

For similar reasons, the Black freedom struggle was able to effect freedom struggles internationally. For example, nonviolent direct action has inspired oppressed groups as diverse as Black South Africans, Arabs of the Middle East, and pro-democracy demonstrators in China to engage in collective actions. The sit-in tactic made famous by the Civil Rights Movement, has been used in liberation movements throughout the third world, in Europe, and in many other foreign countries. The Civil Rights Movement's national anthem "We Shall Overcome" has been interjected into hundreds of liberation movements both nationally and internationally. Because the Civil Rights Movement has been so important to international struggles, activists from around the world have invited civil rights participants to travel abroad. Thus early in Poland's Solidarity movement Bayard Rustin was summoned to Poland by that movement. As he taught the lessons of the Civil Rights Movement, he explained that "I am struck by the complete attentiveness of the predominantly young audience, which sits patiently, awaiting the translations of my words." (Rustin, undated)

Therefore, as we seek to understand the importance of the Black Freedom Struggle, we must conclude the following: the Black Freedom Struggle has provided a model and impetus for social movements that have exploded on the American and international landscapes. This impact has been especially pronounced in the second half of the twentieth century.

What is less obvious is the tremendous impact that the Black Freedom Struggle has had on the scholarly study of social movements. Indeed, the Black freedom struggle has helped trigger a shift in the study of social movements and collective action. The Black movement has provided scholars with profound empirical and theoretical puzzles because it has been so rich organizationally and tactically and because it has generated unprecedented levels of mobilization. Moreover, this movement has been characterized by a complex leadership base, diverse gender roles, and it has revealed the tremendous amount of human agency that usually lies dormant within oppressed groups. The empirical realities of the Civil Rights Movement did not square with the theories used by scholars to explain social movements prior to the 1960s.

Previous theories did not focus on the organized nature of social movements, the social movement organizations that mobilize them, the tactical and strategic choices that make them effective, nor the rationally planned action of leaders and participants who guide them. In the final analysis, theories of social movements lacked a theory that incorporated human agency at the core of their conceptual apparatuses. Those theories conceptualized social movements as spontaneous, largely unstructured, and discontinuous with institutional and organizational behavior. Movement participants were viewed as reacting to various forms of strain and doing so in a non-rational manner. In these frameworks, human agency was conceptualized as reactive, created by uprooted individuals seeking to reestablish a modicum of personal and social stability. In short, social movement theories prior to the Civil Rights Movement operated with a vague, weak vision of agency to explain phenomena that are driven by human action.

The predictions and analytical focus of social movement theories prior to the 1970s stood in sharp contrast to the kind of theories that would be needed to capture the basic dynamics that drove the Civil Rights movement. It became apparent to social movement scholars that if they were to understand the Civil Rights Movement and the multiple movements it spun, the existing theoretical landscape would have to undergo a radical process of reconceptualization.

As a result, the field of social movements has been reconceptualized and this retheoritization will effect research well into the new millennium. To be credible in the current period any theory of social movements must grapple conceptually with the role of rational planning and strategic action, the role of movement leadership, and the nature of the mobilization process. How movements are gendered, how movement dynamics are bathed in cultural creativity, and how the interactions between movements and their opposition determine movement outcomes are important questions. At the center of this entire matrix of factors must be an analysis of the central role that human agency plays in social movements and in the generation of social change.

Thanks, in large part, to the Black freedom struggle, theories of social movements that grapple with real dynamics in concrete social movements are being elaborated. Intellectual work in the next century will determine how successful scholars will be in unraveling the new empirical and theoretical puzzles thrust forth by the Black freedom movement. Although it was not their goal, Black demonstrators of the Civil Rights movement changed an academic discipline.

A remaining question is: Will Black protest continue to be vigorous in the twenty-first century, capable of pushing forward the Black freedom agenda? It is not obvious that Black protest will be as sustainable and as paramount as it has been in previous centuries. To address this issue we need to examine the factors important to past protests and examine how they are situated in the current context.

Social movements are more effective when they can identify a clear-cut enemy. Who or what is the clear-cut enemy of African Americans of the twenty-first century? Is it racism, and if so, who embodies it? Is it capitalism, and if so, how is this enemy to be loosened from its abstract perch and concretized? In fact, we do not currently have a robust concept that grasps the modern form of domination that Blacks currently face. Because the modern enemy has become opaque, slippery, illusive, and covert, the launching of Black protest has become more difficult because of conceptual fuzziness.

Second, during the closing decades of the twentieth century the Black class structure has become more highly differentiated and it is no longer firmly anchored in the Black community. There is some danger, therefore, that the cross-fertilization between different strata within the Black class structure so important to previous protest movements may have become eroded to the extent that it is no longer fully capable of launching and sustaining future Black protest movements.

Third, will the Black community of the twenty-first century possess the institutional strength required for sustaining Black protest? Black colleges have been weakened because of the racial integration of previously all white institutions of higher learning and because many Black colleges are being forced to integrate. The degree of institutional strength of the church has eroded because some of them have migrated to the suburbs in an attempt to attract affluent Blacks. In other instances, the Black Church has been unable to attract young people of the inner city who find more affinity with gangs and the underground economy. Moreover, a great potential power of the Black church is not being realized because its male clergy refuse to empower Black women as preachers and pastors. The key question is whether the Black church remains as close to the Black masses—especially to poor and working classes—as it once was. That closeness determines its strength to facilitate Black protest.

In short, research has shown conclusively that the Black church, Black colleges and other Black community organizations were critical vehicles through which social protest was organized, mobilized and sustained. A truncated class structure was also instrumental to Black protest. It is unclear whether during the twenty-first century these vehicles will continue to be effective tools of Black protest or whether new forces capable of generating protest will step into the vacuum.

In conclusion, I foresee no reason why Black protest should play a lesser role for Black people in the twenty-first century. Social inequality between the races will continue and may even worsen especially for poorer segments of the Black communities. Racism will continue to effect the lives of all people of color. If future changes are to materialize, protest will be required. In 1898 as Du Bois glanced toward the dawn of the twentieth century, he declared that in order for Blacks to achieve freedom they would have to protest continuously and energetically. This will become increasingly true for the twenty-first century. The question is whether organizationally, institutionally, and intellectually the Black community will have the wherewithal to engage in the kind of widespread and effective social protest that African Americans have utilized so magnificently. If previous centuries are our guide, then major surprises on the protest front should be expected early in the new millennium.

REFERENCES

Bobo, L. 1997. "The Color Line, the Dilemma, and the Dream: Race Relations in America at the Close of the Twentieth Century." In *Civil Rights and Social Wrongs: Black-White Relations since World War II,* edited by J. Higham, pp. 31–55. University Park. PA: Penn State University Press.

Conley, Dalton. 1999. *Being Black. Living in the Red: Race, Wealth, and Social Policy in America.* Berkeley: University of California Press.

Rustin, Bayard. no date. *Report on Poland.* New York: A. Philip Randolph Institute.

Oliver, Melvin, and Thomas E. Shapiro. 1995. *Black Wealth/White Wealth: A New Perspective on Racial Inequality.* New York: Routledge.

DISCUSSION QUESTIONS

1. What was the Jim Crow regime? How did Black Americans bring an end to the Jim Crow regime?
2. What does Morris mean by "human agency"?
3. What tactics or strategies developed by the grassroots Civil Rights movement have been employed by other groups?

52

Identity and the Politics of American Indian and Hispanic Women Leaders

BY DIANE-MICHELE PRINDEVILLE

M uch of women's political involvement in the United States traditionally has been motivated by issues pertaining to race, gender,
class, and economic concerns (Gilkes 1994; Tolleson Rinehart
1992). In addition, women of color, who bear the double burden of gender
and racial discrimination, are often economically and politically marginalized.
Consequently, their activism is influenced not only by their experiences of
sexism but also by the effects of oppression and racism on their lives, on the
lives of their families, and on their communities. As Baca Zinn and Thornton
Dill (1994, 4) noted, women of color are subjugated socially, culturally, and
economically because "patterns of hierarchy, domination, and oppression based
on race, class, gender, and sexual orientation are built into the structure of
society."

A growing body of scholarship on women and politics reveals that both
racial/ethnic identity and gender identity inform and influence the political
participation and policy preferences of women of color (see, for example,
Gilkes 1994; Jaimes 1992; Zavella 1988). This research has shown that regardless of the form of participation in which they engage, women of color generally conceptualize politics as a network of human relationships (Baca Zinn and
Thornton Dill 1994; Orleck 1997), that they value participatory democracy

From *Gender & Society* 17(4): 591–608. Copyright © 2003 by Sociologists for Women in
Society. Reprinted by permission of Sage Publications, Inc.

(Kaplan 1997; Kingsolver 1989), and that they tend to view politics as a way of helping others (Hardy-Fanta 1993; McCoy 1992). In effect, they use politics to empower members of their racial/ethnic group and to improve the quality of life in their communities (Cruz Takash 1993; Miller 1992; Pardo 1998).

Both American Indian and Hispanic women have a rich history of political activism. By political activism, I mean participating in activities such as voter registration drives in predominantly minority neighborhoods, organizing rallies to build community cohesion and support for an issue, forming coalitions to achieve a common goal, running for public office, and influencing the public policy-making process. Native women have struggled for policy reforms to attain tribal sovereignty, cultural preservation, and control over their Native lands and natural resources (Chiste 1994; Hoikkala 1995; Jaimes 1992; McCoy 1992). Latinas have long been activists in the labor movement, organizing for workers' rights, equitable pay, safe working conditions, and fair treatment (Kingsolver 1989; Marquez 1995; Segura 1994; Zavella 1988).

While Indian and Hispanic women have traditionally occupied informal positions of leadership in their communities, they are now achieving formal leadership in electoral politics. As tribal leaders, women have continually challenged federal, state, and tribal authorities to formulate and/or reform policy for the benefit of their communities. . . . Latinas have similarly sought to effect change on behalf of their communities by serving in public office. They access resources for their constituents, build and expand social networks, represent women's issues, raise concerns, and provide policy perspectives different from the male norm (Cruz Takash 1993; Hardy-Fanta 1993; Prindeville 2000). These findings reveal patterns both in the activism of the officials and in their conceptualizations of politics that establish the centrality of race/ethnicity, gender, and community for Hispanic women leaders.

To extend this previous research on the role of race/ethnicity and gender identity, and its significance for American Indian and Hispanic women leaders, I have borrowed from Tolleson Rinehart's (1992) model of "gender consciousness." According to Tolleson Rinehart, women who exhibit gender consciousness identify with other women as a group, display a positive affect toward women, and/or demonstrate a sensitivity to women's sociopolitical situations and well-being. I believe that in a similar fashion, the notion of "racial/ethnic consciousness" may be created and used to describe an individual's identification with, positive affect toward, and/or sensitivity to the situation and well-being of the individual's racial/ethnic group. Elaborating on Tolleson Rinehart's work, I create a classification scheme for assessing the role that race/ethnicity and gender play in the political ideology and motives of American Indian and Hispanic women leaders in New Mexico. Specifically, I develop dichotomous measures of racial/ethnic identity and gender identity based on four categories: the individual's propensity to identify with the racial/ethnic group or with women (self-labeling), the individual's sense of racial/ethnic or gender "consciousness," the salience of race/ethnicity or gender for the individual, and whether the individual's political activism is motivated by race/ethnicity or gender. . . .

RESEARCH DESIGN AND METHOD

. . . Using nonrandom purposive sampling, a total of 50 women active in New Mexico politics was selected for participation in this study. I obtained their names using a reputational "snowball" technique in which each woman interviewed was asked the names of other women actively involved in politics in the state. The reliability and validity of my sampling strategy were reinforced when the same women were repeatedly identified as policy leaders by different study participants in numerous organizational settings. . . . These leaders included volunteers and paid staff of grassroots organizations (activists) as well as appointed and elected officials at various levels of government (public officials). Due to their relatively small number, high level of political activity, and aggressive coalition building, and because they constitute a political elite, the 50 women interviewed for this study were frequently acquainted with each other. To protect their identities, and so that no quotation is directly attributable to any individual, pseudonyms have been used throughout.

Twenty-five of the 26 Native women interviewed were American Indian, registered with 1 of 17 sovereign nations. As such, they identified as Bad River Chippewa, Blackfoot, Comanche, Diné (Navajo), Kiowa, or Western Shoshone, or as members of Acoma, Cochiti, Isleta, Laguna, Nambe, Pojoaque, Santa Clara, Santo Domingo, Tesuque, or Zuni pueblos. One leader was Native Hawaiian; I have grouped her with the American Indian women for purposes of this study. The 24 Hispanic women variously self-identified as Chicana, Hispanic, indigenous, Mestiza, Mexican American, and Spanish. More than half of the leaders in this study (26) worked in grassroots organizations concerned with social, economic, and environmental justice, while the other 24 held elected or appointed positions in federal, state, local, or tribal government. . . .

DEMOGRAPHIC CHARACTERISTICS
OF THE LEADERS

The leaders ranged in age from 24 to 64, with a median of 45 years for activists and 50 for officials. Some of the public officials, who started out as activists, entered electoral politics later in their careers. Others waited until their children were grown and/or out of school before they pursued public office.

The New Mexico leaders were exceptionally well educated. Their median level of education was sixteen years or the equivalent of a bachelor's degree, but 40 percent had completed one or more graduate degrees, in contrast to about 6 percent of women in the New Mexico public. The leaders' generally high level of education, however, was not necessarily reflected in their household incomes. For example, the median income for New Mexico households was $24,087, for the activists it was $37,500, but for the public officials it was $52,500. In this respect, the activists were more like the public than the officials whose incomes place them among New Mexico's elite.

Variations in income may be attributed to numerous factors, including the number of income earners in a household, their type of work, hours worked, and level of pay. Despite the fact that similar numbers of activists and officials held college degrees, activists had a considerably lower household income, most likely due to their poorly paid positions in nonprofit organizations. As one activist explained,

> It's hard to find people with the education that I have that are willing to take a low-paying job like this. . . . I know I could find another job that pays much better but I don't want a job that has no real impact on the future of this country or this planet. (Rose)

Determining the social class of the New Mexico leaders, and its influence on their politics, is a difficult task. Household income does not necessarily reflect educational achievement or social class. For many of the 50 leaders, social class is particularly ambiguous. For example, among the Latinas are women whose families are composed of both fourth-generation Spanish landowners and Mexican immigrants, who may be wealthy or low-income. In New Mexico's Hispanic communities, other factors, which may be invisible to outsiders, are frequently more important than income or education for determining one's status within the community. These factors include one's place of birth, whether she is Native born or an immigrant; whether she is light skinned or dark; whether she identifies as "Spanish," "Chicana," or "Mexicana"; whether her family held a Spanish land grant; and whether she speaks Spanish fluently. In similar ways, income and education may be less relevant measures of a Native woman's position in the social hierarchy of her tribe than her clan affiliations, her membership in traditional religious societies, whether she is "full blood" Indian, whether she resides on the reservation, or whether she is fluent in her Native language. For these reasons, I believe that we cannot speak with any degree of certainty about the social class of the New Mexico leaders or of how class considerations might influence their political activism. What we can assess with greater confidence is the role and impact of race/ethnicity and gender on their politics and their concern with achieving social justice and economic equity for their communities. . . .

POLITICS AND THE INTERSECTION OF RACE/ETHNICITY AND GENDER

Racial/Ethnic Identity

Culture and race/ethnicity were highly salient for the 50 New Mexico leaders I interviewed. Many of these women felt strongly about the use of racial/ethnic categories. How they identified themselves was a consciously political decision. For example, Latinas who acknowledged both their indigenous and Spanish heritage called themselves Chicanas or Mestizas. On the other hand, one Hispana who claimed "pure" Spanish heritage was offended when I asked whether she considered herself Chicana or Mexican American. She insisted that she was Caucasian and a descendant of the conquistadores, presumably implying that she had no Native blood despite her family's presence in the region for more than 400 years. In contrast, several of the Indian leaders and Hispanic activists considered themselves to be members of colonized groups who continue to suffer racial, social, and economic oppression at the hands of the dominant Anglo-American culture. A smaller group of women expressed militant or nationalistic views; they identified as "indigenous," "Third World," or "Chicana." The majority of American Indian women identified as members of a particular tribe and used the generic term *Indian* to refer to other Native American peoples. Women of Spanish or Mexican heritage described themselves as Mexican American or Hispanic.

Several women used racial and/or ethnic descriptors to self-identify, often in combination with their gendered roles, such as "a Chicana working mother, a grandmother, a strong woman" or "an independent Comanche woman, an Indian activist." The leaders' responses demonstrate a strong identification with their racial/ethnic groups and cultures. . . .

The New Mexico leaders frequently displayed two or more of these indicators of racial/ethnic identification. This is illustrated by Toni, an advocate of teaching Navajo and other Native languages in the public schools. Toni was motivated and became active in politics by "a sense of empowerment for self and children's sense of pride in self, in their heritage." Her strong sense of racial/ethnic identity was expressed through racial consciousness, racial/ethnic salience, and cultural motivation.

While equal numbers of Native and Hispanic women described themselves by their racial/ethnic groups, three times as many grassroots activists (15) as public officials (5) did. Nearly half of all the New Mexico leaders (23) expressed racial/ethnic consciousness, and a large number of leaders (40) demonstrated the salience of race/ethnicity to their practices of politics, employing their political activism to address specifically the problem of racism and other issues of race, ethnicity, and culture. Finally, cultural preservation provided an incentive for entering politics for nearly all of the Native leaders and grassroots activists (25) and for 19 of the Hispanic activists and public officials. . . .

Identification with their racial/ethnic groups was generally more important for the Native leaders than it was for the Latinas, as well as more important for the grassroots activists than for the public officials. As one Native activist stated,

> The first thing I think about myself is that I'm an Indian. . . . I'm an Indian *person.* . . . In the Mohawk culture, women are pretty strong anyway—pretty influential. . . . It's more important for me to be a Mohawk than to be a woman. (Kari)

The salience of racial/ethnic identity is grounded in the day-to-day experiences of the women interviewed. Camila, a public official, gave a forceful assessment of New Mexico's political and economic environment from the perspective of Indian people:

Right now Native Americans in the state are. . . just now making very small inroads to the legislature. And that's just been in the last 10 to 15 years. That's nothing! . . . That's lamentable for New Mexico. You know, so long as we're just dancing and making our pottery, that's OK. "But you just stay over there! . . . Ya, we can come in and make money off of you but [Indian people] can't have any part of it. Just keep in your own place!" That's the message. . . .

The anger generated in response to racism often fueled the New Mexico leaders' determination to enter into politics or to continue with their community work. As one Indian official noted, "Any time you have a woman of color, color comes out more than the male-female issues." (Camila). Their experiences of racism and of sexism were often entwined. . . .

Like the African American women in Gilkes's study, the New Mexico leaders found positive ways to combat racism in their communities by building social networks among people and by strengthening local institutions. Both Native and Hispanic leaders were active in developing programs that celebrated heritage through language, dance, storytelling, art, and crafts to preserve their communities' cultural traditions. Leaders also participated in cultural exchange programs to better develop cooperative relationships with other racial/ethnic groups in the state. Through their actions, the women expressed their racial/ethnic identities. This finding is consistent with previous research showing that women's collective action to improve the quality of life in their communities is often built on existing networks and serves to unite people across racial and ethnic lines (Baca Zinn and Thornton Dill 1994).

For many women of color, the roles that race/ethnicity and gender play in forming a woman's identity are inseparable and somewhat complex. The New Mexico leaders generally held a multiplicity of, sometimes conflicting, roles and responsibilities. Women frequently expressed feeling torn between cultural gender expectations and their individual needs as women, mothers, workers, and leaders. . . .

Gender Identity

As with the four categories of their racial/ethnic identification, clear expressions of the leaders' gender identities also emerged from the data—self-labeling by gendered roles and/or sex, gender consciousness, gender salience, and gender motivation. . . . Both Native and Hispanic leaders frequently exhibited two or more of these indicators of gender identification.

Several leaders (23) described themselves in relation to their gender as, among other things, a strong woman, single parent, working mother, wife, grandmother, and "supermom." A majority of the leaders (44) demonstrated gender consciousness. Most of the public officials (22), grassroots activists (22), and Native leaders (20) and all of the Latinas identified with other women as a group, displaying gender consciousness. Gender was especially salient for a slightly smaller number of leaders (21) but more so for the public officials (12) than for the grassroots activists (9). Furthermore, twice as many Latinas (14) as Native women (7) addressed women's issues specifically through their political activism by advocating for and/or empowering women. Finally, a larger number of Latina officials (15) than Native leaders or activists (9) were motivated to participate in politics to advance the rights of women and to promote women's issues. Despite their advancing women's issues and fully believing that women should have the "opportunity to participate fully in our society without discrimination," the New Mexico leaders did not necessarily view themselves as feminists.

The Intersection of Race and Gender Identity

While the frequency of racial/ethnic identity was generally greater than gender identity for both the Native and Hispanic women, . . . racial/ethnic identity was more important for a larger percentage of Native than Hispanic leaders as well as more important for a larger percentage of the grassroots activists than public officials. . . . Gender identity was more salient for a greater proportion of the Hispanas than Native women, as well as more salient for a greater proportion of the public officials than the grassroots activists. These findings are consistent with research on Native women's activism, which generally tends to advance the well-being of the community as a group rather than the needs of Indian women in particular (Hoikkala 1995; Jaimes 1992). . . .

For the Native leaders I interviewed, politics is overwhelmingly defined by their efforts to preserve their racial/ethnic and cultural identities in the face of tremendous pressures to assimilate into American mainstream society. While both American Indian and Hispanic people have struggled to overcome the oppressive social and economic conditions that their groups have historically faced in the United States, their experiences of colonization have been different (McClain and Stewart 1995). Furthermore, as Native Americans have a unique legal and political status (they hold dual citizenship as members of sovereign nations and as citizens of the United States) as well as their own tribal political systems, they are less likely than other minority groups to fully assimilate into American political culture.

While the New Mexico leaders, as a group, demonstrated greater racial/ethnic identity than gender identity, this is not to say that public policy relevant to women was unimportant to them (Prindeville 2000). On the contrary, . . . they were very much aware of the social, economic, and political constraints faced by many women simply because of their sex. . . .

DISCUSSION AND CONCLUSION

. . . The findings suggest that there may be greater similarities between public officials and grassroots activists than previously thought. Both sets of leaders, whether Native or Hispanic, sought to empower community members and generally perceived their own role to be one of advocacy. Indeed, 88 percent of those interviewed exhibited gender consciousness—an identification with other women as a group, a positive affect toward women, and a sense of connectedness to women and to their well-being. Gender consciousness significantly influenced their motives for participation in politics and their political ideologies. As a group, the New Mexico leaders valued equality, community participation in shared decision making, and consensus. Their political activism was a means for addressing women's issues by advocating for and/or empowering other women.

In contrast to these findings, the differences observed between the American Indian and Hispanic leaders, based on their racial/ethnic identity rather than gender identity, were more striking. Racial/ethnic identity was highly salient for the majority of the women interviewed. While 80 percent used their political activism to address the problem of racism and other issues relevant to race/ethnicity and culture, nearly 90 percent were motivated to enter politics by their desire to preserve their communities' cultures and traditions. Issues relative to race/ethnicity and culture appeared to be somewhat more important to the Native leaders and grassroots activists. Gender identity seemed generally to be more salient for greater numbers of the Hispanic leaders and public officials. . . .

My findings demonstrate the prevalence of racial/ethnic identity and gender identity—not only as motives for women's political participation but as factors that both inform and shape their political ideologies. While this finding is not altogether new to the research on women of color, it is significant because it reiterates the necessity of having racially and culturally diverse representations of women in politics. Indeed, this research provides further empirical support for the importance of racial/ethnic minority women holding positions of political leadership.

REFERENCES

Baca Zinn, M., and B. Thornton Dill. 1994. Difference and domination. In *Women of color in U.S. society*, edited by M. Baca Zinn and B. Thornton Dill. Philadelphia: Temple University Press.

Chiste, K. B. 1994. Aboriginal women and self-government: Challenging Leviathan. *American Indian Culture and Research Journal* 18:19–43.

Cruz Takash, P. 1993. Breaking barriers to representation: Chicana/Latina elected officials in California. *Urban Anthropology* 22:325–60.

Ford, R. L. 1990. Native American women activists: Past and present. Unpublished manuscript, Southwest Texas State University.

Gilkes, C. T. 1994. "If it wasn't for the women . . .": African American women, community work, and social change. In *Women of color in U.S. society*, edited by M. Baca Zinn and B. Thornton Dill. Philadelphia: Temple University Press.

Hardy-Fanta, C. 1993. *Latina politics, Latino politics: Gender, culture, and political participation in Boston*. Philadelphia: Temple University Press.

Hoikkala, P. H. 1995. Native American women and community work in Phoenix, 1965–1980. CD-ROM. Abstract in *Dissertation Abstracts International* 56-06A:2382:AAI95-33382.

Jaimes, M. A. 1992. American Indian women: At the center of indigenous resistance in North America. In *The state of Native America: Genocide, colonization, and resistance,* edited by M. A. Jaimes. Boston: South End.

Kaplan, T. 1997. *Crazy for democracy: Women in grassroots movements.* New York: Routledge.

Kingsolver, B. 1989. *Holding the line: Women in the great Arizona mine strike of 1983.* Ithaca, NY: ILR Press.

Marquez, B. 1995. Organizing Mexican-American women in the garment industry: La mujer obrera. *Women & Politics* 15:65–87.

McClain, P. D., and J. Stewart Jr. 1995. *"Can we all get along?" Racial and ethnic minorities in American politics.* Boulder, CO: Westview.

McCoy, M. 1992. Gender or ethnicity: What makes a difference? A study of women tribal leaders. *Women & Politics* 12:57–68.

Miller, B. G. 1992. Women and politics: Comparative evidence from the northwest coast. *Ethnology* 31:367–75.

Orleck, A. 1997. Tradition unbound: Radical mothers in international perspective. In *The politics of motherhood: Activist voices from left to right,* edited by A. Jetter, A. Orleck, and D. Taylor. Hanover, NH: University Press of New England.

Pardo, M. 1998. Creating community: Mexican American women in eastside Los Angeles. In *Community activism and feminist politics: Organizing across race, class, and gender,* edited by N. A. Naples. New York: Routledge.

Prindeville, D. M. 2000. Promoting a feminist policy agenda: Indigenous women leaders and closet feminism. *Social Science Journal* 37:637–45.

Prindeville, D. M. and J. G. Bretting. 1998. Indigenous women activists and political participation: The case of environmental justice. *Women & Politics* 19:39–58.

Segura, D. 1994. Inside the work worlds of Chicana and Mexican immigrant women. In *Women of color in U.S. society,* edited by M. Baca Zinn and B. Thornton Dill. Philadelphia: Temple University Press.

Tolleson Rinehart, S. 1992. *Gender consciousness and politics.* New York: Routledge.

Zavella, P. 1988. The politics of race and gender: Organizing Chicana cannery workers in northern California. In *Women and the politics of empowerment,* edited by A. Bookman and S. Morgen. Philadelphia: Temple University Press.

DISCUSSION QUESTIONS

1. What role do race and gender—independently and together—play in the identities of these American Indian and Hispanic women leaders?

2. How do you think the identities and issues that these women share compare to the identities and issues of other political leaders?

53

Signs . . . Signs . . . Turn Visible Again

The Transformative Uses of Biography

BY VINCENT HARDING

W[hen] we look carefully and with insight, it soon becomes evident that few bodies of knowledge are more filled with such living signs of hope than the story of the Black freedom movement in the United States. (Indeed, human struggles for liberation and resurrection, wherever they occur, are often rich in such lively, translucent testimonies, personified in men and women of amazing possibilities.) Interestingly enough, our students seem to have already intuited something of the power of this biographical treasure. . . . In the course of a national assessment of historical knowledge among high school seniors, their response revealed that one of the figures in American history who was best known among the students was Harriet Tubman, the indomitable heroine of the Underground Railroad and the Civil War, who later became a pioneer in the postwar care for the aged.[1]

The scholars who reported the strength of the Tubman/Underground Railroad image in the minds of our students were not sure why it was so real. They suggested that the "inherent drama and conflict" of the subject matter might help account for its presence. . . . At a deeper level it may well be that the students caught the essence of the "drama and conflict," recognizing that in the case of Tubman and others like her they were being confronted by powerful, authentic personalities, rooted in a great human determination to

From Vincent Harding, *Hope and History* (Maryknoll, NY: Orbis Books, 1990), pp. 16–25. Reprinted by permission of Orbis Books.

overcome the forces of darkness within and the systems of domination and destruction all around them.

When we turn to Harriet Tubman's great-grandchildren, to the lives that flowered in the Black-led freedom movement of the post-World War II period, we are flooded with opportunities to provide our students with access to authentic signposts whose biographies may challenge, illuminate, and inspire young people and adults—across lines of race, class, gender, and nationality. There are literally hundreds of well-documented examples and thousands more still waiting for their compelling life stories to receive the documentation that groups of curious and committed students and teachers could provide. The potential subjects of such explorations are living and dead, Black and white, urban and rural, poor and not-so-poor, people of varied religious and nonreligious persuasions, a marvelous cross-section of hope. Each provides us with a vivid opportunity to use biographical study as a teaching tool, to allow the life stories of significant women and men to draw students into their light, and to make it possible for all of us to reassess our own understandings of what makes for a "significant" life.

So we could begin, for instance, with Fannie Lou Hamer, allowing students to hear her magnificent voice on tape and see her in a documentary such as *Eyes on the Prize*, while beginning to ask themselves what it was like to grow up Black (or white) in Mississippi in the 1920s and 1930s. To understand why Mrs. Hamer never went beyond fourth grade, to sense what kind of prison the institution of sharecropping could be, and how a woman in her forties could finally break through the past to become one of the activist-philosopher heroines of the southern freedom movement—these could be compelling tasks for a group of students. . . .

If we continued with the strange light of the Magnolia State, we could turn to Michael (Mickey) Schwerner and his wife, Rita, and ask why a young professional Jewish couple from New York City would want to come to Mississippi and risk their lives working in a Black-led movement for freedom, justice, and the renewal of both the state and the nation. . . .

Keeping some of the fascinating historical lines intact, we might turn our students towards Bernice Johnson Reagon, who grew up as a Black minister's child in Albany, Georgia, and helped organize the Freedom Singers as part of the movement's rich cultural life in that city. Bernice then traveled with the group across the country, raising funds for the movement, telling its story in words and music, finally singing and testifying in New York City, where Mickey and Rita heard the group and were so deeply moved that they decided to give a period of their lives to the southern struggle. Bernice is now the leading spirit of Sweet Honey in the Rock, one of the best-known women's singing groups in the world. At the same time, she is also Dr. Bernice Reagon, musicologist and ethnographer, on the staff of the Smithsonian Institution in the nation's capital, and a recipient of one of the prestigious MacArthur Fellowships. So she stands as a luminous signpost who can be seen, heard, and touched right here and now.

Actually, the list is long enough for years of teaching. Students know something of Martin Luther King, Jr., but often not enough to help them see the richness, strength, and costliness of the life he led. They have usually not been guided to consider the motivations that take a person from a relatively comfortable middle-class existence to the place where finally he was sharply challenging the dominant values, structures, and leadership of a nation he loved so dearly. Perhaps, then, it would be good to let our students see King through the eyes of younger people, like Martin King III. . . .

Of course, even (or especially) as we deal with such relatively young lives, it will be good to be prepared for issues that take us beneath the surface of social and political activism. For Martin III needs to be allowed to recollect the time of his father's assassination and to say, "All through that period . . . I was hurt. I couldn't understand why a man who loves and tried to love so many could be brutally assassinated for no apparent reason."[2] All our students, of all ages, surely need to feel free to go with Marty, probing that ancient question for their own lives, raising their own questions with and about martyred heroes of our times, always in search of their own signs.

On the other hand, we and our students tend to know almost nothing of Malcolm X (many ask, "Who was Malcolm the Tenth?"). Here was one of the greatest signposts of this century, planted originally in the heart of darkness. Here was life so powerful in its self-destruction and then in its redemptive force that its authenticity grasps all who approach him. Here was DeTroit Red, self-hating denizen of criminal depths who rose to become "much more than there was time for him to be,"[3] first Malcolm X, then El-Hajj Malik El-Shabazz. Which students will we assign to learn the meaning of his X and of his new names, to help us and themselves understand the power and meaning of new names in the lives of those humans who seek to become new people? . . .

There are, of course, names that our students may never have heard, like that of Ella Baker, one of the wisest, most courageous, and most influential shapers of the modern freedom movement. Let them meet her—if only through books, films, and the memories of those who knew her—and ask their questions. Let them wonder aloud what she meant when toward the end of her life Miss Baker (as she was called by so many in the movement) kept saying that her autobiography should be titled *Making a Life, Instead of Making a Living*.

Let them uncover the name of June Johnson of Greenwood, Mississippi, who joined the freedom movement there when she was thirteen. Let them reflect on the fact that June was not too young to be brutally beaten in prison the next year as she joined Mrs. Hamer and their fellow movement worker Annelle Ponder in an impromptu challenge to segregation and attempted police intimidation at an interstate bus stop. June continues to work, to speak. Let them hear her.

And perhaps in a different way they will also see and hear Viola Liuzzo, white activist housewife from Detroit, who drove down to Selma, Alabama, in 1965 to work in solidarity with those who were determined to cross the

Edmund Pettus Bridge, marching, working for justice. Let them imagine why she left four children to come south. Perhaps they should be asked to re-enact that night in March, to feel what it was like on that dark highway, to remember how she kept driving her car and humming a freedom song as the Klan members drove up alongside and fired their bullets into her head, because she loved freedom more than life. What signs will become visible on that road?

Of course, students also need to meet the unknown numbers of Klan members who changed their lives, like the one who approached Jesse Jackson during the 1988 presidential primary campaign and confessed that he had been "on the wrong side" in Selma. But now he came to get his picture taken with Jackson, just to reaffirm his resolve never to be on the wrong side of the quest for justice again. To consider how men and women change, to discover that such a change is indeed possible—that is certainly signpost material, especially when the pondering seems to begin with the minds and hearts of students themselves. . . .

Introduce them to Angela Davis and let them ask why this brilliant Black woman who was born in Birmingham, Alabama, whose young friends were killed in a church bombing, eventually became one of the leading members of the Communist Party of the United States. Let them explore the events that made her one of the world's best-known fugitives (was Harriet anyplace near?) and now one of the nation's most articulate advocates and teachers for justice. Diane Nash, who grew up Black in Chicago and then emerged to embody some of the southern movement's most powerful manifestations of nonviolent courage, resistance, and creativity, is still available to our students. Her story is rich, filled with insights concerning the power of spirituality in the transformation of her life. Perhaps she will also tell our students about her continuing determination to organize a "nonviolent army" in America. Will we encourage them to explore such a path? Are there better armies they could join?

Facing that question, it would be good for young people to be introduced to conscientious objectors such as Bayard Rustin, now dead, and James Farmer, still living* but almost without the use of his eyes. The lives of such men convey powerful stories of what it means to be willing to spend years in prison rather than join their lives to the destructive mission of the military. They would help us to see the relationship between such beliefs and the organizing of freedom rides. They would tell us what it meant to have to hide in a coffin to escape a mob, or what it's like to spend almost half a century living and working in the cause of freedom, justice, and hope. Surely such veterans of the struggle for democracy in America would know something and could help us all to understand what it means to be afraid in the midst of danger and yet not submit to the power of fear. . . .

In the same way, let James and Grace Lee Boggs (one a Black native of Alabama, the other a first-generation Chinese-American) open the treasures of their half-century of work for human rights and responsibilities, for the

---

*Eds. note: James Farmer died in 1999 at age 79.

humane transformation of our society. Perhaps their story of life in Detroit over the past four decades will also provide signposts, suggestions for creating outposts of hope in very tough spaces.

It may be that the most important of these long-distance runners is Bob Moses, for he would bring our students many gifts that are sorely needed in our time, beginning with the gift of his story. The story is of his movement from a Depression-time Harlem beginning, to a teaching position at a comfortable private school in New York City, to the profound sense that he must join the great drama unfolding among the 1960s student movement in the South. The story is of Mississippi and its magnificent grassroots Black leaders who risked livelihood and lives to welcome and work with Bob and the small band of nonviolent shock troops who helped open the narrow, blood-bought crevices that eventually became a new beginning for Mississippi and the nation. . . .

Of course, Bob Moses and the others we have named can only begin to touch the surface of the biographical resources available to us. Still, we know that their great value is not in numbers, but in the questions they allow us to raise with ourselves and our students, questions that may open us toward signposts. For somewhere along the line, as students consider such lives, they do try to make sense of them. They do ask what such lives mean, especially as they span decades of commitment to a difficult and often unpopular cause. . . .

What is it that keeps people going for so long a time? In almost every case those women and men who have essentially given their lives to the long work of humane social change answer that question by speaking of some gradual inner development, often unexpected. They tell of the slow nurturing of deep spiritual resources within them, usually connected to regularly practiced disciplines of meditation, prayer, silence, community with others, at times moving outside the conventional paths of America's organized mainstream religions. They speak, like Bob Moses, of constantly discovering how crucial is the work of "internal organizing" for anyone who seeks to work at serious reorganization of the world around us. Clearly these are questions and responses we might ordinarily try to avoid, but when students are in search of signposts, they have a right to expect us at least to explore with them the ways in which other persons have nurtured the resources to keep their lives going.

Moving with our students into such uncertain arenas is the least we can do once they have been awakened to the issues, once they have ventured into the path of self-discovery. For a profound engagement with the lives of authentically human women and men reminds us that such seriously crafted signposts point not only to a way out of the darkness of self-destructive lifestyles and socially constructed traps, but they do more. Such beacon lives and the questions they raise also urge us all toward deep and often untapped levels of our own inner universe, opening issues we might otherwise wish to escape. . . .

To open the way for exploration of such fundamental questions may be one of the greatest privileges of our teaching. To join our students in personally coming to grips with the magnificent signpost lives of our times may be one of the rare privileges of our existence. For it is possible that among the most uniquely humanizing forces we can know and share is the experience of informed gratitude. In that spirit we are able to return to such a sign as was provided by a young Black sit-in heroine in Tallahassee, Florida, thirty years ago. There, Patricia Stephens sat in jail, serving a two-month sentence for her participation in the struggle for democracy, and wrote to a friend: "We are all so very happy that we were (and are) able to do this to help our city, state, and nation. This is something that has to be done over and over again, and we are willing to do it as often as necessary."[4]

With words and lives like Patricia's, the signs and their direction become visible, leaving students and teachers the great privilege of choice, . . . providing constantly new meaning to Bob Moses' invitation to the organizing of our souls, to the transformation of our nation, "over and over again."

NOTES

1. This assessment was conducted in 1986 and reported in Diane Ravitch and Chester Finn, Jr., *What Do Our Seventeen-Year-Olds Know?* (New York: Harper & Row, 1987). See especially pp. 1–5, 61, 76.

2. *Rolling Stone* (April 7, 1988), p. 63.

3. From Robert Hayden's poem "El-Hajj Malik El-Shabazz" in *For Malcolm*, ed. Dudley Randall and Margaret G. Burroughs (Detroit: Broadside Press, 1967), p. 16.

4. Quoted in Harvey Sitkoff, *The Struggle for Equality* (New York: Hill & Wang, 1981), pp. 74–75.

DISCUSSION QUESTIONS

1. What does Harding mean by the concept signpost? Why are signposts important?

2. Why does Harding think high school students remember the story of Harriet Tubman?

3. What motivates people who are not oppressed to work for racial justice?

4. Why do you think people of diverse backgrounds were involved in the Civil Rights Movement?

54

We Are All Americans!

The Latin Americanization of Racial Stratification in the USA

BY EDUARDO BONILLA-SILVA

INTRODUCTION

Racial stratification in the United States has operated along bi-racial lines (White–non-White) for centuries. For demographic (the relative large size of the Black population) and historical reasons (the centrality of Blacks to the national economic development from the 17th to the middle part of the 20th century), the bi-racial order has been anchored on the Black–White experience. Historically, those on the non-White side of the divide have shared similar experiences of colonialism, oppression, exploitation, and racialization. Hence, being non-White has meant having a restricted access to the multiple benefits or "wages of whiteness" (Roediger, 1991) such as good housing, decent jobs, and a good education.

Nevertheless, the post-civil rights era has brought changes in how racial stratification seems to operate. For example, significant gaps in status have emerged between groups that previously shared a common denizen position in the racial order. Asian Americans in particular have almost matched the socioeconomic standing of Whites and, in some areas (e.g., educational attainment), have surpassed them. For example, in selective colleges across the nation, Asian Americans are represented at 3–10 times their national proportion.

From *Race & Society* 5 (2002): 3–17. Reprinted by permission of Elsevier.

Another example of the changes is the high rate of interracial dating and marriage between Latinos and Whites and Asians and Whites. These interracial unions, coupled with the collapse of formal segregation, have created the political space for "multiracial activists" to force the Census Bureau in 2000 to allow respondents to pick all the races they felt apply to them (Daniels, 2002). Yet another instance of the changes in contemporary America is that few Whites endorse in surveys segregationist views. This has been heralded by some as reality as "the end of racism" or pointing to "the declining significance of race." Lastly, Blacks have been surpassed by Latinos as the largest minority group (by 2001, the Census noted that "Hispanics" were 13 percent of the population and Blacks 12 percent).

I propose that all this reshuffling denotes that the bi-racial order typical of the United States, which was the exception in the world-racial system, is evolving into a complex tri-racial stratification system similar to that of many Latin American and Caribbean nations. Specifically, I suggest that the emerging tri-racial system will be comprised of "Whites" at the top, and intermediary group of "honorary Whites"—similar to the coloreds in South Africa during formal apartheid, and a non-White group or the "collective Black" at the bottom. In Figure 1, I sketch what these three groups may look like. I hypothesize that the White group will include "traditional" Whites, new "White" immigrants and, in the near future, assimilated Latinos, some multiracials, and other sub-groups; the intermediate racial group or honorary Whites will comprise most light-skinned Latinos (e.g., most Cubans and segments of the Mexican and Puerto Rican communities), Japanese Americans, Korean Americans, Asian Indians, Chinese Americans, and most Middle Eastern Americans; and, finally, that the collective Black will include Blacks, dark-skinned Latinos, Vietnamese, Cambodians, Laotians, and maybe Filipinos.

As a tri-racial system (or Latin- or Caribbean-like racial order), race conflict will be buffered by the intermediate group, much like class conflict is when the class structure includes a large middle class. Furthermore, color gradations, which have always been important matters of within-group differentiation, will become more salient factors of stratification. Lastly, Americans, like people in complex racial stratification orders, will begin making nationalists appeals ("We are all Americans"), decry their racial past, and claim they are "beyond race"....

WHY WOULD A TRI-RACIAL SYSTEM
BE EMERGING IN THE USA NOW?

Why would race relations in the United States be moving toward a tri-racial regime at this point in history? The reasons are multiple. First, the demography of the nation is changing. Racial minorities are up to 30 percent of the population and, as population projections suggest, may become a numeric majority in the year 2050 (U.S. Bureau of the Census, 1996). And these

---

"Whites"

Whites
New Whites (Russians, Albanians, etc.)
Assimilated white Latinos
Some multiracials
Assimilated (urban) Native Americans
A few Asian-origin people

"Honorary Whites"

Light-skinned Latinos
Japanese Americans
Korean Americans
Asian Indians
Chinese Americans
Middle Eastern Americans
Most multiracials

"Collective Black"

Filipinos
Vietnamese
Hmong
Laotians
Dark-skinned Latinos
Blacks
New West Indian and African immigrants
Reservation-bound Native Americans

FIGURE 1   Preliminary map of tri-racial system in the USA.

projections may be slightly off downward as early releases from the 2000 Census suggest that the Latino population was about 12.5 percent of the population, almost 1 percentage point higher than the highest projection and the proportion of White (77.1 percent White or in combination) was slightly lower than originally expected.

The rapid darkening of America is creating a situation similar to that of many Latin American and Caribbean nations where the White elites realized their countries were becoming "Black" or "Indian" and devised a number of strategies to whiten their population and maintaining White power. Although whitening the population through immigration or by classifying many newcomers as White is a possible solution to the new American demography, a more plausible accommodation to the new racial reality, and one that would still help maintain "White supremacy" is to (1) create an intermediate racial group to buffer racial conflict, (2) allow some newcomers into the White racial strata, and (3) incorporate most immigrants into the collective Black strata.

Second, as part of the tremendous reorganization that transpired in America in the post-civil rights era, a new kinder and gentler White supremacy emerged which Bonilla-Silva has labeled elsewhere as the "new racism" (Bonilla-Silva, 2001). In post-civil rights America the maintenance of systemic White privilege is accomplished socially, economically, and politically through institutional, covert, and apparently non-racial practices. Whether in banks or universities, in stores or housing markets, "smiling discrimination" tends to be the order of the day. This kinder and gentler form of White supremacy has produced an accompanying ideology: the ideology of color-blind racism. This ideology denies the salience of race, scorns those who talk about race, and increasingly proclaims that "We are all Americans."

Third, race relations have become globalized (Lusane, 1997). The once almost all-White Western nations have now "interiorized the other" (Miles, 1993). The new world-systemic need for capital accumulation has led to the incorporation of "dark" foreigners as "guest workers" and even as permanent workers. Thus, today European nations have racial minorities in their midst who are progressively becoming an underclass, have developed an internal "racial structure" to maintain White power, and have a curious racial ideology that combines ethnonationalism with a race-blind ideology similar to the color-blind racism of the United States today.

This new global racial reality will reinforce the trend towards tri-racialism in the United States as versions of color-blind racism will become prevalent in most Western nations (Winant, 2001). Furthermore, as many formerly almost all-White Western countries (e.g., Germany, France, England, etc.) become more and more racially diverse, tri-racial divisions may surface in these societies, too.

Fourth, the convergence of the political and ideological actions of the Republican Party, conservative commentators and activists, and the so-called "multi-racial" movement (Rockquemore & Brunsma, 2002), has created the space for the radical transformation of the way we gather racial data in America. One possible outcome of the Census Bureau categorical back-and-forth on racial and ethnic classifications is either the dilution of racial data or the elimination of race as an official category.

Lastly, the attack on affirmative action, which is part of what Steinberg (1995) has labeled as the "racial retreat," is the clarion call signaling the end of race-based social policy in the United States. Although it is still possible to save a watered-down version of this program, at this point, this seems doubtful. Again, this trend reinforces my thesis because the elimination of race-based social policy is, among other things, predicated on the notion that race no longer affects minorities' status. Hence, as in many countries of the world, the United States may eliminate race by decree and maintain—or even increase—the level of racial inequality. . . .

CONCLUDING REMARKS: TRI-RACIAL ORDER, RACIAL POLITICS, AND THE FUTURE OF WHITE SUPREMACY IN AMERICA

. . . At this early stage of the analysis and given the serious limitations of the data on "Latinos" and "Asians" (most of the data is not parceled out by sub-groups and there is limited information by skin-tone), it is hard to make a conclusive case. Nevertheless, almost all the objective, subjective, and social interaction indicators tend to point in the direction one would expect if a tri-racial system is emerging. For example, objective indicators on income and education show substantive gaps between the groups I labeled "White," "honorary White," and the "collective Black." Not surprisingly, a variety of subjective indicators signal the emergence of *internal* stratification among racial minorities. This has led some minority groups to develop racial atti-tudes similar to those of Whites, and others to develop attitudes closer to those of Blacks. Finally, the findings on the objective and subjective indicators have an interactional correlate. Data on interracial marriage and residential segregation shows that Whites are significantly more likely to live nearby honorary Whites and intermarry with them than live and intermarry with members of the collective Black.

If my prediction is right, what may be the consequences for race relations in the United States? First, racial politics will change dramatically. The "us" versus "them" racial dynamic will lessen as "honorary Whites" grow in size and social importance. This group is likely to buffer racial conflict—or derail it—as intermediate groups do in many Latin American countries. . . .

Second, the ideology of color-blind racism will become even more salient among Whites and honorary Whites and will also impact members of the collective Black. This ideology will help glue the new social system and fur-ther buffer racial conflict.

Third, if the state decides to stop gathering racial statistics, the struggle to document the impact of race in a variety of social venues will become monu-mental. More significantly, because state actions always impact civil society, if the state decides to erase race from above, the *social* recognition of "races" in the polity may become harder. Americans may develop a Latin American- or Caribbean-like "disgust" for even mentioning anything that is race-related.

Fourth, the deep history of Black–White divisions in the United States has been such that the centrality of the Black identity will not dissipate. Research on the "Black elite," for instance, shows they exhibit racial attitudes in line with their racial rather than class group (Dawson, 1994). That identity may be taken up by dark-skinned Latinos as it is being rapidly taken up by most West Indians (Kasinitz, Battle, & Miyares, 2001) and some Latinos (Rodríguez, 2000). . . .

However, even among Blacks, I predict some important changes. Their racial consciousness will become more diffused. For example, Blacks will be more likely to accept many stereotypes about themselves (e.g., "We are more lazy than Whites") and develop a "blunted oppositional consciousness." Furthermore, the external pressure of "multiracials" in White contexts (Rockquemore & Brunsma, 2002) and the internal pressure of "ethnic" Blacks may change the notion of "Blackness" and even the position of some "Blacks" in the system. Colorism may become an even more important factor as a way of making social distinctions among "Blacks" (Keith & Herring, 1991).

Fifth, the new order will force a reshuffling of *all* racial identities. Certain "racial" claims may dissipate (or, in some cases, decline in significance) as mobility will increasingly be seen as based on (1) whiteness or near-whiteness and (2) intermarriage with Whites (this seems to be the case among many Japanese Americans, particularly those who have intermarried). This dissipation of ethnicity will not be limited to "honorary Whites" as members of the "collective Black" strata strive to position themselves higher in the new racial totem pole based on degrees of proximity or closeness to whiteness. Will Vietnamese, Filipinos, and other members of the Asian underclass coalesce with Blacks and dark-skinned Latinos or will they try to distance themselves from them and struggle to emphasize their "Americanness"?

Lastly, the new racial stratification system will be more effective in maintaining "White supremacy." Whites will still be at the top of the social structure but will face fewer race-based challenges and racial inequality will remain and may even widen as is the case throughout Latin America and the Caribbean. And, to avoid confusion about my claim regarding "honorary Whites," let me clarify that their standing and status will be dependent upon Whites' wishes and practices. "Honorary" means they will remain secondary, will still face-discrimination, and will not receive equal treatment in society. For example, although Arab Americans will be regarded as "honorary Whites," their treatment in the post–September 11 era suggests their status as "White" and "American" is tenuous at best. Likewise, albeit substantial segments of the Asian American community may become "honorary White," they will also continue suffering from discrimination and be regarded in many quarters as "perpetual foreigners."

Therein lies the weaknesses of the emerging tri-racial order and the possibilities for challenging it. Members of the "collective Black" must be the backbone of the movement challenging the new order as they are the ones who will remain literally "at the bottom of the well." However, if they want to be successful, they must wage, in coalition with progressive Asian and Latino organizations, a concerted effort to politicize the segments I label "honorary Whites" and make them aware of the *honorary* character of their status. This is the way out of the impending new racial quandary. We need to short-circuit the belief in near-whiteness as the solution to status differences and create a coalition of all "people of color" and their White allies. If the tri-racial, Latin American- or Caribbean-like model of race prevails and "pigmentocracy" crystallizes, most Americans will scramble for the meager wages that near-whiteness will provide to those willing to play the "We are all American" game.

REFERENCES

Bonilla-Silva, E. (2001). *White supremacy and racism in the post-civil rights era.* Boulder, CO: Lynne Rienner Publishers.

Daniels, R. (2002). *More than Black? Multiracial identity and the new racial order.* Philadelphia: Temple University Press.

Dawson, M. C. (1994). *Behind the mule: Race and class in African American Politics.* Princeton: Princeton University Press.

Kasinitz, P., Battle, J., & Miyares, I. (2001). Fade to Black? The children of West Indian immigrants in Southern Florida. In R. G. Rumbaut & A. Portes (Eds.), *Ethnicities: Children of immigrants in America* (pp. 267–300). Berkeley: University of California Press.

Keith, V. M., & Herring, C. (1991). Skin tone and stratification in the Black community. *American Journal of Sociology, 97*(3), 760–778.

Lusane, C. (1997). *Race in the global era: African Americans at the millennium.* Boston: South End Press.

Mills, C. W. (1997). *The racial contract.* Ithaca and London: Cornell University Press.

Rockquemore, K. A., & Brunsma, D. L. (2002). *Beyond Black: Biracial identity in America.* Thousand Oaks, CA: Sage Publications.

Rodríguez, C. E. (2000). *Changing race: Latinos, the census, and the history of ethnicity in the United States.* New York: New York University Press.

Roediger, D. (1991). *The wages of Whiteness: Race and the making of the American working class.* New York: Verso.

Steinberg, S. (1995). *Turning back: The retreat from racial justice in American thought and policy.* Boston: Beacon Press.

U.S. Bureau of the Census. (1996). *Population projections of the United States by age, sex, race, and Hispanic origin: 1995–2050.*

Winant, H. (2001). *The world is a ghetto: Race and democracy since World War II.* New York: Basic Books.

DISCUSSION QUESTIONS

1. What does Bonilla-Silva mean by a tri-racial stratification system? Why are we developing one now?

2. What distinguishes honorary White groups from Black groups?

3. Would White supremacy remain intact in a tri-racial social order?

55

Reducing Inequalities

Doing Anti-Racism: Toward an Egalitarian American Society

BY JACQUELINE JOHNSON, SHARON RUSH AND JOE FEAGIN

We view both structural and individual change as crucial for creating a new society without racism. Prior efforts to destroy structures and institutions that reinforce a system of racism have generally not cut to the heart of the racist prejudice and discrimination still implemented in the lives of most whites. Serious desegregation efforts by U.S. governmental agencies lasted barely a decade, and weak enforcement of most civil rights laws in states and at the federal level is now a national scandal. . . . Likewise, efforts that solely target individual racism do not root out the structural embeddedness of racism. Many programs, for example, that stress a liberal ideology of tolerance or color-blindness encourage people to accept individuals, opinions, and cultures that are different from their own, but require little or no work from those in dominant groups to critique and confront systematically their own privileges and power. . . . In many multicultural programs, whites become just one more "ethnic group" like all the others, rather than the dominant group with great privileges associated with its racial classification (see Van Ausdale and Feagin, 2000).

From *Contemporary Sociology* 29 (January 2000): 95–110. Copyright © 2000 by the American Sociological Association. Reprinted by permission.

A nonracist society cannot be achieved if whites continue to deny the reality of the racist society and of racism within themselves. The painful emotional work of actually undoing individual racism must be accomplished in combination with collective efforts for structural change. . . .

EDUCATION AND RE-EDUCATION
FOR A NONRACIST SOCIETY

. . . Eliminating institutional oppressions like racism is essential to achieve an egalitarian society, but institutions are sustained and administered by individuals. Therefore, it is important to emphasize that institutional and individual racism are co-dependent. Racist systems construct as "natural" the tendency of individuals to use skin color as a basis for assigning others to in-groups and out-groups (Tajfel 1982). . . . If these anti-human impulses were not overtly and covertly taught to children, a society would not embed them in its realities. . . .

Educational systems, such as public schools, were seen as primary sites where racial misconceptions could be refuted, resulting in greater racial accord in all social arenas. However, because most attempts at actual desegregation have been controlled by whites, the attempts to deal with institutional racism or re-education were mostly tentative baby steps. Most white desegregationists did not seem to understand, or perhaps did not wish to understand, that racism was deeply embedded in white minds and institutions, including the educational institutions at hand. . . .

Changing the way that the public schools are supported is one way to attack segregation. The current system, where public schools are funded largely by local taxes, guarantees social inequality, segregation, and continued racism across large areas of cities and states. This system reinforces white notions of entitlement and privilege by linking wealth routinely to educational quality. What was once considered overtly racist "white flight" to avoid desegregated schools and neighborhoods has now been reconceptualized, as the socially conscious attempts of white parents to build a "better" life for their children. But, often what makes this life "better" is their limited contact with people of color. . . .

We need new educational approaches and structures if we are to move to the nonracist society. Schools would need to be nonracist, nonsexist, nonclassist, democratic, and egalitarian in structure and process as well as in fundamental values. Children would be taught both through example and ideology that all people—black or white, Anglo or Latino, Asian or non-Asian, boy or girl, man or woman, rich or poor—have a right to grow and develop to their fullest potential. In a nonracist society, all children would be supplied with learning environments of equal quality so that their abilities could be developed. . . .

"DOING" ANTIRACISM: SOCIAL
AND PERSONAL ACTIVISM

... To some degree, most Americans of color are forced routinely to engage in anti-racism work, at least in regard to their own group. These Americans of color may need to expand their activities to include the discrimination faced by other groups of color. But the most challenging task is to move significant numbers of whites into anti-racist actions and activism. This means that whites must move out of their present comfort zones to confront personally the painful and usually emotional work of doing anti-racism every day. We also envision the widespread formation of cross-racial coalitions with others who are devoted to doing anti-racism. Overall, we visualize many white individuals actively, consciously, and consistently working to eliminate racism by rejecting systems of privilege-maintenance in favor of human dignity, mutual respect, and liberty.

Organizational Efforts

African-Americans and other Americans of color have long led the struggles against racism in the United States. They continue to lead that struggle. As we see it, the goal must be to continue that struggle and to recruit more whites to the nonracist cause. Some members of the dominant group, albeit a very small percentage, are moving already toward the ideal nonracist society. Over the last several decades nonracist whites, with other nonracist Americans, have participated in a number of grassroots organizations working against racial oppression. . . .

A next step in a broad nonracist strategy for the United States would be to expand the number of these nonracist organizations and to connect them into a national and international network of all peoples working against systemic racism in this and other societies across the globe. . . .

To effect a genuine move away from racism and toward the nonracist egalitarian utopia, these well-intentioned whites must understand their own racism and that of the society. They must combat institutional racism—the racism built into every facet of American life. Although they may be opposed to discrimination, too many liberal whites are unaware of the demon of white privilege in their own lives. The next step is for them to acknowledge that their own white privilege contributes to the persistence of racism.

Individuals Undoing Racism

... Whiteness and white racism must be carefully learned and maintained over lifetimes. Individual selves and psychologies are shaped by structural realities. Thus, the question arises: What would the nonracist individual in the nonracist society be like? If we can begin to construct such a person, then perhaps we can take real steps toward the utopian society. The emerging nonracist person, like all human beings, is not without faults. Many Americans of color have already moved well down the road to nonracist or nonracist attitudes and actions. It appears that a pressing need today is to create a multitude of whites who can be started along that road. Unlike most whites today, however, a white person committed to the nonracist utopia would be filled with a very deep and lasting *respect* for all human beings as equals, including those who are physically or culturally different.

Respect, not just tolerance, is the necessary emotional orientation. Because the primary goals of the nonracist utopian society are to eliminate unnecessary suffering and to create the rights to life, liberty, and human happiness so eloquently asserted in the U.S. Declaration of Independence, people therein would be motivated to treat each other equally with dignity and respect. Within this positive energy cycle, everyone will enjoy the equal rights to be free from unnaturally imposed suffering and to be happy. A skeptic might suggest that this sounds good, but ask whether individual Americans, especially individual whites, are up to this difficult task. How do we bring changes in those centrally responsible for racist oppression?

Clearly, being willing to talk candidly about individual and societal racism is one essential step for whites to take in moving toward the nonracist society. Honestly discussing with Americans of color the realities of racism increases the possibility that whites will move beyond their misunderstandings and fears and begin to put good intentions into the hard work of dismantling racism. This effort requires whites to actively join the struggle by working with African Americans and other people of color side by side, day in and day out. . . .

Moving to the ideal nonracist society will require much work on the cognitive and emotional aspects of contemporary racism. This is perhaps the most difficult task for white Americans. However, whites' identification with oppression on the other side of the color line can develop through at least three different stages: sympathy, empathy, and what might be called transformative insight. The initial stage, sympathy, is important but limited. It usually involves the willingness to set aside some of the racist stereotyping and hostility taught in white communities and the development of a friendly interest in what is happening to the racial other. Empathy is a much more advanced stage, in that it requires the ability to reject distancing stereotypes and a heightened and sustained capacity to see and feel the pain of the racial others. Empathy involves the capacity to sense deeply the character of another's pain and to act on that sensitivity.

Empathizing with victims of racial discrimination is an important and valuable but limited emotional skill. Such empathic feelings are limited because they stem largely from perceiving the reality of anti-black discrimination. The empathic person's energy may be directed mainly at ending those practices. However, this outward focus on blacks' pain may incline empathic whites to avoid the inward reflection necessary to understand the role that their own white privilege plays in maintaining patterns of racism.

Actually crossing the color line provides an opportunity for a more informed understanding of the dynamics of racism that even very liberal whites have not gained from academic studies, civil rights activities, or limited social contacts with black Americans. This third stage of white development we call transformative insight. Transformative insight is more likely to develop in loving and caring interracial relationships. Interpersonal love characterizes most relationships in which people care deeply about each other: parent/child, husband/wife, friend/friend, and so forth. . . . Transformative insight, or transformative love, is most likely to develop in people who are in loving relationships that challenge institutional norms about power distribution. As a result, whites in such a position come to understand much more about the way in which the racialized hierarchy bequeaths power and privilege. The transformative insight includes a clear understanding of the broad range of privileges that comes from being white in a racist society, privileges a person has whether she or he wants them or not. . . .

A well-intentioned white person has to dig deep to uncover knowledge of this privilege, even though it is obvious to black Americans and other Americans of color. Unfortunately, most whites do not have a caring or loving relationship with even one black person. For that reason, increasing real respectful relations across the color line is essential to the long-term battle against white racism. . . .

Unlearning racism and the essential emotional work to eliminate individual racism are essential steps in becoming an effective nonracist activist. Nonracist activists cannot interact merely within limited social circles. They must actively work to break down existing structures that maintain and reproduce inequality. It is not enough to acknowledge difference or to be tolerant of others. The new anti-racists must acknowledge the real pain and the white privilege embedded in the existing system, must be willing to give up much privilege, and must work actively to create more egalitarian structures. This is tough work—emotionally, spiritually, and physically—but it is ultimately crucial for the construction of a nonracist utopia.

REFERENCES

Carr, Leslie. 1997. *"Color-Blind" Racism.* Thousand Oaks, CA: Sage.

Tajfel, Henri. 1982. *Social Identity and Intergroup Relations.* Cambridge: Cambridge University Press.

Van Ausdale, Debra and Joe Feagin, 2000. *The First R.* Lanham, MD: Rowman & Littlefield.

DISCUSSION QUESTIONS

1. How can schools contribute to building a nonracist society?
2. Why do Johnson, Rush, and Feagin think White People have to understand their own racism and that of the society?
3. Can you identify a group on your own campus engaged in antiracist work?

InfoTrac College Edition:
Bonus Reading

http://infotrac.thomsonlearning.com

You can use the InfoTrac College Edition feature to find an additional reading pertinent to this section. You can also search on the author's name or keywords to locate the following reading:

Bruce Nissen. 2000. "Living Wage Campaigns from a 'Social Movement' Perspective: The Miami Case." *Labor Studies Journal,* 25 (Fall): 29–50.

Around the nation there are local campaigns to push for living wage ordinances. Many of those currently working in low wage jobs are people of color, including immigrants. In this article on the Miami living wage campaign, Bruce Nissen examines the potential and limitations of living wage campaigns to build enduring social movements that unite organized labor with community partners. Nissen explores whether this movement has the components that scholars identify as essential for a social movement: political opportunities, mobilizing structures, and framing processes. His article shows the difficulties of building a genuine social movement that involves egalitarian multilateral alliances with different groups.

STUDENT EXERCISES

1. Social protests are important ways that people disrupt business as usual and demand that others pay attention. Have you ever participated in a protest? If so, what motivated you to take such action? Was this protest an individual act or part of an organized effort? What was your reaction to the experience? If you have not participated in a protest, why not? You can use your answers to these questions as the basis for a group discussion.

2. Imagine that your college or university announced that they were no longer giving scholarships to students for either economic need or to enhance diversity. Only students with excellent high school records and high SAT or ACT scores would now be eligible for scholarships. Others would have to rely on loans and jobs. Write a brief paper describing the implications of such a policy for your school. What does this assignment suggest about the work of building non-racist institutions in a color-blind era?

Index